W A.

OF
WEST
UNTRY

D1460564

Geological Highlights
of the
West Country

Geological Highlights of the West Country

A Nature Conservancy Handbook

W. A. MACFADYEN, M.C., M.A.(CAMB.), SC.D.,
PH.D., F.G.S., F.R.G.S.

Formerly Chief Geologist,
The Nature Conservancy

With contributions by Dr A. W. G. Kingsbury

LONDON   BUTTERWORTHS

ENGLAND
Butterworth & Co (Publishers) Ltd
London: 88 Kingsway, W.C.2

AUSTRALIA
Butterworth & Co (Australia) Ltd
Sydney: 20 Loftus Street
Melbourne: 343 Little Collins Street
Brisbane: 240 Queen Street

CANADA
Butterworth & Co (Canada) Ltd
Toronto: 14 Curity Avenue, 374

NEW ZEALAND
Butterworth & Co (New Zealand) Ltd
Wellington: 49/51 Ballance Street
Auckland: 35 High Street

SOUTH AFRICA
Butterworth & Co (South Africa) Ltd
Durban: 33/35 Beach Grove

First published 1970

ISBN 0 408 70002 5

Printed in England at the Pitman Press, Bath

Contents

Preface

To a geologist England's West Country is classic ground. Not only are its rocks of great intrinsic interest, but it was here in the earliest years of the nineteenth century that William Smith first deduced the natural 'laws' on which stratigraphical geology is firmly based. The general application of these has led to an enhanced understanding of the Earth—an understanding that has benefited, through its numerous and continuing economic applications, not merely scientists but mankind in general. Today in south-western England the geologist can examine the strata that fired the imagination of his early predecessors and he can find for himself the evidence that led to their outstanding discoveries.

The Nature Conservancy is responsible for the maintenance of features of geological interest and many of those in the West Country have been selected for conservation. In this book Dr W. A. Macfadyen, the Conservancy's first geologist, has described the most important geological sites in the counties of Cornwall, Devon, Dorset, Somerset and Gloucester. By treating each site individually it has been possible to give fuller coverage than would have been the case if the whole countryside had to be described. The author has sought to take into account the essence of every published description, from the earliest mention of the scientific interest up to the present day; he has added new material for a number of sites from his own studies in the field.

This volume is designed to have a wider appeal than to professional geologists and students; it has been written for all who are interested in the former conditions and inhabitants of what is now one of the most visited regions of Britain.

M. E. D. POORE,
Director,
The Nature Conservancy

November, 1969

'After the *stones* of a restrained peculiar use let us next consider those of no use at all, at least that are put to none; which yet possibly may not altogether be unworthy our *admiration*.'

Robert Plot, *The Natural History of Stafford-shire*, 1686

Introduction

In 1945 the Geological Sub-Committee of the Society for the Promotion of Nature Reserves published a list of some 400 geological sites in England and Wales. A few were mainly of physiographic importance but all were considered worthy of preservation because of their outstanding scientific interest. When the Nature Conservancy was established in 1949, it notified them all to the local authorities concerned as 'Sites of Special Scientific Interest' under Section 23 of the National Parks and Access to the Countryside Act, 1949, and later declared a few of the most important as National Nature Reserves under Sections 16–19 of the Act. Thus, a start was made to implement the Conservancy's Royal Charter responsibility for 'the maintenance of physical features of scientific interest'. The list of sites has now been more than doubled by further recommendations from geologists of repute all over the country, and Scotland also is now well represented.

It may be recalled that the fundamental principles of stratigraphical geology were discovered in England by William Smith when living in the West Country between 1791 and 1805. Thus, there is a special obligation to conserve the classic sections or other available exposures of the strata on which the science was founded. It is the aim of the Nature Conservancy to preserve unobscured for present and future generations these and other sections in Great Britain that are of outstanding value to geology.

Since 1957 the Cave Research Group has helped to increase considerably the number of caves Notified by the Nature Conservancy for preservation in order to represent more adequately the natural history and other scientific interests. These include geology and physiography, hydrology, meteorology, physics and chemistry, archaeology and palaeontology of cave deposits, and the fauna and flora now living in the caves. Some of the small animals are of special interest because they spend their whole life cycles in complete darkness. From another point of view, a wealth of stalactites and stalagmites often very beautifully decorates the underground chambers and passages.

This first geological Handbook of the Nature Conservancy describes the notified geological sites and caves only within the Conservancy's South-West Region of England. If one may borrow words recently used to introduce another new publication, the objective here also is to provide an account 'learned enough to satisfy the most exacting scholar and yet lucid enough to be read with pleasure by any intelligent member of the public'. With this in mind simple language has been used wherever that seemed reasonably possible, and a glossary is provided for unavoidable technical terms. The Handbook aims to give a relatively complete account of the geological and kindred interests (so far as they are known) of each site for the information of students, and also to make clear the claims to preservation. The early scientific history has been given when known, and in some cases there is added a conspectus of the local geological circumstances of which the site forms an example. The fossil fauna and flora have been reviewed for certain sites, and include the more rarely encountered groups of organisms ranging from mammals and giant saurians down to the microscopic fossils. Apart from their intrinsic interest, these help one to appreciate the fossil ecology of the strata in which they are found. The individual peculiarities of the caves have been noted, together with the available records of the living fauna and flora, which are often as yet incompletely known. Finally a select bibliography has been included.

Of the 90 named sites described in this Handbook, 79 are of varied geological interest. There are 11 caves, and three other groups of caves are included within geological sites. Some of the sites include stretches of coast up to 17 miles long, each one exposing many different features and sections of outstanding interest. Since reference is made to literature spanning more than two centuries, it is inevitable that occasional obsolete fossil names will have remained uncorrected and various old errors and misidentifications may have been repeated. It has naturally not been possible completely to revise the palaeontological records and nomenclature; that would present a formidable task for many specialists, involving the re-examination of all the specimens, some of which, indeed, are no longer available. Nevertheless, in the case of a few sites revised lists of certain groups of fossils have been made for this volume by the kindness of specialists.

Dr Kingsbury has included his unpublished records of certain minerals, etc. in his accounts of Botallack, Cheesewring, Clicker Tor and Cligga Head in Cornwall; and Haytor and Smallacombe Iron Mines and the two Meldon Quarries in Devon. Those familiar with the literature will notice a modicum of new material contributed by the present writer to the accounts of St Erth, Blashenwell, Chesil Beach and Axmouth–Lyme Regis Undercliffs, with occasional details in some others.

EXPLANATION OF SYMBOLS, ETC.

The row of figures above the description of each site indicates, in sequence, the National Grid reference; the sheet number of the Popular and Seventh editions of the Ordance Survey one inch to the mile map; the sheet number of the Geological Survey one-inch map; the sheet number of the Ordnance Survey six-inch map of the county in question, or of the new gridded six-inch sheet, the symbol below it indicating that the Geological Survey have (G) or have not yet (–) got available for reference the six-inch geological sheet. All these figures are explained at the head of the first site under each county.

The main site interests are indicated at the head of each: A = Archaeology; B = Biology of caves; C = Caves; Ig = Igneous rocks; M = Minerals; Met = Metamorphic rocks; P = Palaeontology or Palaeobotany; Ph = Physiography; R = Raised Beach; S = Stratigraphy; T = Tectonics; U = Unconformity.

PSD signifies that the site is included in the Palaeontographical Society's *Directory of British Fossiliferous Localities, 1954.*

IGC signifies that the site was visited during an excursion of the XVIII Session of the International Geological Congress in 1948.

Finally, the year given under the Interest symbols is that of the latest inspection by a Nature Conservancy geologist. A year in parentheses in the case of caves signifies that only the exterior of the entrance has been so inspected.

AUTHORITY FOR NOMENCLATURE

Three British Museum (Natural History) Handbooks, 1959, *British Cainozoic Fossils;* 1962, *British Mesozoic Fossils;* and 1964, *British Palaeozoic Fossils,* have been followed for modern nomenclature of zones, stratigraphical equivalents and such revised fossil names as are mentioned therein.

When a species has been definitely identified in the literature, this has been accepted as correct, and where necessary the name has been revised in accordance with the B.M. Handbooks. The original name is shown in square brackets to facilitate correlation with the old literature, as *Liostrea hisingeri* [=*Ostrea liassica*], or *Spirillina* [*Orbis*] *infima.*

Where generic names alone, such as *Rhynchonella* or *Terebratula*, have been recorded in the literature without species, these names can of course only be so given. When a species not mentioned in the B.M. Handbooks is named in the old literature, it is in general given as published there, for a revised name is impossible without specialist assistance.

The authors of the fossil names have generally been omitted as unnecessary in a work such as this.

LITERATURE

For the writing of this Handbook it has been the aim to consult, so far as practicable, the whole of the original literature concerned with each site, for every such publication has its own unique contribution to make. Only a selection of the more important and more recent papers have, however, been given in the bibliography.

ACKNOWLEDGMENTS

For much help in the preparation of this Handbook, I am greatly indebted to the following friends, particularly to the late Dr A. W. G. Kingsbury of the University of Oxford, who, as indicated in the text, kindly rewrote the accounts of seven sites in Cornwall and Devon involving rare mineral species, and thereby conferred upon them his authority.

Dr F. W. Anderson, Dr R. Casey and Dr F. B. A. Welch of H.M. Geological Survey (lately renamed the Institute of Geological Sciences).
Mr A. M. ApSimon and Dr R. J. G. Savage of the University of Bristol.
Miss Muriel A. Arber of Cambridge.
Mr D. F. W. Baden-Powell and the late Dr A. W. G. Kingsbury of the University of Oxford.
Dr H. W. Ball, Dr A. J. Charig, the late Dr L. R. Cox, F.R.S., Mr S. P. Dance, Dr C. Patterson, Dr A. J. Sutcliffe, the late Dr H. Dighton Thomas, Mr H. A. Toombs and Dr E. I. White, F.R.S., of the British Museum (Natural History).
Miss Marjorie E. J. Chandler of Bridport.
The late Mr Henry Dewey of Newton Abbot.
Professor D. T. Donovan of University College, University of London.
Dr A. Hallam of the University of Edinburgh.
Dr J. M. Hancock of King's College, University of London.
Dr J. D. H. Hooper of the Devon Spelaeological Society.
Dr J. Wilfred Jackson of Buxton.
Dr P. E. Kent, F.R.S. of British Petroleum Ltd.
Professor C. Kidson of the University of Wales, Aberystwyth.
The late Professor W. B. R. King, F.R.S., and Dr R. G. West, of the University of Cambridge.
Dr A. D. Lacaille, F.S.A., of London.
The late Dr W. D. Lang, F.R.S., of Charmouth.
Dr J. D. Lawson of the University of Birmingham.
Mr L. A. Pritchard of The Nature Conservancy.
The late Mr Linsdall Richardson of Worcester.
The late Mr C. W. Taylor of Barnstaple.
Dr G. T. Warwick of the Cave Research Group of Great Britain, University of Birmingham.

To the Librarians of the following libraries I owe a debt of gratitude for their help and for the facilities so freely placed at my disposal: British Museum (Natural History), H.M. Geological Survey and Museum, Geological Society of London, The Nature Conservancy, Royal Geographical Society, Royal Society, and Society of Antiquaries, all in London; and Dorset County Archives, Dorchester, and Wells Natural History Society, Wells.

Finally I have to thank two members of the Maps Section of The Nature Conservancy Headquarters: Mr Robin Fenton for supervising the illustrations, and Mr Nicholas Rous for drawing all the diagrams.

CODE FOR VISITORS TO GEOLOGICAL SITES

In the interests of everyone concerned, the owners of the land, other geologists and students, the Nature Conservancy, and not least yourself, *please* read this code carefully before making any field excursion, particularly to a geological Nature Reserve or Site of Special Scientific Interest, *and then put it into practice*!

1. A Site of Special Scientific Interest (S.S.S.I.) is one that has been notified as such by the Nature Conservancy to the Local Planning Authority under Section 23 of the National Parks and Access to the Countryside Act, 1949. It remains in the complete possession of its individual (and generally private) owner, and its notification to the Local Planning Authority in no way diminishes the owner's rights over his land. In particular, it gives no right of access or to the collection of specimens by anyone.

National Nature Reserves either are in the possession of the Nature Conservancy or are managed by that body by agreement with the owner.

2. Any individual or party wishing to visit an S.S.S.I. has the obligation to seek prior permission of the landowner. The landowner has no obligation to permit such visits, although he has been requested as a courtesy to allow access to those who wish to study the site. In practice owners courteously permit such visits to the great majority of sites. In the case of a closed National Nature Reserve prior permission of the Nature Conservancy must be sought.

3. If the name of the landowner is not known to the person wishing to visit an S.S.S.I. in the south-west region, it may generally be obtained by application to the Nature Conservancy: Geological Section, Oak Cottage, Brimpton, Reading, Berks., RG7 4RJ, Tel. Woolhampton 3266; or South West Region, Furzebrook Research Station, Wareham, Dorset, Tel. Corfe Castle 361; or London Headquarters, 19 Belgrave Square, S.W.1, Tel. 01–235–3241. Since ownership may change, the information may not always be up to date. But intending visitors must make their own arrangements with owners; the Nature Conservancy cannot undertake this for them.

4. Visitors are asked to conduct themselves with scrupulous care, meticulously carrying out the provisions of the Country Code, e.g. ensuring that all gates found shut are left shut; leaving no litter; doing no damage to gates, hedges, fences or walls, etc., or to growing crops by trampling; and avoiding the disturbing of stock, horses, cattle or sheep, especially in the breeding season, or of poultry, etc.

Whenever foot-and-mouth disease or fowl pest has been notified anywhere in the area, special care must be taken to obtain permission to visit a site, otherwise, rightly or wrongly, farmers may attribute to visitors the spreading of the disease to their farms. A case of this has occurred.

5. Failure to observe these proper courtesies has led in the past to the annoyance of the site owner and his subsequent refusal to permit access. Certain farmers have erected extra barbed wire fences, and even put a bull in their fields to keep geologists out, which testifies to their resentment.

6. In certain coast sections care must be taken to avoid getting trapped by the incoming tide; and to beware of falling stones and cliff falls, particularly in incompetent strata of Mesozoic, Tertiary and Quaternary age.

7. Be careful to prevent stones or boulders rolling down steep hill-sides or over cliffs; they may cause damage to unseen people or stock below, or may end up in fields which will be cut for hay, and have to be removed by the justly annoyed farmer, or may damage his cutters.

8. Do not discard unwanted specimens or trimmings in fields; they may cause damage to grazing stock by being accidentally swallowed; and they may damage grass cutters.

9. Good sections, particularly in Nature Reserves and S.S.S.I.s should never be defaced by hammering, and the sites should be disturbed as little as possible; for others will want to see the sections in good condition just as you do.

10. Finally, leaders of parties are most earnestly asked to make these points unmistakably clear to every member of their party, and themselves to ensure that proper discipline is maintained; for the leader who secures permission to visit a site is responsible to the landowner for the good behaviour of his party.

CORNWALL

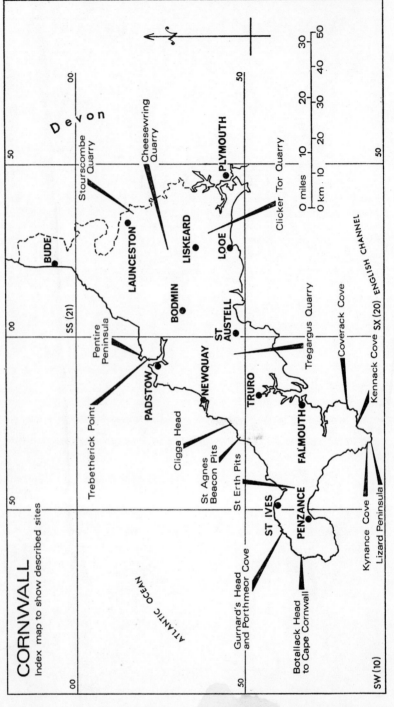

Fig. I

CORNWALL
Index map to show described sites

Devon

PLYMOUTH

Cheesewring Quarry

Stourscombe Quarry

LISKEARD

LOOE

Clicker Tor Quarry

BUDE

LAUNCESTON

BODMIN

ST AUSTELL

Tregargus Quarry

Coverack Cove

Pentire Peninsula

NEWQUAY

TRURO

FALMOUTH

Kennack Cove SX (20) ENGLISH CHANNEL

Trebetherick Point

PADSTOW

Cligga Head

St Agnes Beacon Pits

St Erth Pits

ST IVES

PENZANCE

Kynance Cove

Lizard Peninsula

ATLANTIC OCEAN

Gurnard's Head and Porthmeor Cove

Botallack Head to Cape Cornwall

SW (10)

SS (21)

O miles   10   20   30
O km   10   20   30   40   50

National Grid Reference	One-inch Sheet Pop. and 7th ed. Ordnance Survey	One-inch Sheet Geol. Survey	Six-inch Sheet Ordnance Survey	Main Site Interests
SW(10)362330	189	351	67SW, 73NW (G)	Ig, M, Met IGC, 1965

These cliffs provide some of the finest exposures of contact metamorphism and metasomatism in Cornwall and lie within the aureole of the Land's End granite mass of Armorican age (late Carboniferous to post-Carboniferous) emplaced some 270–300 million years ago.

In many places Mylor (Downtonian) Slates (killas) and greenstone sills have been sheared together to produce banded and foliated epidiorites in contact with the slate. The subsequent intrusion of the granite gave rise to thermal and pyrometasomatic alteration (with varying effects) of the slates, the more massive greenstones and the epidiorites, producing tough hornfelses of very varied types. In addition many bands of tourmaline were formed, and considerable masses, veins and assemblages of typical 'skarn' and other minerals such as garnet, axinite, epidote, pyroxenes, amphiboles, magnetite, etc. Many mineral veins occur and outcrop in the cliffs, and tongues and veins of granite can be seen cutting the altered rocks in many places.

The alteration of the greenstones and epidiorites is well shown above Botallack Head; a little to the south, at the foot of the cliffs, are three veins of prehnite with axinite, etc., and at the Crowns Rock there is a mass of garnet rock. In the cliffs adjoining the ruined engine-houses of the Crowns Mine, and in the rocks below Chycornish Carn, well-formed dodecahedra of almandine garnet occur.

At Kenidjack Castle there is a remarkable suite of cordierite-, cummingtonite-, and anthophyllite-hornfelses, unique in Britain (Tilley, 1935).

In Priest's Cove, adjoining Cape Cornwall, weathered tourmaline-granite is transgressive through highly contorted chiastolite-slate with many veins of aplite, pegmatite and quartz, into a purplish slate hornfels, some with cordierite.

Botallack was the centre of a once famous sett, or group, of mines, all now abandoned, which included the ancient mines of Wheal* Cock, just north of Botallack Head, and Wheal Owles, a little to the south. Mining here was active in mediaeval times and reached its zenith in the latter half of the 19th

* The word *Wheal* is stated to come from the Celtic *huel* meaning a mine.

century. Botallack Mine itself was mainly a tin producer, mining for this metal having started as early as 1721. An attempt was made to re-open it about 1908 but it was unsuccessful and the mine closed again in 1914. Some of the workings extended as far as 600 yards out under the sea.

Between 1836 and 1895 this group of mines was claimed to have produced tin ores sold for £829 604; copper ores sold for £220 701; and arsenic ores sold for £6481. Between 1901 and 1905 the working of shallow levels yielded 268 tons of black tin, sold for £16 792 (Collins, 1912).

The various sections of Botallack Mine, together with Wheal Cock to the north and the Wheal Owles group to the south, and the cliffs in between them, have produced a very large number of rare and interesting minerals, including unusual crystallisations of calcite and siderite. Atacamite and many other compounds of copper; native copper and bismuth and compounds of arsenic, bismuth, cobalt, iron, lead, manganese, silver, tin, uranium and zinc have also been recorded. More recently a number of additional rare minerals, several of them new to Britain, have been found by the writer (A.W.G.K.) in this area of cliffs and at the various mines, including the rare beryllium minerals phenakite (at Wheal Owles) and herderite and helvine (at Wheal Cock).

A. W. G. Kingsbury

1907, Reid and Flett; 1912, Collins; 1930, Tilley and Flett; 1935, Tilley; 1954, 1958, 1961, Kingsbury.

Cheesewring Quarry, 5 miles N of Liskeard

SX(20)259724	186	337	28NW	Ig, M
			(G)	1965

This large quarry was until recently (1960) working the sound, unweathered grey granite (the 'normal' granite) of the south-eastern margin of the Bodmin Moor mass, of Armorican age.

The granite is of medium grain and composed of porphyritic orthoclase crystals about an inch long (often Carlsbad twins), abundant quartz, less plagioclase, two micas and some tourmaline. A durable rock obtainable in large blocks, it was much used in the past for public works in Plymouth, and also in London for the Guards Memorial in Waterloo Place, for the pillars of Westminster and Waterloo Bridges and in the construction of docks. The granite was transported from the quarry by railway.

The quarry is traversed by two sets of nearly vertical joints or fissure planes, at right angles; one set runs N–S, and the joints are closed; the other set runs

E–W and the joints are 'open' generally less than an inch wide, but sometimes as much as three inches. These E–W joints are parallel to the adjoining lodes in the well-known Phoenix Mine, and have been mineralised during more than one stage. They carry an interesting assemblage of quartz, chlorite, tourmaline (two generations), orthoclase of adularia habit, fluorite, red sphalerite, pyrite, torbernite, anatase, bertrandite, and, more rarely, cassiterite and wolframite. Small pegmatitic druses and cavities occur in many parts of the quarry and these also carry orthoclase, quartz, fluorite, gilbertite, and, rarely, apatite and bertrandite. Autunite occurs in joints in one part of the quarry, and andalusite, sillimanite and possibly cordierite have been recorded in hornfelsed xenoliths of slate.

Phenakite was found here in 1905 by Sir Arthur Russell for the first time in Britain; the few crystals were found in quarry debris. The mineral has more recently been discovered *in situ* by the writer (A.W.G.K.) together with some further rare beryllium and other minerals, in two different environments in the quarry, each showing a different crystal habit.

The quarry face now extends almost up to the well-known Cheesewring of weathered granite looking like piled slabs or blocks, often discussed in the literature. For nearly a century this was regarded as partly at least of human handiwork instead of a purely natural feature, although Speed in 1611 had described it, as the Wring-cheese, quite non-committally.

A. W. G. Kingsbury

1611, Speed; 1758, Borlase; 1887, Teall; 1903, Hull; 1911, Reid *et al.*; 1911, Bowman; 1911, Russell; 1927, Ghosh; 1964, Barton.

Clicker Tor Quarry, Menheniot, 3 miles SE of Liskeard				
SX(20)289613	186	348	36SE (G)	Ig, M IGC, 1965

A large and partly disused quarry in a boss of dark-green augite-picrite, which forms a small hill close to Menheniot railway station; it affords a very fine exposure of this unusual type of intrusive ultrabasic igneous rock. Its weathered skin has the appearance of rusty cast-iron.

The Clicker Tor rock consists mainly of serpentine pseudomorphs after olivine, together with augite, a very small amount of plagioclase feldspar and some secondary hornblende (tremolite). Minerals found in the quarry include fibrous calcite, fibrous wollastonite, datolite, prehnite, asbestos, magnetite and dark-green precious serpentine.

The picrite has been intruded into Upper Devonian sediments, and is closely related to many other contemporaneous greenstones, some being intrusions but most extrusive spilitic lavas, which occur in this district and elsewhere in Cornwall and Devon, particularly in the Upper Devonian. Many of these rocks are characteristic of the Devonian spilitic province which extends into central Europe, and the picrites are the typical ultrabasic facies of the suite.

A most interesting feature of this intrusion seems to have been recognised only recently, during a field trip from Oxford in 1960 (personal communication from Dr G. M. Brown and Dr A. W. G. Kingsbury, University of Oxford). As in other similar intrusions, the Clicker Tor picrite has a characteristic envelope of gabbro, which was well exposed in 1960 on the northern side of the quarry. This gabbro envelope apparently acted as an intrusion lubricant, without which the picrite, probably in a semi-solid condition, would have suffered much brecciation or deformation.

A distinctive rock of unusual type, the picrite is readily recognised as the source of certain far-travelled boulders and human artifacts. Studied since 1813 and worked at least from that date for road metal, it was still in work in 1962. About 1900 it was yielding ballast for the Great Western Railway (Evens, 1958), and some was used in Somerset on the Cheddar Valley branch line. A small quantity of the ballast seems adventitiously to have reached a field near the Ebbor Rocks (q.v.), where it was found in 1904 and led to much speculation as to how the rock arrived there, its use as railway ballast nearby apparently then not being appreciated (see p. 244).

A. W. G. Kingsbury

1813, Townsend; 1821, Sedgwick; 1888, Teall; 1907, Ussher; 1958, Evens.

Cligga Head, 1 mile W of Perranporth

SW(10)738535	190	346	47NE, 48NW (G)	Ig, Met, M IGC, 1965

A magnificent 300-feet-high sea-cliff section exposing the remnant of the eastern side of a granite boss of Armorican date intruded into Devonian slates (killas) which it has metamorphosed. It forms one of the finest exposures of greisen-bordered mineral veins in Britain and perhaps in Europe. The site has attacted the attention of geologists certainly since 1813 when mining was active.

Mining here for tin (and later for tungsten) probably goes back to pre-historic times and continued sporadically until 1945, so that the rocks are honeycombed with old workings.

The granite is riddled through with quartz veins, many containing tourmaline, cassiterite, wolframite and other minerals, and ranging in thickness from more than a foot to less than a quarter of an inch. Each vein is bordered by bands of greisen. Elsewhere, and between the greisen bands, the granite is largely soft and kaolinised. The veins, curved into the shape of an anticline and syncline, occupy divisional planes in the granite, roughly parallel to the contact, and form a large stockwork.

Nearby are the extensive workings of the old Perran St George copper mine; many of the lodes outcropped in the cliffs and afford excellent illustrations of the very early methods of cliff mining.

The large number of minerals found and recorded from here include: andalusite, arsenopyrite, bismuthinite, bornite, cassiterite, chalcocite, chalcopyrite, chlorite, copper, garnet, gilbertite, hematite, isostannite, kaolinite, lithia-mica, molybdenite, muscovite, olivenite, pharmacosiderite, scorodite, stannite, topaz, torbernite, tourmaline, varlamoffite, wolframite.

Other more recent discoveries are: botallackite, paratacamite, and connellite (Kingsbury, 1954); atacamite (Kingsbury and Hartley, 1956) and the first and so far only known British occurrence of the rare beryllium mineral euclase (Kingsbury, 1958).

A. W. G. Kingsbury

1813, Townsend (p. 330); 1903, Scrivenor; 1906, Reid and Scrivenor; 1954, 1958, Kingsbury; 1956, Kingsbury and Hartley.

Gurnard's Head and Porthmeor Cove, $5\frac{1}{2}$ miles NW of Penzance

---

SW(10)432387	189	351	61A (on 67NE) (G)	Met 1965

Coast exposures of some of the best sections in Cornwall for studying contact alteration of greenstone and killas, within the metamorphic aureole of the Land's End Granite, intruded during the Armorican movements.

At Gurnard's Head alternating greenstone and altered Mylor Slates (Downtonian) pitch towards the sea at high angles. At Porthmeor Cove excellent shore sections show the junction between granite and the rocks into which it was intruded, veins of granite penetrating the greenstone and slate, with bands of greisen and schorl rock, respectively, cutting the granite and all the rocks. The greenstone has been converted into a splintery dark hornfels.

Spotted slates occupy the larger part of the metamorphic aureole, but close to the junction they also pass into dark hornfels, one band containing small perfect crystals of garnet. The outstanding interest of these sections was recognised and they were well described in detail by Forbes in 1819.

Records of the old Gurnard's Head Mine are very incomplete. Started before 1821 under the name Treen Copper Mine, in the years 1834–35 and 1842, 24 tons of copper ore were sold for £277, but no very profitable lode was struck (Jenkin, 1961). In 1843 another Company, North United and Gurnett's Head Mine, started work, but it seems to have closed in 1847. Further working appears to have taken place again later, for there is a record of 25 tons of copper ore yielded in 1853.

1822, Forbes; 1907, Reid and Flett; 1961, Jenkin.

Lizard Peninsula 12 miles SSW of Falmouth

SW(10)7012	189, 190	359	84SE (G)	Ig, Met IGC, 1965

Viewed from a distance, this peninsula appears flat, the effect of the 'Pliocene' marine planation. It has an average height of about 300 feet above the sea, and is one of the best examples of an elevated platform in Great Britain. It is bounded by cliffs up to 200 feet high, whose origin is controversial.

The Lizard Peninsula consists of a great plutonic intrusive complex, represented by serpentine, gabbro and gneiss, surrounded by an aureole of metamorphosed sediments—mica-schists, greenschists, quartz-granulite, etc.—and hornblende-schists that were originally igneous rocks. The exposures are nearly all in the sea cliffs.

Of all these rocks, termed the Lizard Series, there is no direct evidence of age, but they used to be considered as probably Precambrian* and among

* This was the old Geological Survey view as given in Flett's 1946 Memoir. Dr Hendriks (1937, 1959), however, on the basis of certain fossils that she had by then discovered, considered the Old Lizard Head Series to be metamorphosed Lower Devonian sediments, and the area to be one of nappes, whose movements and the intrusion of the serpentine were latest Devonian or early Carboniferous.

The isotopic age of the Lizard rocks has now been measured by Dodson (1961), who obtained a mean value of 360 million years, indicating from Middle to Upper Devonian according to the Geological Society Phanerozoic time-scale, 1964. Miller and Green (1961) made further measurements which were in good agreement with Dodson's, but they found the Kennack Gneiss (q.v.) to be still older. This is contrary to the geological evidence. They therefore regarded the Kennack Gneiss as correctly dated and the Old Lizard Head Series isotopic age as anomalous, owing to loss of argon associated with slight retrogressive metamorphism. The Old Lizard Head Series metamorphism would thus appear to be pre-Devonian.

the very oldest rocks in Britain. All are in a variable but often intense state of metamorphism.

The succession is given as follows:

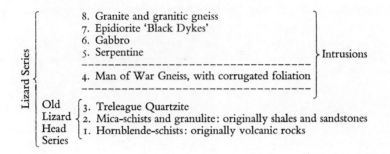

Numbers 1 and 2 above are in a very high state of metamorphism and the intrusive rocks (Nos. 4–8) must have been folded during their consolidation to produce the fluxion effects seen in them.

There were two periods of metamorphism: the first probably due to the intrusion of the Man of War Gneiss; and the second, of far greater intensity, probably due to the intrusion of the serpentine, gabbro, granite and granitic gneiss.

The highly variable individual reactions of the intrusions to atmospheric weathering and to marine erosion are responsible for the strikingly different aspects of their outcrops. The granite-gneiss is inconspicuous; it weathers relatively rapidly and forms low ground with good soil and a rich growth of vegetation. The gabbro is exceedingly tough and strong, highly resistant to marine erosion; but it weathers readily, so that its surface is deeply rotted, in places leaving residual boulders. The serpentine is highly resistant to weathering, forming a poor and barren soil, but it is a soft rock easily eroded by the sea.

Some of the most informative sections in this S.S.S.I. are found at the following localities.

Pistil Ogo (SW698116): Old Lizard Head Series.

Old Lizard Head (695115): Man of War Gneiss invading mica-schists.

Venton Hill Point (694120): Andalusite mica-schists.

East side of Pen Olver (713117): Sillimanite mica-schist among dark hornblende-schists.

Housel Bay (708119) to Church Cove (715127): Epidotic hornblende-schists.

The Balk (715128): Thrust junction of serpentine over schists.

Three separate sites of special scientific interest, Coverack Cove, Kennack

Cove and Kynance Cove, further illustrate the Lizard intrusives. Of the very extensive literature the following references include all four Lizard S.S.S.I.s.

1822, Sedgwick; 1877, Bonney; 1913, 1946, Flett; 1913, Flett and Hill; 1937, 1959, Hendriks; 1961, Dodson; 1961, Miller and Green; 1964, Barton.

Coverack Cove, 6½ miles NE of Lizard Village

SW(10)785188	190	359	81SE (G)	Ig, Met IGC, 1951

Easily accessible and well-exposed foreshore sections of great merit, giving an epitome of the igneous history of the Lizard district, with its plutonic intrusions of the Lizard Series altered by metamorphism.

The country rock of bastite serpentine is broken through by a profusion of dykes, the later ones cutting the earlier, as follows:

(Latest) 4. 'Black Dykes' of epidiorite.
3. Gabbro-pegmatite, with crystals of diallage up to seven inches across.
2. Grey-green or coal-black olivine-gabbro, in places sheared to a schist by contemporary movements and enclosing many blocks of the serpentine.
1. The remarkable red- or green-spotted troctolite ('troutstone'), an olivine labradorite rock.

Below the village is a Raised Beach covered by a thick deposit of Head. Scattered over the fields and down above lie great residual boulders of gabbro, known locally as 'crusairs'; these form the more resistant parts of the gabbro mass, which have weathered out *in situ*, like the Fyfield Down sarsens and the granite tors of the south-west.

Kennack Cove, 3½ miles NE of Lizard Village

SW(10)737166	190	359	85NW and SW (G)	Ig, Met IGC, 1965

Remarkable geological shore sections of the Lizard Series intrusives, a beauty spot, and the principal sandy shore in the Lizard district, including the 'towans' (blown sand) and a sandy beach largely of shell debris, with abundant Recent foraminifera.

Here are found the best exposures of the Kennack Gneiss, the most abnormal rock of the Lizard, a mixture of a dark basic and a pale acid ingredient. The dark component is doleritic and probably the last emission of the magma of the 'black dykes', and the pale component is the precursor of the granite-gneiss. This Kennack Gneiss forms dykes, sills and irregular injections, often of large size, in the country rock of bastite serpentine, which, to the east, makes striking cliffs.

Radiometric measurements on biotite from the Kennack Gneiss by the potassium–argon method date the intrusion between 384 and 397 million years ago (Miller and Green, 1961). This indicates a Lower Devonian age, within the Caledonian orogeny.

A famous exposure in a cliff 40 yards south of Kennack Gate (at the end of the Kuggar road) shows a gabbro dyke in the serpentine, stepped by three little faults, and cut across by a 'black dyke' (epidiorite).

Hereabouts also are found veins of pseudophite, talc and asbestos, and fluxion bands of chromite-serpentine.

Kynance Cove, 1½ miles NNW of Lizard Point

SW(10)685133	189	359	84SE	Ig, Met
			(G)	IGC, 1951

'Among the magnificent scenes exhibited on the coast of Cornwall, Kynan's Cove has long been justly celebrated' (Sedgwick, 1822). Various rock forms and colours show the effect of erosion upon different kinds of the Lizard serpentine and a number of small dykes cutting it.

The country rock of the central and western part of the cove is typical tremolite serpentine, a fine-grained, streaky, banded, rather fissile rock; that of the east side is bastite serpentine, a coarsely crystalline rock with large shining plates of bastite. These two varieties are separated by a well-defined fault, with accompanying brecciation.

There are some small intrusions or dykes of hornblende schist, gneiss, granite, gabbro and epidiorite, which are largely responsible for the topographic diversity; and a vein of pseudophite.

27

Pentire Peninsula, 3 miles N of Padstow

SW(10)930808	185	336	13ASE	Ig
			(G)	IGC, 1965

Strikingly exposed in the 250-foot cliffs at Pentire Head is perhaps the most notable feature of Cornish geology, the extensive and massive development of spilite (pillow lava) in the Upper Devonian. It is stated here to overlie grey shales with goniatites characteristic of Wedekind's Zone II, the lowest zone of the Famennian (Middleton, 1960).

A vast sheet of lava covered the sea-floor for hundreds of square miles and even now its remnants form a not inconsiderable part of Devon and Cornwall. The spilites are characteristic volcanic rocks, submarine lava flows, probably from fissures, poured out remote from shore-lines into seas of considerable depth, an accompaniment of long-continued subsidence. The highly vesicular material consists of a series of alternating concentric bands of solid and amygdaloidal rock, each 'pillow' being, in truth, a thick-walled bubble of lava. It is notable for abundant soda-feldspar and frequent albitisation. At Pentire there is also a fine exposure of adinole (albite rock).

Pentire Head was described and figured by Whitley in 1849. He wrote that the pillows, in section, were 'much like the ends of bales of cloth piled one on another'.

1849, Whitley; 1910, Reid, Barrow and Dewey; 1911, Dewey and Flett; 1914, 1948, Dewey; 1960, Middleton; 1964, Barton.

St Agnes Beacon Pits, 1 mile W of St Agnes Church; 7 miles NNE of Camborne

SW(10)705510	190	346	47SE	S
			(G)	1965

Pits, some disused and overgrown, others still in work, in a rapidly varying 'orphan' series of sands and clays. They lie between 350 and 450 feet above O.D. (Ordnance Datum), forming a collar a mile in diameter and up to 500 yards wide, contouring the north-east, north and west slopes below St Agnes Beacon, which rises to 629 feet above O.D. The beds reach to within 300–400 yards of the north Cornwall coast on each side of St Agnes Head. They are mapped as covering no more than 270 acres.

These beds were, perhaps, first described by William Borlase in 1758. He gave a detailed section of 21 feet thickness of strata down to bedrock. He clearly implied that they were of marine origin, and the cause of the elevation of the land to its present level he ascribed to a great event that could be 'no other than the universal deluge'.

Hawkins (1832) found the total thickness of the beds to be 24 feet in the New Downs Pit, north of the Beacon.

Each measured section is peculiar to itself, and the whole thickness is now nowhere exposed, but in general the succession shows:

Recent to Pleistocene	3–9 ft	Surface soil and Head: blackish sandy soil with rather angular stones, cobble-sized and larger, derived from the hill immediately above, and sometimes termed *cobb*.
? Pleistocene or Pliocene	7–24 ft	Brown sand, fairly evenly bedded.
	c. 10 ft	Blue-grey clay, partly sandy; 'candle clay' and fireclay.
	c. 10 ft	White, yellow, brown, red or other coloured sand, some of it coarse, with pebbles at the base.

The deposits are practically unfossiliferous* and lime free.

The quartz sands, feldspathic and kaolin-bearing, much of them with ferruginous cement, are similar to those of St Erth (q.v.), containing glauconite and abundant grains of tourmaline, magnetite, ilmenite, andalusite, kyanite, zircon, rutile and topaz. The pebbles and even large boulders that are found are all those of the local rocks.

The beds rest upon a smooth, waterworn, and presumed marine erosion platform. In 1875, at Wheal Coit, north of the Beacon, mining operations exposed a buried cliff, facing north, and a stack. The cliff was 16 feet high, 15 feet being below the present surface. The face was almost perpendicular and bore strong evidence of having been shingle-worn for a considerable time, being eroded into caves and hollows. Old miners then reported that, some 150 yards to the south, a buried cliff with many hollows stood facing east (Davies and Kitto, 1878).

The country rock is largely of the Ladock Beds, Lower Devonian metamorphosed sandstone, close to the eastern boundary of the metamorphic aureole of the St Agnes Beacon Granite. But on the western side it is of the weathered granite. Both sandstone and granite are traversed by veins of tin ore trending east-north-east.

Certain pits are still (1959) in work for moulding sand. In the past their pure 'candle clay' was used all over Cornwall to hold the miners' candles on

* De la Beche recorded 'traces of plants that have the appearance of Fucoids' and Davies and Kitto refer to indistinct markings somewhat resembling seaweed on a piece of shale about four inches square. A sample of the sandy clay collected in 1951 when washed yielded me no microfauna.

to their hard hats as well as to the rocks underground. This was no small matter, for it is recorded that in 1827 no less than 1 344 000 pounds of candles valued at £35 000 was used in the mines of Cornwall and Devon. The clay had once been used, prior to 1758, for making tobacco pipes; and by 1878 it sold at 7–8s. per ton at the pits. It is stated to be thickest where it lies in a basin on the granite.

A bed of white sand, used for cleaning and for sanding local cottage floors, is reported by the present owner to have been dug here by women, working in well-like pits with vertical walls, from about A.D. 1400.

In the absence of direct evidence of age, these famous and controversial deposits are considered to be connected with the Pliocene planation of the county, with a sea level of some 430 feet above present O.D., partly from their elevation and topographic situation and partly by analogy with the St Erth Beds, 13 miles to the south-west.

Small outcrops believed to be of the same date occur within 20 miles, at Crouza Common (SW7819), and near Lelant (SW5538).

1758, Borlase; 1832, Hawkins; 1839, De la Beche; 1878, Davies and Kitto; 1879, Ussher (separate publication); 1890, Reid; 1906, Reid and Scrivenor; 1916, Dewey; 1922, Milner; 1923, Boswell; 1960, Fryer; 1964, Barton.

St Erth Pits, 6 miles NE of Penzance

SW(10)556350	189	351	69NW (G)	P, S PSD, 1965

Disused pits in an isolated sand and clay deposit unique in the British Isles, and mapped as covering no more than 50 acres, largely overgrown with brambles. There is no present exposure of the clay. The age is not exactly known, but is probably Pliocene or early Pleistocene. The highly fossiliferous clay has yielded a very large marine fauna, including many species described as new and not known elsewhere. There is a large literature.

The deposits lie on gentle slopes a quarter of a mile east of St Erth village and east of the low ground stretching between St Ives Bay on the north and Mount's Bay on the south. Whitley (in Wood, 1885) levelled the fossiliferous blue clay as 98 feet above mean tide-mark at Hayle, the ground surface at the pit being 15 feet higher. The beds rest upon country rock of Mylor Slates (killas) of Downtonian age, between a dyke of quartz porphyry (elvan) on the north and a greenstone sill on the south-east.

The succession varies rapidly from point to point, but shows, in general:

Surface soil.

Pleistocene or Pliocene {
    3–7 ft Head, clayey and sandy loam with angular fragments of killas, quartz and other rocks.
    c. 7 ft Yellow clay without fossils, passing down into blue shelly clay with many fossils.
    ? 2 ft (or more) Fine quartz sand, in places loamy, coarse and gravelly at base.
}

In 1959 a section seen on rising ground in the south pit showed only some six feet of brown sands, earthy on top and all but the lowest foot containing angular pebbles of quartz and elvan.

The pits were first opened for what was known as 'Cornish Red' moulding sand, and worked from about 1832 to 1872, and apparently later for sand until fairly recently. The yellow upper part of the clay was used as puddling clay for a Penzance dock in 1881; a pit was reopened for a geological examination on 26 August, 1884; the clay body was worked from about 1886 to 1890; and again briefly for pottery clay about 1931.

Nicholas Whitley discovered fossil shells in the lower, blue part of the clay in 1881, and read his paper on it in the same year; but it was apparently not published until 1887. He described it as a boulder clay, an identification not accepted by other geologists until Mitchell (1965) revived the idea, suggesting that it may have been a block of the sea-floor dredged up by an ice-sheet crossing St George's Channel, which 'thrust the marine clay mixed with other materials into a pre-existing deposit of sand at St Erth'.

The deposits were investigated in 1887 with a Royal Society grant.

Of the quartz sand, which is glauconitic, many grains are well rounded, suggesting a wind-drifted character. The commonest accessory minerals are tourmaline, topaz, magnetite, ilmenite, leucoxene, kyanite, andalusite, zircon, rutile, staurolite and chert, and there is a ferruginous cement. The kyanite and staurolite are considered to have come from the old Armorican land mass to the south. The absence or rarity of garnet and certain other locally occurring minerals is noteworthy (Milner, 1922; Boswell, 1923).

The blue clay has yielded to prolonged search a recorded fauna of some 502 species. These include 240 gastropod species (89 described as new), 62 bivalves (eight new), 173 foraminifera (six new) and fish remains (vertebrae, otoliths and teeth, including one of the shark *Galerus vulgaris*), crustaceans, ostracods (hitherto undescribed), pteropods, *Balanus*, echinoids, holothurian plates, polyzoa (*Melicerita*, *Salicornaria* and *Lepralia*), tunicate bodies (*Leptoclinum*), calcareous sponge spicules (*Leuconia*, *Leucandria*), annelids, an alcyonarian and *Melobesia*.

Macro-fossils are often sparingly distributed, and only about five per cent of the forms are plentiful, the commonest being three gastropods, *Nassa semireticosa*, *Bittium reticulatum* var. *trinodosa* and *Turritella erthensis*, all

31

peculiar to St Erth. Many of the gastropods are minute, but well preserved; but the bivalves are mostly fragmentary. The tunicate bodies and calcareous sponge spicules were the first fossil records of these animals. The fauna is distinctively different from anything yet discovered elsewhere in Britain, and at least 104 new forms have been described from it, some 20 per cent of the whole fauna. It poses problems not yet solved of age, sea depth and temperature, affinities and the disposition of land and sea.

Watts (in Mitchell, 1965) has recently made the very interesting discovery of pollen in marine clay taken from inside a *Turritella* shell collected from St Erth in 1881. From this pollen he identified four genera of conifers (*Pinus, Abies, Picea, Tsuga*), bushes (Hazel, *Coryllus*) and woody shrubs (Ericales, mostly Crowsfoot, *Empetrum*, with rare ? *Rhododendron*); also salt marsh plants (Chenopodiaceae, Gramineae, Crucifereae, Cyperaceae and Thrift, *Armeria*); rare spores of a moss, *Sphagnum*, and of a fern; and pollen of Bottle Brush Grass, *Hystrix*.

Thus two communities are represented, one salt marsh, the other heath and coniferous trees. The latter association is typical of the later stages of an interglacial period. It is clearly of Pleistocene age, and Watts considers it might well belong to the Cromerian or Hoxnian Interglacial.

The abundance of molluscan vegetable feeders such as *Rissoa*, predators such as *Nassa*, and shallow water dwellers like *Hydrobia*, but the absence of typical shore dwellers like *Patella*, indicate a sea depth in the laminarian zone, not more than 15 fathoms, for the deposition of the blue shelly clay (Bell, 1886).

However, Clement Reid (1890) argued cogently in favour of a more probable depth of about 40 fathoms, with a sea level standing some 300–400 feet above that of the present day. This would have turned Cornwall into a scattered archipelago comparable with the present-day Scilly Isles, with submerged shoals reaching the laminarian zone.

Reid found the land below about 400 feet planed and smoothed by the Atlantic waves. This would fit in with the St Agnes Beacon sands and clays (q.v.) at exactly the level needed, a conclusion supported by the petrological similarity of the sands of St Erth and St Agnes (Milner, 1922).

Reid correlated the St Erth deposits with similar small and scattered fossiliferous strata in northern France, assigned by French geologists to an Astian to Plaisancian date in the Pliocene, approximating to the Coralline Crag of East Anglia. However, there is great difficulty in dating this isolated clay containing so many species not known from elsewhere, and because facies plays so great a part.

The fauna of the St Erth Clay has been thought to indicate a climate warmer than at present and comparable with the present Mediterranean. However, Bate and Rowe (1881: *Brit. Assoc. Rept*, 199) observed that crustacea from the deeper waters of the entrance to the English Channel, dredged between Falmouth and Plymouth, especially off the Dodman, included species common in the Mediterranean, as well as others more at home in the Arctic.

The St Erth foraminifera have not been revised since they were originally described by Millett, whose specimens are preserved in the British Museum (Natural History). I have to thank the keeper of Zoology and Dr C. G. Adams for permission to study them.

Millett likened the fauna to that of shallow seas in a sub-tropical climate and its affinities with the Italian Tertiaries rather than the Crags of England. But I find that the great majority of the foraminifera, about 150 species, some 85 per cent of Millett's recorded list, are known to me in the Recent and Post-glacial British faunas.

A number of oddities present in Millett's collection as mostly single specimens are best put in a suspense account and regarded as strays unless they can be confirmed by fresh material. However, one striking species is represented by many specimens of a large form identified by Millett as *Rotalia punctato-granosa* Seguenza and confirmed in another specimen of the clay.

Of the important indicator *Gümbelina globulosa* there are three specimens; this is a well-known species derived from the Chalk.

From a small sample of the clay kindly made available to me in 1941 by Mr A. G. Brighton, from the Herries Collection in the Sedgwick Museum, Cambridge, and authenticated as from St Erth by the abundance of *Faujasina carinata*, I have obtained about 72 species, including a dozen other than those in Millett's lists, and all well known to me from the British area; and one species of *Heronallenia* that I regard as new to science.

Two of Millett's records, confirmed in other samples, are of special interest, *Bolivina robusta* Brady and *Faujasina carinata* d'Orbigny; both occur plentifully at St Erth. *Bolivina robusta* is known almost world-wide from Recent and Tertiary deposits back to the Eocene, but it seems to be recorded from the British area only in Recent material off Plymouth and round the Eddystone Lighthouse, 50 miles to the east of St Erth, in about 30 fathoms. *Faujasina carinata* was described in 1839 erroneously as from the Chalk of Maestricht. It was next recorded by Millett from St Erth in 1887, the only known locality where it is abundant. Since then it has been significantly recorded from England by Macfadyen, from the Corton Sands in Suffolk (1932: *Geol. Mag.* 487), and, as *F. subrotunda*, by Funnell from the Lower Icenian at Ludham, Norfolk (1961: *Trans. Norfolk and Norwich Nat. Soc.* **19,** 347); from the Normandy coast by Roger and Freneix (1946: *Bull. Soc. geol. France* ser. 5, **16,** 103); and from the Netherlands by ten Dam and Reinhold (1941: *Geol. Stichting Mededeel.* (C) **V,** No. 1, 55). *F. carinata* is uniquely promising for correlation and dating if its range can be established elsewhere.

My provisional conclusion is that, omitting some of the rare oddities claimed, the St Erth foraminifera might well have lived in the English Channel under no warmer conditions than those of the present day. But this does not explain the presence of the new species that are almost peculiar to

St Erth, or of *Faujasina*, not now known as living in the British seas or elsewhere. For comparison the Swansea Docks foraminiferal fauna of Post-glacial (Boreal) age contains 19 per cent of the indigenous species of typically warmer-water habitat (Macfadyen, 1942: *Geol. Mag.* 140).

Professor G. F. Mitchell re-exposed the fossiliferous clay at St Erth in 1966 and the results are in process of description.

1885, Wood; 1886, Kendall and Bell; 1887, Whitley; 1887–1902, Millett; 1890, Reid; 1898, Bell; 1907, Reid and Flett; 1914–1924, Harmer; 1922, Milner; 1923, Boswell; 1964, Barton; 1965, Mitchell.

Stourscombe Quarry, 1 mile SE of Launceston

SX(20)344839	186	337	17NW (G)	P 1965

Two small disused quarries in the Upper Devonian Stourscombe Beds, a group of nodular and thin bedded cherts and slates, with a rich and well-preserved fauna of ammonoids (clymenids) and trilobites largely found in the nodules.

Two faunal divisions can be made out in the westerly quarry, the upper characterised by *Wocklumeria*, and the lower characterised by *Parawocklumeria* but *Wocklumeria* absent.

Thirty-seven species or varieties of ammonoids were identified from here, of which 10 were described as new, with this site as their type locality; and seven species of trilobites, as follows.

Ammonoids: *Gonioclymenia* (2 species), *Wocklumeria sphaeroides*, *Epiwocklumeria* (1), *Postglatziella* (1), *Kosmoclymenia* (6), *Cyrtoclymenia* (3), *Cymaclymenia* (7), *Parawocklumeria* (3), *Kenseyoceras* (3), *Imitoceras* (5), *Discoclymenia* (3), *Sporadoceras* (2).
Trilobites: *Phacops* (6), *Chaunoproetus* (1).

The fauna correlates the beds closely with the subzone of *P. paradoxa* (Wedekind), which forms the upper part of the *Wocklumeria* Zone, the highest part of the Upper Devonian (Famennian) of the Rhineland.

1960, Selwood; 1963a, House.

Trebetherick Point, Padstow Bay

SW(10)925779   185   336   18SE   S, A, R
                               (G)   1965

A coast exposure of raised beach and later deposits, the best and probably the most important Pleistocene occurrence in Cornwall. It was first described in detail by Ussher in 1879, in his separately published paper.

The deposits rest on a wave-cut platform at 10 feet above high-water mark, backed by an ancient cliff in the country rock of purple and green slates. These are now known to be of Frasnian, Upper Devonian age, from the occurrence of the goniatite *Gephuroceras* [*Manticoceras*].

The very variable sections reach a maximum thickness of about 50 feet. They include parts of the following succession (generalised after Arkell, 1943).

Recent ⎰ ? Sub-Boreal, Bronze Age	1½–6 ft	Blown sands with many land-snail shells; at base a kitchen midden with mussel shells, etc., and flint artifacts, mainly flakes of Mesolithic, Tardenois style.
Atlantic	0–1 ft	Submerged forest and old soil. Late Mesolithic or early Neolithic.
Würm	0–1 ft	Pebbly bed and Younger Head (solifluction).
Eemian Interglacial	0–12 ft	Trebetherick Boulder Gravel; boulders up to one foot in diameter.
Pleistocene Riss	0–6 ft	Main Head (solifluction).
Hoxnian Interglacial	3–45 ft	⎰ False bedded blown sand and sand rock, with small pebbles and shell fragments. 10-ft beach platform with raised beach conglomerate and occasional huge rounded greenstone boulders up to 6 ft in diameter —ice-borne erratics.

The above correlation, after Arkell, is tentative.

A submerged forest lies at six feet below high-water mark in the adjacent Daymer Bay. It is a peaty deposit, which has yielded oak, yew, hazel, landshells and horns of Red Deer.

1858, Henwood; 1879, Ussher; 1943, Arkell; 1956, 1963a, House; 1964, Barton.

Tregargus Quarry, 1 mile NNE of St Stephen-in-Brannel

SW(10)949540	190	347	50NW (G)	Ig, M, Met IGC, 1965

A very deep, actively working (1962) quarry of chinastone or 'Cornish Stone', complete with its primitive grinding mill driven by stream water (Fig. 2).

Fig. 2. Tregargus Quarry, St Stephen-in-Brannel, Cornwall; looking north-west (1905). A chinastone quarry showing the vertical jointing of the granite. (Crown copyright Geological Survey photograph A.213. Reproduced by permission of The Controller, H.M. Stationery Office.)

Quarries in the Cornish Stone are usually very deep, with sheer sides formed by the vertical joints of which there are two sets nearly at right angles, and there is also a horizontal set. This jointing system is responsible for the general appearance of these quarries. The joints may contain quartz, tourmaline and fluorite, but they are generally not more than a fraction of an inch thick, only widening out in the more altered sections.

In the St Austell Granite mass of Armorican (Permo-Carboniferous) date four types have been recognised (after Exley, 1958).

1. Porphyritic biotite–muscovite granite, outcropping over some 16 square miles on the east side of the mass.

2. Porphyritic early lithionite granite, outcropping over 17·5 square miles, mainly in the centre, a small part lying on the west side.

3. Non-porphyritic lithionite and gilbertite granite, outcropping over 2·2 square miles in the west centre.

4. Non-porphyritic fluorite granite (Cornish Stone) outcropping as two enclaves within No. 3, the larger of about 0·3 square mile, west of Nanpean, and the smaller of about 0·1 square mile, at Tregargus.

The compositions of these four types of granite conform to the principles of magmatic differentiation. Granite No. 1 was intruded first and appears to be separated from No. 2 by an intrusive junction. Granites 2, 3 and 4 form one intrusion, and they crystallised in that order, the Cornish Stone being the last to consolidate from a residual magma in which very little ferromagnesian material was left. It is on this lack that the value of the stone to the potter depends.

The Cornish Stone or fluorite granite is a white or pale-buff medium granite altered by hydration and the introduction of lithium and fluorine. There are four varieties, known as Hard Purple, Mild Purple, Hard White and Soft White; these have been progressively altered, in that order, by volatiles of the original magma.

The alteration took place in stages (after Exley, 1958).

1. Late magmatic processes:
   (*a*) tourmalinisation by boron,
   (*b*) greisening by fluorine—the alteration of potash feldspar to secondary mica and quartz, with enrichment with K, Sr, Ba, Ti, V.
2. Post-magmatic processes; after consolidation and jointing of the outer granite, pressure in the still unconsolidated magma injected solutions which caused:
   (*c*) tourmalinisation, secondary tourmaline lining the joints,
   (*d*) kaolinisation, by siliceous hydrothermal solutions devoid of B, F, CO_2, S and Cl.

The Hard Purple is the least altered; the Mild Purple and the Hard White show the effects of increasing greisening; and the Soft White is due to a

37

final kaolinization. A fifth type, Buff Stone, occurs near the ground surface, and the alteration of the feldspars and the coloration appear to be due to downward percolation of acid moorland waters.

The analysis of a sample from Tregargus mill is given by Richardson (1923) as:

	(%, by wt.)
Quartz	33·1
Orthoclase (potash feldspar)	32·9
Plagioclase (lime-soda feldspar)	17·4
Colourless micas (gilbertite)	11·2
Topaz	2·8
Fluorite	1·6
Tourmaline	0·6
Iron ores	0·2
Apatite	0·2

Cornish Stone thus contains four principal minerals. In addition to the above there have been recorded kaolinite and an economically troublesome clay mineral, probably montmorillonite.

Other minerals recorded from Tregargus Quarry include autunite, lithia mica, and löllingite.

Cornish Stone is finely ground for mixing with china clay (kaolin) and other materials. A high-grade glazing material, it vitrifies at about 1200°C to a colourless glass. It was in use in the Staffordshire potteries by 1759.

China clay and china stone were first found by William Cookworthy, a Cornishman and Quaker apothecary, about 1750, in Tregoning Hill, some eight miles east of Penzance. Here they were commonly in use for mending the tin furnaces and the hearths of steam engines; perhaps this suggested to him their potentiality for making porcelain. Later he found great quantities of these materials near St Stephen, where they were more easily wrought and also gave him a better product. Cookworthy was the first manufacturer of porcelain similar to the Chinese from English materials, and had established his first china factory at Plymouth by 1768 (Worth, 1876).

1876, Worth; 1909, Ussher, Barrow and MacAlister; 1914, Howe; 1923, Richardson; 1929, Leech; 1958, Exley; 1961, Keeling.

DEVON

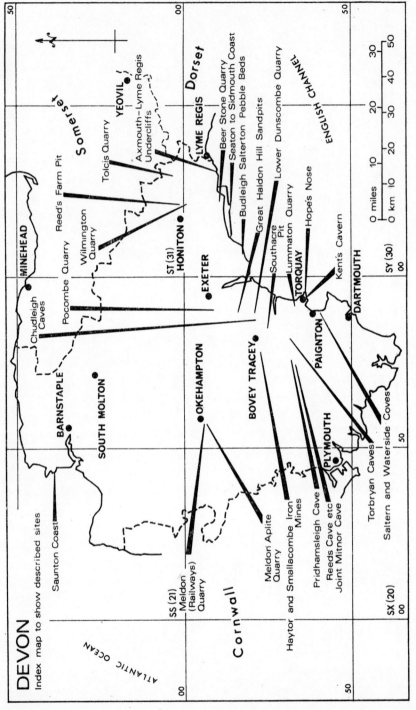

DEVON
Index map to show described sites

ATLANTIC OCEAN

Saunton Coast

MINEHEAD

Somerset

Pocombe Quarry

Reed's Farm Pit

Wilmington Quarry

Tolcis Quarry

YEOVIL

Axmouth–Lyme Regis Undercliffs

Dorset

LYME REGIS

Beer Stone Quarry

Seaton to Sidmouth Coast

Budleigh Salterton Pebble Beds

Great Haldon Hill Sandpits

Lower Dunscombe Quarry

Hopes Nose

ENGLISH CHANNEL

Southacre Pit

Lummaton Quarry

Kent's Cavern

Chudleigh Caves

BARNSTAPLE

SOUTH MOLTON

OKEHAMPTON

BOVEY TRACEY

EXETER

ST (31)
HONITON

Meldon Aplite Quarry

Meldon (Railways) Quarry

SS (21)

Cornwall

Haytor and Smallacombe Iron Mines

Pridhamsleigh Cave

Reeds Cave etc
Joint Mitnor Cave

PLYMOUTH

Torbryan Caves

Saltern and Waterside Coves

TORQUAY

PAIGNTON

DARTMOUTH

SX (20)
00

SY (30)
00

50

0 miles   10

0 km   10   20   30

10   20   30   40   50

50

00

50

00

50

Fig. 3

Axmouth–Lyme Regis Undercliffs, 4 miles SW of Lyme Regis

National Grid Reference	One-inch Sheet Pop. and 7th ed. Ordnance Survey	One-inch Sheet Geol. Survey	Six-inch Sheet Ordnance Survey	Main Site Interests
SY(30)3090	177	326	83SE, 84NE, 84SE, 84SW (–)	T, S, P, U PSD, IGC, 1966

A thickly wooded coastal strip of very broken country, this National Nature Reserve is five miles long and from 200 to 700 yards in width; the inland-bounding cliffs, mainly capped by white chalk, range from 300 to 500 feet in height, broken at intervals by deep ravines or 'goyles'. 'The great landslip alone would stamp any coast with a distinctive individuality, and the contrasts afforded by the juxtaposition of red Trias marl, yellow Greensand, and white Chalk realise a field of rare geological interest and a picture of surpassing beauty' (Rowe and Sherborn, 1903). This picture is now largely masked by vegetation.

For geologists the Reserve has two major interests: the classic landslips to which the terrain owes its form, one of them 'the most remarkable example ever recorded to have occurred within this island' and the best documented; and the fine natural sections in Cretaceous, Lias, Rhaetic and Trias strata, some of them highly fossiliferous, and unmatched elsewhere in Devon.

The strata are first described, to give a background to the landslips, but the succession is interrupted, no one section exposing the whole. The Reserve shows a considerable unconformity, the Cretaceous overstepping an eroded surface of lower strata progressively from east to west. As far as about Bindon Cliffs it rests upon Lower Lias; thence for a little way upon the Rhaetic; and finally up to the River Axe directly upon the Trias. Unfortunately, there is no large-scale geological map.

In general, the unslipped Cretaceous strata are nearly flat-lying, although near the coast there is a slight seaward dip of about 5°. However, the Lias and underlying strata have a slight south-easterly dip. At Pinhay (or Pinney) is a fault downthrowing about 40 feet to the west. East of Pinhay are several small folds exposed along the shore in the Lower Lias (see Fig. 4).

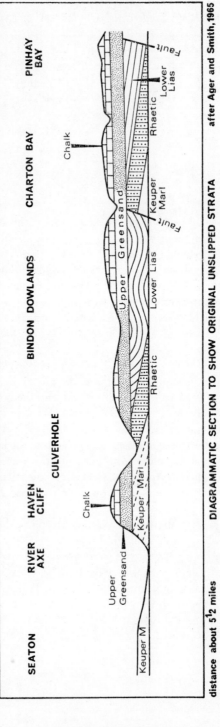

after Ager and Smith, 1965

DIAGRAMMATIC SECTION TO SHOW ORIGINAL UNSLIPPED STRATA

Fig. 4. Seaton–Pinhay Bay: diagrammatic coast section to show the original unslipped strata; Chalk and Upper Greensand unconformably overlie Lower Lias, Rhaetic and Keuper (Trias).

The complete section is as follows (after the authorities named below).

	Recent to Tertiary			Clay with flints (discontinuous) Zone	
Upper Cretaceous	Upper Chalk	Senonian	50 ft	*cortestudinarium*	Hard nodular chalk, with prominent bands of black flint; siliceous and phosphatic nodules; 12 ft of 'Chalk Rock' at base; *planus* zone with many small fossils.
		Turonian	40 ft	*planus*	
	Middle Chalk	Turonian	71 ft	*lata* (=*gracilis*)	Hard and soft nodular chalk with flints; *Terebratulina lata* abundant and a rich fauna of echinoids.
			60 ft	*labiatus* (=*cuvieri*)	Hard nodular chalk with scattered flints, and a rich fauna including ammonites such as *Neocardioceras* and *Metoicoceras*.
	Lower Chalk	Cenomanian	up to 3 ft	*subglobosus varians*	Arenaceous limestones with glauconite and conglomerate, in places with well-preserved ammonites, as *Mantelliceras*, *Hyphoplites*, also *Orbirhynchia wiesti* and *O. multicostata*; and, at the base, *Scaphites equalis* and *Ceriopora ramulosa*.
Lower Cretaceous	Upper Greensand	Upper Albian	? c. 90 ft	*dispar* (=*Pecten asper*)	Chert Beds yellowish sandstones with chert and sands, in part glauconitic; with *Exogyra conica* etc., also with the holotype, of *Salenia dux*, and with *Nucleolites lacunosus* and *Ochetes morrisi* (Wright, 1967: *Proc. Geol. Ass.* **78**, 19).
			? c. 65 ft	*inflatum* (=*rostrata*)	Greensands, soft Foxmould overlying concretionary Cowstones, with many echinoids and other fossils.
	Gault	Middle Albian	c. 25 ft	*dentatus* (=*interruptus*)	Dark loamy and glauconitic silt, base sometimes conglomeratic; fossils scarce.

–––––––––––––––––––––– Major unconformity ––––––––––––––––––––––

Lower Jurassic	Lower Lias c. 100 ft		? ft	*obtusum*	Black marl.
			13 ft	*semicostatum*	Blue Lias; alternating calcareous shales and thin limestones; with ammonites, including giant Arietitids, *Gryphaea arcuata*, *Calcirhynchia calcarea*, crinoids, etc.
			37 ft	*bucklandi*	
			25 ft	*angulata*	
			20 ft	*planorbis*	
			c. 10 ft	Pre-*planorbis* Beds	

Upper Rhaetic	26 ft Langport Beds, White Lias: Sun Bed (2 in) at top, riddled with U-shaped borings, overlying white limestone, fossiliferous.
	6 ft Cotham Beds: Mostly pale clays, shales and cream-coloured limestone, including Cotham Marble of false-Cotham type up to 2½ in thick; Ostracods (*Darwinula*), fish-scales, *Cardium cloacinum*; *Estheria* Bed.

(*continued overleaf*)

Lower Rhaetic	17 ft	Westbury Beds (*contorta* Beds): Black selenitic shales with some thin limestones; many fossils, *Rhaetavicula contorta, Schizodus, Chlamys valoniensis*, fish remains; Basal Bone Bed (2 in) with fish remains.
	10 ft	Tea-Green Marls: pale grey or greenish marls with eroded upper surface.
	20 ft	Pale green and creamy marls with hard bands of marly limestone and dark green and black clays.
Trias; Keuper	15 ft	Grey, green and red marls, with cuboidal jointing, and including a hard layer of pale grey or buff-coloured banded marl with dark clayey streaks.
	More than 1000 ft	Red Marls: very thick, with cuboidal jointing, and with greenish-grey bands and patches especially in the upper part; gypsum also occurs and geodes lined with calcite.

The best sections are found as follows.

Upper Cretaceous: Chalk. Pinhay Cliff, SY315410; Chapel Rock, 312906; Dowlands, 285896; with many blocks of Cenomanian limestone on the shore between Humble Point and Pinhay Bay.

Lower Cretaceous: Upper Greensand. Pinhay Cliff 315410; Dowlands, 285896; Gault. Culverhole, 275893.

Jurassic: Lower Lias. Pinhay Bay, 320908 to Seven Rock Point, 327909; Culverhole, 275893.

Upper Rhaetic: White Lias. Pinhay Bay, 320908 ('the best exposure in the county'); and Charton Bay, 300900.

Main Rhaetic sequence. Culverhole, 275893.

Trias. Haven Cliff, 265897, and back of Charton Bay, 300900.

LANDSLIPPING

Landslipping has taken place here, from Ware Cliffs (SY325915) in the east to Bindon Cliff (275896) in the west, from time immemorial and still continues, for movements in the Whitlands sector, and to a smaller extent in the Bindon sector, were active in 1961 and 1962. If, as is the tradition, the slipped Chapel Rock (312906) at Pinhay was used as a place of worship during the religious persecutions in the reign of Mary Tudor (1553–58), its date of slipping must have been earlier. Other slips have been more definitely recorded, at Whitlands (305903) about 1775; near Dowlands (285895) about 1790–1800; from Pinhay to Ware in 1828; the great slip between Bindon and Dowlands in 1839: at Whitlands again in 1840; and now in 1961–62.

The activating cause is abnormally heavy rainfall over a period, which leads to the face of the cliff, to a variable depth, becoming unstable. Such high rainfalls here are known in 1774, 1839 and, very locally round Pinhay, in 1960–61. At Pinhay the total for 1960 was 54·88 inches as compared with an

44

average of about 35 inches. The permeable Chalk overlies porous Greensand, part of which is the soft loose sand of the Foxmould; that in turn rests upon the clayey Gault and the slightly inclined impervious Lower Lias clays, or, in the more westerly part of the Reserve, on the Rhaetic or Trias marls. Prolonged heavy rain is stated partly to wash out some of the loose Foxmould, and partly to convert it into a quicksand. The overlying Chalk and the Chert Beds of the Greensand, whose weight has been greatly increased by saturation with the rainwater, have their foundations sapped and weakened; they founder and tend to slide seawards over the water-lubricated clays below, leaving a confused ruin of mounds and hollows. This was the explanation given by Buckland and Conybeare in 1840.

The most famous slip was that between Bindon and Dowlands on Christmas Day 1839. Two eminent geologists, Buckland and Conybeare, were fortunately in the neighbourhood and at once caused a detailed survey to be made, collected the evidence of eye-witnesses, and published accounts in 1840, one of them with detailed illustrations based on the survey.

This slip was completed from first to last within about 48 hours, the movements having been gradual and continuous. 'There was no displacement at the farmhouses of Dowlands and Bendon (*sic*), each within three quarters of a mile of the chasm; they had not a tile displaced nor a pane of glass shivered.' The survey showed that some 20 acres of arable land subsided in rifted fragments into a great chasm; and nearly 15 acres more were cut off as an insulated fragment 'Goat Island' beyond the chasm, having subsided about 60 feet; but that 'island' was also rifted by broad and deep fissures and subsidences of up to 10 feet. The chasm, mainly floored by Chalk, which significantly dipped fairly strongly landwards, was measured as about 4000 feet long, from 200 to 400 feet wide and from 130 to 210 feet deep, measured from the top of the old cliff; and so it and the 'island' remain, little altered, to-day, though now much obscured by woodland. The subsided rocks were calculated at not much less than 150 million cubic feet, weighing nearly eight million tons.

Simultaneously with the subsidence a reef, also nearly 4000 feet long, rose off-shore, some 300–500 feet from the original high-tide mark. The surface of this reef had apparently lain some 10 feet below low-tide level, and the highest part of the reef reached about 40 feet above high-tide level, giving a total rise, including the tidal range, of some 60 feet in all. The raised material was Greensand, cherty sandstones over Foxmould, with layers of Cowstones; and the beds were tilted to dip 30–40° landwards. This reef was soon eroded by the sea.

A similar off-shore reef arose in connection with the Whitlands landslip of 8 February, 1840, when semi-fluid Foxmould was observed being squirted upwards as the reef was rising.

In March 1961, again, a subsidence of some 20 feet in a section of the cliff top at Whitlands coincided with a rise of some 10 feet in a 1200-foot stretch of the modern beach below; this ridge included Lias.

This striking feature of uprising, on a very considerable scale, of off-shore reefs or beaches at the same time as the downslipping of the cliffs immediately behind them, with other features, suggests that rotational shear slipping plays a controlling part in the mechanism of these movements, the reefs, etc., forming the uprising toe of the shear. But it is impossible to ignore the substance of the old explanation, which for a century has been generally accepted as satisfactory. Provisionally it is here suggested that the strata of the 1839 slip which disappeared underground to form the chasm took part in a great rotational shear slip, whose toe was seen in the off-shore raised reef; but that 'Goat Island' suffered a simple vertical displacement without appreciable slipping. Thus elements of both explanations were involved. (See Fig. 5.)

1840, Conybeare and Buckland (landslip survey); 1900, Jukes-Browne (Lower Cretaceous); 1903, Rowe and Sherborn (Upper Cretaceous); 1906c, Richardson (Rhaetic); 1924, Lang (Lias); 1940, Arber (landslip); 1948, Dewey (stratigraphy); 1960, Hallam (White Lias); 1962, Smith and Drummond; 1965, Ager and Smith; 1965, Smith; 1969, Torrens (No. 1).

Note. Geological work is not lightly to be undertaken in this Reserve, being both difficult and potentially dangerous; it should not be attempted by one person alone.

The ground is covered with thickets of vegetation, which restrict the view and make it very difficult to locate one's position on the map. As well as the dangers of being trapped on the shore by the tide, and crumbling cliffs, there are other hazards peculiar to the place, such as many vertical or precipitous slopes, sometimes concealed by vegetation. In March 1961 and again in 1962 slips made Pinhay Bay inaccessible by the old steps near the fault, and it could then only be reached by walking along the beach. In 1963 in the undercliff was seen an open fissure in the Chalk, two feet wide and plumbed to a depth of 11 feet; in places it was bridged over by turf and growing vegetation which there made it invisible. The Reserve is isolated, difficult of access, and often deserted, so that help could not be quickly obtained in case of need. Rowe and Sherborn (1903) give warning of the not uncommon presence of adders.

Beer Stone Quarry, 1 mile W of Beer, 2 miles WSW of Seaton

SY(30)215894	177	326	83SW	S
			(-)	1966

A very large quarry, with face some 80 feet high, where the famous Beer Freestone has been worked for centuries. The strata dip about 4° east. The succession shows (after Jukes-Browne, 1903):

46

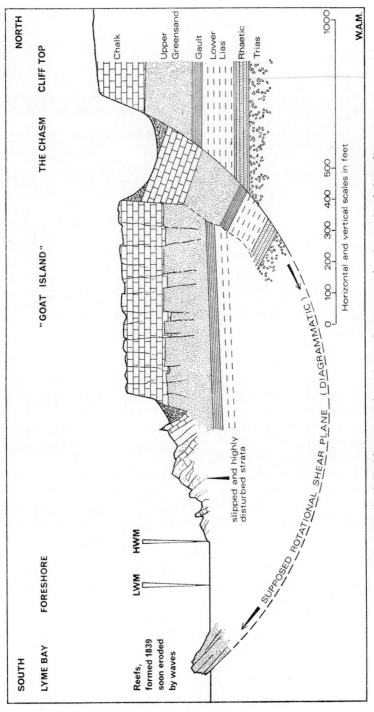

Fig. 5. Dowlands landslip: true scale section to show supposed main rotational shear slip.

Zone

lata (=*gracilis*) 50 ft	c. 9 ft Soil over broken chalk with few flints, and one conspicuous layer of black flints.
	24 ft Chalk with many layers of flints.
	6 ft Chalk with few flints.
	8 ft Soft white chalk with scattered flints and some flint layers.
	2½ ft Soft smooth chalk passing down into rough nodular chalk; with *Echinoconus subrotundus, Orbirhynchia* [*Rhynchonella*] *cuvieri.*
labiatus (=*cuvieri*) 30 ft	2 ft Very hard yellowish limestone, passing down into hard shelly brecciated chalk; *O. cuvieri, Gibbithyris* [*Terebratula*] *semiglobosa.*
	11½ ft Rough yellowish nodular chalk, largely of comminuted shell fragments; *Inoceramus mytiloides, O. cuvieri.*
	3 ft Nodular chalk, the 'roof course' or 'cockly bed' of the quarrymen.
	13 ft Beer Stone: crystalline, granular limestone, the upper 8 ft thick-bedded, the lower 5 ft harder and thinner-bedded; mainly composed of the comminuted shells of *Inoceramus;* identifiable fossils are scarce, with *Inoceramus mytiloides, O. cuvieri, G. semiglobosa, Conulus castanea, Hemiaster minimus, Nautilus, Ptychodus mammilaris, Lamna appendiculata.*

Upper Cretaceous / Middle Chalk / Turonian (bracket label to the left)

This section rests upon Cenomanian Limestone, the upper surface of which is exposed in the floor of the quarry, where there have been seen about five feet of hard rough white chalk with quartz and glauconite at the base.

The Beer Freestone has been worked here both by open quarry and by mining in adits. The adits form a large cavern under the hill to the north, and De la Beche (1826) wrote that it afforded a miniature representation of the Maestricht quarries. The nearly even roof is supported by large square pillars. In 1825 the cavern extended about 170 yards in from the entrance. By 1965 it was stated to be about a quarter of a mile in extent, but the adits were no longer worked.

The stone is sawn out *in situ*, where it is fairly soft, into blocks of up to eight tons weight; it hardens on exposure to the air, and has been tested to bear a stress of 2106 pounds per square inch. Because of its uniform texture and close grain it is admirably suited for carving, especially for indoor decoration, but it is also used for outside work. Possibly worked since the Romans, it was used in Exeter Cathedral in Norman times and later, the cathedral archives from 1427 to 1434 referring to this quarry. It was also used in the churches of Ottery St Mary, Honiton and Axminster, and in the Roman Catholic cathedral at Norwich.

Chalk from the *lata* Zone has been burnt here for lime.

The name 'Beer' appears to come from the Norse *byr*, meaning an abode or farmstead; compare the English words 'byre' and 'bower'.

In 1822 it was written that the village of Beer was 'notorious for the daring community of smugglers who inhabit it and form as it were a peculiar race distinguished by many singular customs' (Conybeare and Phillips).

48

1826b, De la Beche; 1903, Jukes-Browne; 1911, Woodward and Ussher; 1965, Ager and Smith.

Budleigh Salterton Pebble Bed, 3 miles E of Exmouth

SY(30)050813	176	339	103NE	S, P
			(–)	1962

A fine red coastal cliff rising from 100 feet in the western outskirts of the town to over 400 feet a mile further west. Beneath a little irregular gravel the nearly vertical cliff face exposes a dip-section dipping about 5° NNE, as follows:

Superficial gravels, partly derived from weathering of the exposed Middle Bunter Pebble Bed

Trias ⎰ Upper Bunter: c. 100 ft (seen). Coarse, red, current-bedded sandstones.
Trias ⎨ Middle Bunter: 70–80 ft Budleigh Salterton Pebble Bed.
Trias ⎩ (Lower Bunter: absent)
Permian   c. 500 ft Red marls, mottled with small green spots; in some places these contain vanadiferous nodules.

The interest centres in the sparingly fossiliferous pebbles of the Budleigh Salterton Pebble Bed, which rises gently from beneath the shingle beach near the town and passes out of the top of the cliffs one mile to the west. Locally known as 'Budleigh Pebbles' or 'Popples', these are largely of flattened ovoid form, of all sizes but mostly large, some reaching a foot long. Some 99 per cent are of sandstone or quartzite, of several varieties and colours, light grey to reddish or purple, or speckled. There are some pebbles of tourmalinised grit, and of quartz; perhaps one per cent are igneous pebbles, of black volcanic rocks, with a few of decomposed granite and porphyry. A strongly current-bedded formation, the pebbles are set in a dark red, grey or buff matrix of sand, and there are irregular intercalations of sand as lenses and thin seams. Three small faults are seen in the cliff face, with throws of about three, eight and 10 feet, respectively.

The broken-up pebbles were used as the best source of local road metal during at least the nineteenth century.

The Pebble Bed appears to have been first mentioned by Polwhele, the historian of Devon, in 1797; he was impressed by the wasting of the cliffs, and he described the pebbles.

The first description by a geologist is probably that of the Geneva-born de Luc,* who visited it in 1806. He gave a clear and detailed account of the bed and its location, noting the overlying gravel of the same kind of pebbles that are found on the beach, and the faults; found 'the impression of a large bivalve deeply channelled' in one of the siliceous pebbles; and discussed the cliff erosion, the pebbles remaining to form the present beach.

H. J. Carter (in Davidson, 1881) described the bed and recorded that in 1835 he had picked up a pebble full of impressions of a small *Orthis* later known as *O. budleighensis*. The proportion of fossiliferous pebbles is very small (Shrubsole found only two in a fortnight), but Vicary made a considerable collection, from which 40 forms were described or recorded by Salter in 1864 comprising:

> Algae, 2 species: *Vexillum* ?, and *Daedalus*?
> Trilobites, 4: *Homalonotus* (2), *Calymene*, *Phacops*.
> Brachiopods, 14: *Lingula* (4), *Orthis* (3), *Spirifer* (2), *Rhynchonella* (2).
> Bivalves, 14 plus: *Modiolopsis* (5).
> Gastropods, 2: *Bellerophon* (not described).
> Cephalopods, 2 or 3: *Orthoceras* (not described).

There were included 19 new species, with three new genera, the bivalves *Pseudaxinus* and *Hippomya*, and a form described as a crustacean, *Myocaria*, may really be one side of a bivalve.

Salter found this fauna quite new to British geology, but parallelled in the 'Lower Silurian' (now Ordovician) of the *Grès de May* and the *Grès Armoricain* of Normandy. Other of the fossils suggested a Devonian date. It was later recognised that each fossiliferous pebble contained either Ordovician or Devonian fossils; there was no mixing of the faunas. In explanation of this is the land barrier in Ordovician and Devonian times, postulated by Barrande, a sector of which stretched across the present Devon and Cornwall. This barrier separated the Scandinavian province, which included almost the whole of Britain, from the Mid-European province, which included northern France, and was the provenance of the Budleigh fossils.

Later palaeontologists, Davidson, Edgell and Tromelin, increased the new species described to 41. Davidson (1881) monographed 45 species of the brachiopods and classed them as follows.

---

* Jean André de Luc, elected F.R.S. in 1773, toured south-west England by chaise, on horseback and on foot, often alone or with but one servant, in the late eighteenth and earliest years of the nineteenth century up to 1806, when he was aged seventy-nine. A follower of Werner and his geological sect styled Neptunists, at a time when the foundations of the science were being argued, his theoretical views are largely unacceptable; no doubt owing to these and to his polemics against Hutton and Playfair, his geological work seems to have been rejected *in toto*. Nevertheless he had a most acute, ingenious and inquiring mind (witness, his previous meteorological discoveries and inventions) and many of his early descriptions of the geology of sites in the West Country are first-rate and of enduring importance on both scientific and historical grounds.

Lower Devonian: 33 species, the commonest being *Spirifer verneuilii*, followed by *Rhynchonella inaurita* and *Productus vicaryi*, with *Athyris* (3), *Rhynchonella* (9), *Orthis* (3), *Strophonema* (8), *Discina* (3).

Ordovician
- Caradocian, *Grès de May:* 8 species: commonest *Orthis budleighensis* and 4 other *Orthis* species.
- Basal Llandeilo, *Grès Armoricain:* 4 species (*Lingula lesueuri*, *L. ? hawkei*, *L. salteri* and *Dinobolus brimonti*).

Of Davidson's list of Devonian species only *Spirifer verneuilii* and *S. speciosa* were known to occur *in situ* in Britain. Much of the remainder of the fauna has been incompletely identified, some forms only generically, with many queries. To obtain an accurate list would require a re-investigation of the material, of some 80 species or more.

It is generally agreed that the origin of the fossiliferous Budleigh pebbles, which were deposited up to some 20 miles north of Budleigh Salterton, must be sought from the south, in the Mid-European province of those days, and that they were transported by a large north-flowing river. A mineral species, staurolite, found by H. H. Thomas in the sand of the bed supports a southern origin.

The nearest point on the Cherbourg Peninsula lies 87 miles to the southeast, and Brittany lies 121 miles due south. The transport of the large Budleigh pebbles a matter of 100 miles by river may seem difficult to envisage, but present-day parallels are known, e.g. the River Po in Italy and the Helmand in Afghanistan, and testify to the possibility (Jukes-Browne, 1911, pp. 232-4).

An earlier suggestion that the pebbles might have travelled a lesser distance, from a now-vanished Armorican land mass located somewhere within the present English Channel, seems to be negatived by recent work on the geology of the English Channel (King, 1954).

1797, Polwhele (Vol. 1, p. 61); 1811, de Luc (Vol. 3, pp. 73–78); 1864, Salter; 1864, Vicary and Salter; 1870, 1881, Davidson; 1874, Edgell; 1877, Tromelin; 1877, 1913, Ussher; 1902, Thomas; 1903, Shrubsole; 1911, Jukes-Browne; 1954, King; 1960, Stubblefield.

Chudleigh Caves, Chudleigh, 5 miles NNE of Newton Abbot

SX(20)865787	188	339	101SE	C, P
			(–)	1967

A series of caves in Middle Devonian limestone. They include:

Pixie's Hole, located in Chudleigh Rock on the side of the steep little valley of the Kate Brook; it is a vertical network with 870 feet of passages, and four openings, one of which, excavated by Buckland in 1823, yielded a prehistoric hearth with charcoal, pottery and flint knives; and also, under stalagmite, mammalian remains, including Hyaena, Deer and Bear.

Chudleigh Cavern, a simpler cave.

Cow Hole, a more usual linear cave, yielded remains of Bear and a few flint flakes, including one identified as an Azilian blade.

Another fissure yielded at least two species of Lemming, and remains of some 38 species of birds.

A Bat population, mainly the Greater Horseshoe Bat (*Rhinolophus ferrum-equinum*), is under study, and large-scale ringing experiments have been made on them since 1947.

1873a, Pengelly; 1949a, 1949b, Shaw; 1953, 1962, Cullingford, Ed.

Great Haldon Hill Sand Pits, 5½ miles SSW of Exeter

SX(20)897839	176	339	92SW	S, P
			(–)	1966

Capping the flat-topped Great and Little Haldon Hills up to a height of about 800 feet above O.D. are found Tertiary coarse flint gravels which overlie the Lower Cretaceous Haldon Beds, sandy representatives of the Upper Gault, and the most westerly outliers of the Cretaceous in England. They form a westerly extension of the Blackdown Beds 20 miles to the north-east.

The Haldon Hills were visited by de Luc in 1806. He found sands and sandstones with marine fossils, and recognised their close similarity to those of the Blackdown Hills, which he investigated, recording many marine fossils and describing the whetstone mining then active there.

The best available exposures now to be seen are in cuttings made along new forest roads, and the succession is somewhat as follows.

(GENERALISED ACCOUNT)

Supposed Bagshot Beds (uppermost Lower Eocene) 30–40 ft	Coarse gravels of large and partly worn chalk flints, greensand cherts, and Palaeozoic pebbles from the Permian breccia; with matrix and occasional seams of granitic sand and whitish clay. The sand contains abundant tourmaline and plentiful topaz. The beds are decalcified; the only fossils are derived Cretaceous forms such as *Echinocorys* preserved in flint.

52

WOODLANDS SECTION AT SX903842; JUKES-BROWNE, 1900

<table>
<tr><td rowspan="2">Lower Cretaceous</td><td rowspan="2">Haldon Beds: Upper Greensand; (Upper Albian) c. 65 ft</td><td>6–10 ft Fine yellowish sand, with large cherts.</td></tr>
<tr><td>20 ft Yellow and grey sand with layers of small quartz pebbles and a thin seam of grey micaceous clay.</td></tr>
<tr><td></td><td></td><td>1 ft Coarse pebbly sand with many shell fragments.</td></tr>
<tr><td></td><td></td><td>19 ft Fine grey glauconitic sand with siliceous concretions passing down into fine sand with some broken shells.</td></tr>
<tr><td></td><td></td><td>1 ft Grey glauconitic sandstone.</td></tr>
<tr><td></td><td></td><td>1½ ft Laminated grey and brown sand with broken shells.</td></tr>
<tr><td></td><td></td><td>6 ft Fine yellow-green glauconitic sand, with sandstone lumps passing down into</td></tr>
<tr><td></td><td></td><td>5 ft Greenish-grey glauconitic sand with lenticles of glauconitic sandstone with chalcedonised fossils.</td></tr>
<tr><td></td><td></td><td>2 ft Conglomerate of pebbles derived from the Permian breccia, resting upon uneven surface of</td></tr>
<tr><td>Permian</td><td></td><td>30–40 ft (seen). Red breccia.</td></tr>
</table>

Like the overlying tertiary gravels these Haldon sands contain abundant tourmaline and plentiful topaz.

Fossils, which seem to occur in local pockets, are not easily found. Although they have been known since the early nineteenth century the section was not described, and now even the site of the old pit in which most of them were found is uncertain, although it was near the Woodlands section given above.

From these Haldon Beds have been recorded some 122 species, including molluscs, especially bivalves, and polyzoa, and the foraminifer *Orbitolina*, and a remarkable fauna of some 22 species of corals. This, with compound corals resembling littoral forms, and thick-shelled mollusca, is suggestive of shoal water. The horizon of the shell-bed is thought to be in the upper part of the lowest 40 feet of the beds.

The corals, preserved in chalcedony, include a genus (*Haldonia*) and a dozen species described by Duncan (1879) as new. Some are as large as the simple tropical corals of the present day, and others are encrusting forms indicating the conditions of a fringing reef.

Unlike the corals of the Cambridge Greensand area, these from Haldon (they are not known from Blackdown) are the northern expression of the French and Central European equivalent deposits.

A radiometric age determination by the potassium–argon method on the glauconite has yielded a date 91 million years ago for the Upper Greensand of Telegraph Hill (SX910840) Haldon Hills (Evernden *et al.*, 1961: *Geochim. Cosmochim. Acta* **23,** 78; the strata are ascribed to the basal Cenomanian).

1811, de Luc; 1879, Duncan; 1882, Downes; 1900, Jukes-Browne; 1913, Ussher; 1923, Boswell.

Haytor and Smallacombe Iron Mines, 2½ miles WSW of Bovey Tracey

SX(20)775770	188	338	100SE (G)	M, Met 1965

Disused iron mines, formerly worked for magnetite, 'limonite' (brown hematite) and umber, with an interesting suite of associated minerals, in metamorphosed Culm Measures (Carboniferous) shales and sandstones, 500 yards east of the margin of the Dartmoor Granite. 'Lode-stones' from here were known in the sixteenth century or earlier.

At the Haytor Mine there were four beds of magnetite; three were exposed in the adit and were, respectively, 10, 14 and six feet thick (including four feet of waste), and the fourth bed, about three feet thick, outcropped about 300 yards to the north-east. The mine was worked opencast early in the nineteenth century, but later was exploited by an adit and levels.

The ore has been traced for over half a mile towards the south-east, to Smallacombe. It consisted of remarkably pure magnetite associated with actinolite, and among other minerals present were quartz, chalcedony, garnet, axinite, pyrite, some chalcopyrite, arsenical minerals, siderite, and some haytorite, a name given by Tripe in 1827 to pseudomorphs of quartz after crystallised datolite (a silicate of boron and calcium). The whole formed a typical metamorphic assemblage which had, however, been silicified at a later date.

At Smallacombe Mine 'limonite' and magnetite were both present. The former occurred in the form of irregular beds of nodules in decomposed slates and grit, and the fact that they were slightly manganiferous suggests that these nodules were probably some form of manganiferous clay-iron-stone in the shales, now highly oxidised. The magnetite, no doubt the south-easterly continuation of the Haytor Beds, occurred at a lower horizon, south-west of the 'limonite' beds; it was associated with a thick mass of greenstone, and with much garnet-rock and hornblende (actinolite).

There is no doubt that these deposits are metamorphic in origin, and the assemblages are characteristic of many similar occurrences elsewhere in Devon and Cornwall, where bedded sedimentary iron deposits and/or greenstones have been metamorphosed by the granite.

From 1858 to 1861 Haytor Iron Mine sold 5717 tons of iron ore, chiefly brown hematite, for £2601 (Collins, 1912).

A. W. G. Kingsbury

1828, Kingston; 1875, Foster; 1912, MacAlister; 1912, Collins; 1919, Cantrill, Sherlock and Dewey.

Hope's Nose, Torquay

SX(20)948635   188   350   116SE   P, R, S, T
                                (G)   PSD, 1963

A coastal promontory of Devonian strata with striking tectonic features and a Raised Beach.

The succession shows:

Devonian	Upper (Frasnian) to Upper Middle (Givetian)	Thin-bedded limestones, including two bands of tuff with *Maenioceras terebratum, Hypothyridina* [*Rhynchonella*] *cuboides* (*sensu stricto*) and *Stringocephalus burtini*.
	————————unconformity: erosion surface————————	
	Lower Middle (Couvinian =Eifelian)	Massive limestones; a quarry in the northern part of the promontory yields an abundant and well-preserved coral fauna, with *Calceola sandalina, Mesophyllum damnoniense, Cystiphyllum vesiculosum, Heliolites porosus, Alveolites, Thamnopora,* and Stromatoporoids.

There are faulted Lower Devonian slates, grits, and quartzite found to the west, and on the southern coast.

Interesting tectonic features are well displayed in transverse sections. The main folding, with east–west axis, is of Armorican date, and this is cut by later thrusting. Three small thrust planes are mapped, outcropping roughly north–south.

New Red Sandstone Neptunian dykes in the limestone of the old quarry have recently been recognised (Richter, 1966).

Fragments of a well-preserved shelly Raised Beach, some 17 feet thick in all, are found on the southern tip of the promontory.

The section shows (Orme, 1960):

Hillwash and limestone debris.

Pleistocene Eemian Interglacial	2–3 ft	Blown sand.
	12 ft	Fine sand passing down into coarse, current-bedded sand cemented with calcite; debris of flint, limestone, slate, dolerite, quartz and sandstone, with comminuted shell fragments throughout. Well-preserved shells are found only towards the base. Seventeen species of marine mollusca and crab have been recorded, the commonest being *Ostrea edulis, Mytilus edulis, Patella vulgata* and *Cardium edule.*
	1 ft–1 ft 4 in	Basement deposit of pebbles and boulders chiefly of limestone and slate, some larger than a foot across, in a matrix of sand.

55

The beach rests upon a wave-cut bench at 30–35 ft above O.D.

1835, Austen; 1853, Milne-Edwards and Haime; 1933, Lloyd; 1928, Shannon; 1960, Orme; 1961, Dineley; 1963, Elliott; 1966, Richter.

Kent's Cavern, Torquay

SX(20)935641	188	350	116SE	A, C, P
			(G)	1962

A famous large bone-cave, a ramification of water-worn passages and chambers, now dry. Sometimes called Kent's Hole, it is privately owned and commercialised, and the interior is floodlit to display to visitors the spectacular stalactite and stalagmite formations. It is scheduled and protected as an Ancient Monument.

The cave is situated in a hill a mile east of the centre of Torquay, on the west side of a well-wooded glen known as Ilsham Valley and less than half a mile from the present sea coast lying to the east. The country rock is the upper, thick-bedded part of the Torquay Limestone group of about Middle Devonian, Givetian, age, dipping 20° north-west. Rain-water percolating through joints in the rock over a period of time, perhaps to be measured in millions of years, dissolved out the chambers and passages nearly 700 feet in length and extending over an area of some two and a half acres. The two entrances now open lie about 180 feet above the present mean tide level; three other entrances have been stopped up again after excavation.

According to Ogilvie (1941), the cave was known as Kent's Hole as far back as 1659; the name probably derived from the Cornish word *kent* or *cant*, a border or headland, so that it meant the Hole in the Headland. But according to Kennard (1945), Kent's Hole was merely the local name derived from an owner in the seventeenth century.

The cave may have been known to man continuously since Palaeolithic times. The earliest modern date is stated to be scratched on a large boss of stalagmite: 'William Petre 1571'. It is thought unlikely that Man ever used the cave as his permanent residence, but as the temperature nowadays remains at about 53°F all the year round, it may have attracted early Man by the modicum of warmth it provided, although for some of the time it will have been colder than it is today.

In 1824 Thomas Northmore discovered in it bones of extinct mammals, and in 1825 Buckland found a flint knife-blade, the earliest record of a human artifact from any cave. More flint implements were then found by McEnery and others, and these showed for the first time that early Man had co-existed with the extinct Pleistocene animals, a fact that at the time led to much heat and controversy in that it contradicted the Biblical chronology. In 1867 a fragmentary human jaw with four teeth was discovered here embedded in stalagmite, and finally it was proved beyond any doubt that man-made flint implements were found associated with prehistoric animals, at depth, sealed under undisturbed stalagmite of great age.

It was largely due to the first excavation of this cave that men of science fully grasped the idea of animal life having existed on earth for periods of thousands of years before there was any written history.

Father McEnery was the first to explore parts of the cavern systematically, beginning in 1825, and he accumulated thousands of teeth and bones of Horse, Hyaena, Bear, Rhinoceros and Elephant; he wrote that by 1826 40 elephant teeth had been recovered. These were animals (other than Horse) which very few people then believed ever to have lived in England, or which if their presence was admitted, were thought to have been drowned in the 'Deluge'.

Excavations have been made in Kent's Cavern over 140 years, and there still remains unexcavated material, but no complete section. McEnery died in 1841, and Pengelly carried out long-continued excavations under the auspices of the British Association from 1864 to 1880. Unfortunately, neither McEnery nor Pengelly properly described and published his results, and the collected material has been dispersed and much of it has apparently been lost. From 1926 onwards the Torquay Natural History Society continued to investigate the cave, again under the aegis of the British Association, with grants from the Royal Society and the Society of Antiquaries. Accounts of this work are given in the Annual Reports of the British Association.

A very large literature has grown up, Kennard (1945) recording a selected 85 titles of 'real importance'. He also valiantly assembled a lengthy tally of the remaining manuscript and published data and collections.

The cave was a Hyaena den for long periods, and then most of the other bones were the residues of the Hyaenas' meals. Some carcases must have been dragged in piecemeal, for the larger were too big to have got in whole through the available entrances. Practically every large bone and antler found in the cave earth shows marks of Hyaena teeth.

The cave deposits are very irregularly developed, but the table overleaf shows the general succession. The dating of the beds underlying the Black Mould is as yet quite uncertain. The tentative ages assigned below were given by Dr A. J. Sutcliffe (*in. lit.* 9 November, 1964), to whom I am greatly indebted.

Animals other than those noted in the table whose remains have been recorded from Kent's Cavern are: Straight-tusked Elephant, Wild Boar, cf. Fallow Deer, Lynx, Hare, Mountain Hare, Pine Marten, Lemming, Water Vole, Field Vole, Bank Vole, Golden Eagle, Mallard, and Salmon; a total of some 36 species, including Man.

57

Recent	Recent, Mediaeval, Romano-British, Iron Age, Bronze Age, Neolithic	Up to 1 ft Black Mould: carbonaceous material with the remains of fires; found only in large chambers near the entrances; yielded mediaeval material and Roman ware; rings, a spoon, combs, spearhead and blade, spindle whorls, and pottery of La Tène type, Iron Age; bronze celts, spearheads and potsherds of Bronze Age type; and stone celts and flint arrow-heads of Neolithic type; also bones of domestic animals and of Seal, Fox, Badger, Brown Bear, and Long-horned Ox, and some Human bones and teeth.

Pleistocene	? Würm Glaciation	c. 3 ft Upper (Granular) Stalagmite, with remains of Hyaena, Fox, Brown Bear, Horse, Mammoth, Woolly Rhinoceros; also a few flint flakes, charcoal, and fragments of a Human upper jaw.
		0–30 ft Cave Earth: unstratified reddish loam with angular lumps of limestone and grit. Found locally near the main entrance, at the top of the Cave Earth, is the Black Band: hearths with remains of fires and burnt bones; with many artifacts, Magdalenian Reindeer antler harpoons, and bone needles.
		The Cave Earth has yielded a vast quantity of bones and teeth of extinct animals, including Hyaena, Cave Bear, Cave Lion, Sabre-toothed Tiger (*Machaerodus latidens*), Grizzly Bear, Brown Bear, Bison, Reindeer, Irish Giant Deer, Red Deer, Horse, Mammoth, Woolly Rhinoceros, Wolf, Fox, Badger, Beaver, Glutton, Cave Pika, and a fragment of Human jaw-bone.
		Artifacts include a hand-axe of early Mousterian or Acheulian type; other artifacts of early Solutrian type, and a bone pin of later date; a peculiar assemblage of tools of both primitive man and early modern man.
	? Eemian Interglacial	0–12 ft Lower (Crystalline) Stalagmite.
	? Riss Glaciation and ? Hoxnian Interglacial	Breccia of red grit with a few fragments of limestone, in a red sandy matrix, mostly cemented hard; up to 50 per cent of the material is of mostly unbroken bones and teeth of Cave Bear,* and fewer of Cave Lion; few rude artifacts referred to Abbevillian (=Chellean), usually in rolled condition.

A specimen of Hippopotamus remains supposed to have come from Kent's Cavern has recently been fluorine tested, the results proving, Dr Sutcliffe tells me, that it has not come from this cave.

The series of flint artifacts allegedly from Abbevillian onwards, and ending with Magdalenian tools made from Reindeer antlers, complete a Palaeolithic suite which cannot be surpassed by any other cave in the world. According to Rogers (1956), some 1200 flints are still preserved, of which 60 can be classified as implements.

---

* Dr Sutcliffe tells me that Cave Bear remains are very rare in this country, and known only from Kent's Cavern, and from the Hoxnian Interglacial deposits at Swanscombe, Kent. Records from other sites appear to be incorrect.

The diversity of the cave deposits, including the stalagmite floors, shows that they must have been assembled over long periods of time, and under different circumstances, climatic and otherwise, which are as yet inadequately understood, with insufficient evidence for dating them reliably. These problems remain for the future.

1797, Maton; 1869, Pengelly; 1941, Ogilvy; 1945, Kennard; 1948, Dewey; 1949, Oakley; 1953, Vachell; 1956, Rogers.

Lower Dunscombe Quarry, Chudleigh, 5 miles NNE of Newton Abbot

SX(20)886790	188	339	102SW	P, S
			(-)	1962

A disused quarry in Upper Devonian limestones, in a small triangular faulted inlier, 400 yards across, within Culm Measures (Carboniferous) exposing (after Anniss, 1933):

Upper Devonian / Frasnian

6 ft (seen). Highly ferruginised thin-bedded limestone with red shaly partings; with a rich goniatite fauna of the *cordatum* Zone of the *Manticoceras* Stufe, with *Manticoceras cordatum* (the zone fossil), *M.* cf. *intumescens*, *Beloceras sagittarium*, *Tornoceras* (T.) sp.; also *Phacops*, *Coccosteus*, etc.

3–4 ft Thin-bedded cream to liver-coloured calcareous mudstones, with many tiny fragments of trilobites, *Phacops latifrons*, and *Proetus batillus*, and a bivalve.

— Massive pale grey crystalline limestone, with *Hypothyridina* [*Rhynchonella*] *cuboides* and eleven other brachiopods, two bivalves, a gastropod, *Stromatopora*, and crinoids.

House (1963a) redetermined the goniatites as above, finding no conclusive evidence from them of any other Frasnian horizon than the *cordatum* zone.

Of the overlying Famennian, however, House found evidence of the *sandbergeri* Zone of the *Platyclymenia* Stufe in a specimen of *Sporadoceras* cf. *contiguum* from Lower Dunscombe (exact location not stated); a museum specimen of *Cymaclymenia cordata* which presumably came from the field above the quarry, where clymenids were noted by Clement Reid, represents the *Clymenia* Stufe of a higher Famennian horizon. In the Chudleigh area both Famennian and Frasnian are of the schwellen facies and are significantly condensed.

The goniatites correlate these beds with equivalent strata in the Eifel district of Germany and the European Devonian succession.

Dineley and Rhodes (1956) investigated two samples from this site, one of massive crystalline limestone with fragmentary fossils at the foot of the western face of the old orchard quarry; and the second of massive shelly and crystalline limestone 10 feet above the first sample. Both yielded a prolific and generally well-preserved fauna of conodonts, including the Upper Devonian genera *Bryantodus*, *Ancyrodella* and *Icriodus*. Seven other genera were identified and members of four other conodont groups.

1880, Roemer; 1889, Kayser; 1889–1898, Whidborne; 1913, Ussher; 1933, Anniss; 1956, Dineley and Rhodes; 1963a, House.

Lummaton Quarry, Torquay

SX(20)913665	188	350, 339	116NW (G)	P, S PSD, 1962

Large and partly disused quarries in an isolated faulted hill of clearly exposed light-grey Devonian limestones, mainly massive, and largely dolomitised, when they are sometimes red; slickensiding is visible. The bedding cannot be seen but the dip is probably to the north-west. Some 200 feet thickness has been calculated in the northern part of the exposure. On the west the limestones are flanked by red Permian clays.

Traced from south-east to north-west, the massive and thrust limestones probably range from the Couvinian stage (=Eifelian) through the Givetian to the base of the Frasnian, that is, from the Middle Devonian to the lowest part of the Upper Devonian.

A very large fauna (monographed by Whidborne, less the corals and stromatoporoids) has been found in different patches of the unaltered limestones, which show varying communities. In places the rock is largely composed of stromatoporoids up to four feet across; and there are locally abundant other fossils including crinoid fragments, corals, polyzoa, many brachiopods, bivalves, gastropods, cephalopods, ostracods and trilobites.

The main Lummaton Shell Bed, exposed at the top of the eastern face of the small western quarry, forms only a small part of the Devonian limestone of the quarries. Elliott (1961) found in it a distinctive polyzoan facies, whereas only a foot below (and where the rock is not dolomitised) the facies is of stromatoporoids and crinoids. From it he obtained a sample of 300 determinable fossils, mainly rounded or globular brachiopods only 10–15 mm in diameter; 13 per cent were terebratuloids, 33 per cent spiriferoids and 37 per cent rhynchonelloids. Fenestellids were prominent in the polyzoan debris and the alga *Palaeosporella lummatonensis* was described as new. Elliott concluded that the Shell Bed is an accumulation due to current action, and of Upper Givetian age, the top of the Middle Devonian.

In an investigation of the ammonoids House (1963) recognised from the Shell Bed the zone fossil *Maenioceras terebratum*, and *M.* cf. *decheni, Agoniatites fulguralis, A.* cf. *costulatus* and *Tornoceras* spp. The bed is correlated with the *terebratum* Zone at the top of the Givetian stage of the Middle Devonian of the European succession and is thus in accord with the evidence of the brachiopods.

Copper (1965) has described unusual structures in some English Devonian Atrypidae, including from this quarry *Spinatrypa trigonella* (Davidson) from the Shell Bed, and directly below many large specimens of a recently described genus *Mimatrypa desquamata* (Sowerby).

In recent years mammalian remains have been found here in fissures in the limestone during quarrying. Sutcliffe (1962) records remains of 12 animals, identified as Hyaena, Horse, Rhinoceros, Red Deer, Reindeer, Ox or Bison, and Hare. They may have fallen into small open fissures, most in late Pleistocene times, although they are not all necessarily of the same date.

1889–98, Whidborne; 1906, Jukes-Browne; 1913, Ussher; 1933, Lloyd; 1961, 1963, Elliott; 1962, Sutcliffe; 1963a, House; 1965, Copper.

Meldon Aplite Quarry, 2 miles SW of Okehampton

SX(20)568921	175	324	76SE	Ig, M, Met
			(–)	IGC, 1962

This site comprises two adjoining quarries—one on each side of the Red-a-Ven Brook—which work the so-called Meldon Aplite dyke. This dyke, which is unique in the south-west, is really an unusual lepidolite-soda microgranite which has been intruded into tuffs and sediments of Lower Culm (Lower Carboniferous) age. These comprise the Meldon Calcareous Group (of shales and impure limestones) and the older Meldon Shale and Quartzite Group, which includes a volcanic group of agglomerates and tuffs. These rocks have been much folded, faulted and overturned before the intrusion of the granite—which has also metamorphosed them—so that they now dip steeply to the north-west.

The aplite is a white, fine-grained, holocrystalline rock, but here and there it contains patches and lenses of pegmatite, some of coarse grain. In many places it has been altered by the introduction of late-stage minerals, and it contains xenoliths of country rock, also much altered. The country rocks have been greatly affected both thermally and metasomatically by the Dartmoor Granite, and contain much garnet, axinite, pyroxenes, pyrrhotite

61

and other metamorphic minerals. The specific gravity of the aplite is 2·66 and it has a very high crushing strength measured as 1549 tons per square foot.

Essentially the aplite consists of quartz, albite, orthoclase and muscovite, but it contains many accessory and later-formed minerals such as lepidolite, apatite, topaz, fluorite, axinite, petalite, pink montmorillonite, prehnite, heulandite and tourmaline (pink, green, parti-coloured and blue). Many of these occur well crystallised in druses, especially in the pegmatite lenses. Other minerals, some new to Britain, have been found here recently by the writer (A.W.G.K., unpublished) and include a number that are unusual or very rare, containing beryllium; lithium, caesium and other alkali metals; and beryl, arsenopyrite, löllingite, zoisite and scheelite.

A short distance to the south-east are old mining trials known as Red-a-Ven Brook Mine, which were referred to in 1878 by Warington Smyth. They were opened in mineralised bands in the upturned Meldon Calcareous Group, and the small dumps contain excellent examples of the typical pyrometasomatic assemblages, with sulphides such as chalcopyrite, much löllingite, some arsenopyrite, sphalerite, pyrrhotite, galena, in 'skarn' matrices of garnet, idocrase, hedenbergite and other pyroxenes, calcite, wollastonite, fluorite, scheelite, etc., found by the writer (A.W.G.K., unpublished). The first British occurrence of the rare mineral helvine (a complex silicate of beryllium, manganese, iron and zinc) was also discovered here (Kingsbury, 1961).

Here also the mineral malayaite, $CaSnSiO_5$, a tin-bearing analogue of sphene, is present in wollastonite hornfels containing datolite (Dearman, 1966: *Proc. Geol. Ass.* **77**, 230).

A. W. G. Kingsbury

1839, De la Beche; 1912, Reid; 1920, Worth; 1959, Dearman and Butcher; 1961, Kingsbury.

Meldon (Railways) Quarry, 2 miles SW of Okehampton

SX(20)568927	175	324	76SE	M, Met
			(–)	IGC, 1962

The principal reason for the notification of this quarry is the array of minerals developed here, on the north-west corner of the metamorphic aureole of the Dartmoor Granite of Armorican date (from late Carboniferous to post-Carboniferous). The country rock is of Lower Culm Measures (Lower Carboniferous) of the Meldon Calcareous Group provisionally assigned to a

Viséen Age and the older Meldon Shale and Quartzite Group, assigned to the Viséen and Tournaisian; the latter includes volcanic tuffs and agglomerates. Two types of basic dyke intrusions are confined to the Lower Culm.

The structure is exceedingly complex, the Meldon district comprising major recumbent zigzag folds overturned to the south, with minor rucking; the whole is broken up by strike faulting, normal and reversed, and shearing.

The vast quarry is actively worked by British Railways, mainly for ballast; it has a working face over half a mile long on several levels, and affords magnificent sections of the altered Lower Culm sediments, tuffs, and two intrusive basic dykes, all of which have been thermally and metasomatically metamorphosed by the Dartmoor Granite.

As work progresses towards the granite, the faces of the quarry constantly change, and expose a great variety of calc-silicate and other tough hornfelses, chiastolite slate, etc. Occasional bedded manganese deposits have been metamorphosed into assemblages of unusual manganese silicates.

A great variety of minerals have been recorded and found by the writer (A.W.G.K.), including arsenopyrite, axinite, bismuth, bismuthinite, bustamite, calcite, chabazite, chalcolite, chalcopyrite, chiastolite, datolite, diopside, fluorite, galena, garnet (grossular, andradite and spessartite), heulandite, hornblende, idocrase, laumontite, lepidolite, löllingite, molybdenite, prehnite, pyrite, pyrrhotite, rhodonite, rhodochrosite, scheelite, sphene, siderite, tephroite, tourmaline and wollastonite; also bavenite (Dearman and Claringbull, 1960).

A. W. G. Kingsbury

1839, De la Beche; 1912, Reid; 1920, Worth; 1959, Dearman; 1959, Dearman and Butcher; 1968, Edmonds *et al.*

Pocombe Quarry, Exeter				
SX(20)899914	176	325	80SW	Ig
			(−)	1967

A large disused quarry occupied as a caravan site. It is the type locality for the Pocombe variety of the Permian 'Exeter Traps' or lavas, a ciminite or iddingsite-trachyandesite, from 30 to upwards of 70 feet thick. These rocks form a prominent topographic feature, and as building stone they have attracted attention since Roman times. They have figured in the scientific literature since the end of the eighteenth century.

Two main varieties with transitional forms are distinguished. In the lower parts of the flow the lava is pink, finely vesicular, and shows prominent reddish-brown pseudomorphs after iddingsite (an alteration product of olivine). Vesicles are filled with carbonates and crypto-crystalline silica. The compact facies is a dark bluish-grey rock with brown pseudomorphs after iddingsite in an aphanitic groundmass. The titanium mineral pseudobrookite is a prominent accessory. The rocks are veined with dolomite.

The latest surface manifestations of the Dartmoor cycle of igneous activity, these lavas form products of a later stage of magmatic differentiation than any other known rocks of the region. In places they lie on upturned and denuded Culm Measures, and elsewhere on Permian sands, with which they are contemporaneous.

Four radiometric age determinations by the potassium–argon method on biotite from the 'Exeter Trap' at Killerton Park, seven miles to the north-east, at SS975015, yielded a mean value of about 279 million years old (Miller et al., 1962: Geophys. J. **6**, 394) equivalent to a basal Permian age according to the Geological Society Phanerozoic time-scale 1964.

1835, De la Beche; 1902, Ussher; 1932, Tidmarsh.

Pridhamsleigh Cave, Buckfastleigh, 10 miles WNW of Torquay

SX(20)750678	188	338	114NE (G)	C, B 1963

The second largest natural cave system in Devon, dissolved out of Middle Devonian limestone, it is a complex phreatic network with over 3000 feet of passages so far known. At the end is a pool of limited surface area but 95 feet deep, a drowned 'pot' which opens out below the surface. Stalactites were seen in it 40 feet below the water level, and connecting tunnels 70 feet down indicate a previously lower level of the water table, possibly during the Last (Würm) glaciation, when the sea level was 200–300 feet lower than at present. The cave has a wealth of remarkably fine stalactite formations, and also helictites.

Of the living fauna inhabiting the cave, the bat population is being studied, mainly the Greater Horseshoe Bat (*Rhinolophus ferrum-equinum*). The invertebrates are of much interest but as yet are only partially known. They include some 13 named species, including two true troglobites, which spend their entire life-cycle in darkness, the collembollan (springtail) *Lepidocyrtus cavernarum* and the rare amphipod *Niphargus glenniei*, a tiny blind shrimp, whose type station it is.

1937, Yeldham; 1953, 1962, Cullingford, Ed.

64

Reed's Cave—Baker's Pit System, with Joint Mitnor Bone Cave, Buckfastleigh, 10 miles WNW of Torquay

SX(20)743665	188	349	114SE	C, B, P
			(G)	1967

A large cave system with over 7000 feet of passages, including Baker's Pit Cave (discovered by quarrymen in 1847), Reed's Cave, Disappointment Cave, Rift Cave, Joint Mitnor Cave and Spider's Hole. Access is gained from two quarries, Baker's Pit and Higher Kiln Quarry, both near the Church. These worked the limestone during the latter part of the eighteenth century and early part of the nineteenth century, but were abandoned about 100 years ago.

Of early notices, Polwhele (1797) mentioned a vast cavern in what was probably part, later destroyed, of what is now known as Reed's Cave, named after Mr Edgar Reed, a spelaeologist now living at Buckfastleigh. De Luc (1811) visited Buckfastleigh in 1806, gave some notes on the geology, and mentioned quarries working 'marble' of the same kind as he had seen in Chudleigh Rock. He was given a description of large caverns in the calcareous strata, but these do not appear to have been the Buckfastleigh caves. The later caving records are given by Sutcliffe (1960).

The country rock is Middle Devonian Limestone, of which a thickness of some 60 feet is exposed in Higher Kiln Quarry, where small faults are visible in the face. The Limestone here includes some interbedded volcanic ash, and is cut by a lamprophyre dyke.

The cave system, which now lies between about 270 and 200 feet above O.D., is considered to have been formed by solution of the limestone at a time when this lay below the water table, in the phreatic zone. This could have been taking place as late as the date of the Ambersham Terrace of the Cromerian (First) Interglacial. Since then there has been very little vadose enlargement, i.e. enlargement while the water table has lain below the cave system. Cracked flowstone floors and some brecciated water-laid sediments are thought to be manifestations of periglacial frost action during the rigours of a later glaciation, in fact apparently of the Würm Glaciation.

Stalactite formations are found in most parts of the system, the best being in Reed's Cave, where there are fine floor and wall formations, curtains, very unusual stalactites formed of single crystals, small helictites and pool deposits, although the formations in the earlier known part of the cave have been badly damaged.

The caves are the home of cave-dwelling invertebrates, including spiders, springtails, small crustacea, moths and other insects still being studied. Of special interest is the blind shrimp *Niphargus glenniei*, a breeding colony of which has recently been established conveniently for study in Joint Mitnor Cave, the copepod *Acanthocyclops bisetosus*, and (found on the bats) the mites *Eugamus lunulatulus* and *Spinturnix euryalis*, and the tick *Ixodes vespertilionis*. Rift Cave is now set aside as a reserve for the bats, the Greater Horseshoe Bat, the Lesser Horseshoe Bat (*R. hipposideros*) and Natterer's Bat (*Myotis nattereri*) (which have been studied by the Hoopers) together with the cave-dwelling invertebrates which live on the bat droppings. In Spider's Hole lives the spider *Meta menardi* in considerable numbers (Figs. 6 and 7).

Joint Mitnor Cave was discovered only in 1939, and is named after Messrs. W. Joint, W. Mitchell and F. R. Northey. The deposit was excavated from 1939 to 1941 by Mr A. H. Ogilvie, who obtained 4307 specimens of mammal bones and teeth, now known to comprise at least 127 individual animals, of 16 species. Most of this material is preserved in the Torquay Museum.

This has proved to be the richest assemblage of interglacial mammalian remains yet found in a British cave; it was buried in a cone of earth and stones at the foot of a shaft (now blocked) into which the animals and talus had fallen. It is considered to be dated to the warmest part of the Eemian (Last) Interglacial, perhaps some 100 000 years ago. This dating is deduced partly from consideration of the relation of the cave system to the terraces of the River Dart found nearby, and partly from the composition of the mammalian fauna. The fauna of the Ogilvie Collection studied by Sutcliffe (1960) forms a homogeneous assemblage of mammals, some characteristic of a warm climate, with others of no climatic significance. Although the bones and teeth are very numerous, the number of animals represented is much smaller, and amounts to the following approximate tally of individuals: Bison (44), Cave Hyaena (15), Fallow Deer (11), Fox (9), Wolf (8), Narrow-nosed Rhinoceros (7), Red Deer (7), Hippopotamus (6), Straight-tusked Elephant (6), Cave Lion (5), Badger (2), Hare (2), (? Brown) Bear (1), Giant Deer (1), Pig (1), Wild Cat (1). There is also a fauna of small rodents in course of description.

Characteristic features of this assemblage are as follows.

(a) Relative abundance of Hippopotamus, Narrow-nosed Rhinoceros (*Dicerorhinus hemitoechus*) and Hyaenas; the Hyaenas were all adults, and only one or two of the other bones had been gnawed, showing that the cave had never been a Hyaena den.

(b) Striking absence of Horse and of species commonly found in British cave deposits accumulated during the cold phases of the Ice Age, such as Woolly Rhinoceros, Woolly Mammoth, Reindeer and Glutton.

(c) The undetermined Fallow Deer are slightly smaller than the *Dama clactoniana* of Swanscombe (Hoxnian Interglacial).

(d) An appreciable proportion of the species still survive or survived until recently in Britain.

Fig. 6. Rift Cave, Reed's Cave, Buckfastleigh, Devon. Greater Horseshoe Bats, *Rhinolophus ferrum-equinum* (Schreber) at roost; a portable receiver detects the ultrasonic signals emitted by the bats as they wake up. (Photo, J. H. D. Hooper.)

Fig. 7. Spider's Hole, Reed's Cave, Buckfastleigh, Devon. The spider *Meta menardi* (Latreille) and its egg sac (approximately natural size). (Photo, A. E. McR. Pearce.)

Higher Kiln Quarry was purchased in 1961 by the Society for the Promotion of Nature Reserves, and thereafter leased to the Devon Trust for Nature Conservation Ltd. Now known as the William Pengelly Cave Research Centre, it is being developed for research and teaching. The Joint Mitnor Cave, preserved in safety behind a massive locked door, has been set up as a demonstration site with exposed sections of the bone deposit and of the basal water-laid sediments of mingled clay and clastic material. Within the old quarry two old stone-built farm buildings are being renovated to provide a museum.

1797, Polwhele; 1811, de Luc; 1873b, Pengelly; 1939, 1947, 1950, Hooper; 1953, 1962, Cullingford, Ed.; 1960, 1965, Sutcliffe.

Reed's Farm Pit, Wilmington, 3 miles E of Honiton

ST(31)213003	177	326	71NW	S, P
			(–)	PSD, 1966

This is sometimes called Hutchin's Pit, and is referred to by Woodward and Ussher (1911) and in PSD as near the lane from the village to Hayne's Farm.

Soil and flints

		Up to 2½ ft Soft bedded white chalk.
Middle Chalk; Turonian		2½ ft Hard rubbly glauconitic chalk with scattered quartz grains and a layer of very hard chalk with green coated nodules at the top; *Inoceramus mytiloides*.
		9 in Soft glauconitic marly chalk.

Upper Cretaceous

Chalk; Lower Cenomanian

c. 2 ft Hard glauconitic and quartziferous limestone with worn surface at the top, encrusted with brown phosphate; base not well marked; *Scaphites equalis*, *Turrilites costatus*, *Pycnodonte* [*Ostrea*] *vesicularis*, *Holaster subglobosus*, *Discoidea subuculus*.

6 ft Rough yellowish calcareous sandstone; with a shell-bed containing a very rich fauna, crowded with casts of *Trigonia*, and including Lower Cenomanian ammonites *Forbesiceras largilliertianum*, *F.* sp. cf. *beaumontianum*, *Hyphoplites crassofalcatus*, *Mantelliceras* spp., and other forms; also *Chlamys* [*Pecten*] *aspera*, *Lima semiornata*, *Holaster subglobosus*, *H. altus*, *H. carinatus*, *Pseudodiadema variolare*, *Rhynchonella dimidiata*; and a box crab (*Calappa cranium* C. W. Wright) was described from here, the first record of this genus from the Cretaceous (1945: *Geol. Mag.* 128).

The sandstone weathers to sand with lumps of sandstone. It appears to pass down into the Upper Greensand of Albian age.

It is now a degraded and overgrown sand pit on the roadside, half a mile north-east of the main Wilmington Quarry (q.v.), to which it is complementary. It exposes the lowest part of the calcareous sands not seen in the Wilmington Quarry.

At the northern end of Reed's Farm Pit the beds are cut by a fault throwing not more than eight feet, and the section on p. 69, which is here combined with later observations, was observed on the downthrow side in 1897 (Jukes-Browne, 1898), when the strata were more clearly exposed.

1898, 1903, Jukes-Browne; 1911, Woodward and Ussher; 1961, Smith.

Saltern and Waterside Coves, Goodrington, 3½ miles SSW of Torquay

SX(20)895587	188	350	122SW (G)	P, S, T, PSD, IGC, 1962

Waterside Cove is the name used by geologists for a little otherwise unnamed cove just beyond the headland, and about 100 yards north of Saltern Cove; it is referred to as Saltern Small Cove by House (1963).

This is a highly disturbed coast section, with the different rocks bounded by thrusts or faulted junctions. The following strata are present.

New Red Beds: Permian	Conglomerate.
Upper Devonian: Lower Famennian, Upper Frasnian	c. 500 ft Purplish-red cleaved shales with thin limestones (Famennian); on the south side of Waterside Cove these pass down into Upper Frasnian red mudstones with large concretions of paler hard calcareous mudstone—the Goniatite Bed (see below). There are intrusions of dolerite (now decayed) associated with basic spilitic tuff.
Middle Devonian	Shattered crystalline limestones, in places ferruginised or dolomitised.
Lower Devonian, Emsian	Staddon Grits, with sparse fauna.

At Waterside Cove New Red Beds rest unconformably upon Lower Devonian, and these are separated from the Upper Devonian by a reversed fault.

The main interest of the site lies in the Goniatite Bed in the Upper Devonian, some 40 feet below the uppermost bed of the ostracod *Entomis* shale. The Goniatite Bed has yielded some 29 forms, including trilobites (*Phacops*), brachiopods (*Athyris, Spirifer*), bivalves (*Mytilus, Buchiola*), a gastropod

(*Pleurotomaria*), cephalopods (*Orthoceras, Bactrites*), crinoid stems and a fish palate; and also some 17 species of goniatites.

Discovered here by Lee in 1877, the goniatites proved the presence of Upper Devonian strata in South Devon for the first time. They serve to correlate the bed with the *holzapfeli* Zone of the *Manticoceras* Stufe of the Upper Frasnian (House, 1963a), rather than with the German Büdesheimer Schiefer of the underlying *cordatum* Zone, as hitherto believed.

More than 500 specimens of ammonoids have been obtained over the years from this Saltern Cove bed, and House has redetermined them as follows: *Archoseras angulatum; A. varicosum; A. ussheri,* sp. nov; *Manticoceras cordatum; M.* aff. *serratum; M.* sp. nov.; *M.* cf. *adorfense; M.* cf. *retrorsum; Crickites holzapfeli; Tornoceras* (*T.*) aff. *linguum; T.* (*T.*) aff. *uniangulare; T.* (*T.*), sp. nov.; *Aulatornoceras auris; A.* sp. nov.; *A.* aff. *paucistriatum;* gen. et spp. nov.

Crickites holzapfeli, recognised from here for the first time by House, is not known in the *cordatum* Zone, and the abundance of *Archoceras* and distinctive species of *Tornoceras* and *Aulatornoceras* shows that the fauna belongs to the *holzapfeli* Zone.

Famennian. At Saltern Cove rocks of ostracod slate facies overlie dated Frasnian. There is no evidence from Famennian ammonoids, but ostracod dating of slates at Anstey's Cove, Torquay, indicates *Clymenia* or *Wocklumeria* Stufe. Ostracod slate facies hereabouts was probably established at the close of the Frasnian and continued into the Upper Famennian. Volcanic ash occurs within the ostracod-dated slate of Anstey's Cove.

1877, Lee; 1927, Anniss; 1933, Lloyd; 1963a, House; 1963, Elliott.

Saunton Coast, 8 miles NW of Barnstaple

SS(21)432384	163	OS27	8NW, SW	R, S
			(–)	PSD, 1962

A Pleistocene Raised Beach with far-travelled boulders can be recognised intermittently over ten miles of coast in North Devon, from Morthoe, west of Ilfracombe, to Westward Ho! near Bideford. It is well exemplified at Saunton, in cliff sections in front of Down End House.

The erratic boulders have been noticed and described since 1837; they include foliated red or pink granite, grey trachy-andesite, greyish gneiss, granulite, grit, altered dolerite, and agglomerate. Some are not of local materials and have been brought by ice, probably from western Scotland. One red granite boulder resembles a rock from near Gruinard Bay, Ross-shire.

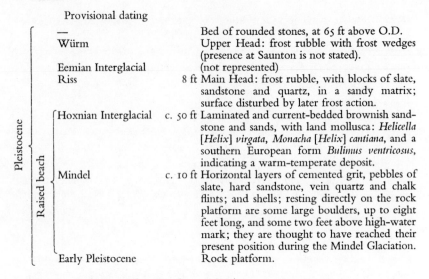

Provisional dating

Pleistocene / Raised beach	—	Bed of rounded stones, at 65 ft above O.D.
	Würm	Upper Head: frost rubble with frost wedges (presence at Saunton is not stated).
	Eemian Interglacial	(not represented)
	Riss	8 ft Main Head: frost rubble, with blocks of slate, sandstone and quartz, in a sandy matrix; surface disturbed by later frost action.
	Hoxnian Interglacial	c. 50 ft Laminated and current-bedded brownish sandstone and sands, with land mollusca: *Helicella* [*Helix*] *virgata*, *Monacha* [*Helix*] *cantiana*, and a southern European form *Bulimus ventricosus*, indicating a warm-temperate deposit.
	Mindel	c. 10 ft Horizontal layers of cemented grit, pebbles of slate, hard sandstone, vein quartz and chalk flints; and shells; resting directly on the rock platform are some large boulders, up to eight feet long, and some two feet above high-water mark; they are thought to have reached their present position during the Mindel Glaciation.
	Early Pleistocene	Rock platform.

The above dating (after Stephens, 1960) is as yet uncertain.

The Raised Beach rests on a platform cut across the upturned edges of highly inclined Pilton Beds, fossiliferous grey flags and shales with calcareous bands, of Upper Devonian, Famennian, age, passing up into lowest Carboniferous.

1840, Sedgwick and Murchison; 1887, Hughes; 1892, Prestwich; 1910, 1913, Dewey; 1956, Taylor; 1960, Arber, with later discussion by N. Stephens (1961: *Proc. Geol. Ass.* **72**, 469–471).

Seaton to Sidmouth Coast

SY(30)2088	176	326	83SW, SE, 94NW, NE (–)	S, P PSD, 1963

A nine-mile stretch of coast extending from Seaton Hole, SY236896, to Windgate Cliff, SY104858, exhibits striking white and red cliffs of great scenic beauty; they include the rugged white, 300-feet-high Beer Head, and the red, sheer 500-foot Windgate Cliff, west of Sidmouth.

The Upper Chalk, the highly fossiliferous Middle Chalk, and the Lower Chalk (Cenomanian Limestone) succession rests unconformably upon Upper Greensand, which in turn unconformably overlies the red marls of the Triassic Upper Keuper.

The section exposed is as follows (Upper and Middle Chalk after Rowe and Sherborn, 1903; Base of Middle Chalk and Lower Chalk after Smith, 1961; Lower Cretaceous after Jukes-Browne, 1900).

		Superficial Zone	Thickness at Beer	
				Flint gravel capping.
Upper Cretaceous	Upper Chalk	Senonian: Coniacian	*cortestudinarium*	c. 30 ft	Chalk with several bands of marl, and of yellow, hard siliceous, nodular chalk; flints in irregular courses, some flints spongiform. The zone fossil *Micraster cortestudinarium* is here very rare; *Micraster praecursor, Holaster placenta, Cidaris serrifera, Cardiaster cotteaui, Rhynchonella reedensis,* etc.

Upper Cretaceous / Upper Chalk

Senonian: Coniacian — *cortestudinarium* — c. 30 ft

Chalk with several bands of marl, and of yellow, hard siliceous, nodular chalk; flints in irregular courses, some flints spongiform. The zone fossil *Micraster cortestudinarium* is here very rare; *Micraster praecursor, Holaster placenta, Cidaris serrifera, Cardiaster cotteaui, Rhynchonella reedensis,* etc.

Turonian — *planus* — 60 ft

Hard greyish marly nodular chalk; flint lines strong and regular; scattered phosphatic and glauconitic nodules. The zone fossil *Sternotaxis* [*Holaster*] *planus* very abundant; *Micraster praecursor, M. corbovis, M. leskei, Echinocorys vulgaris* var. *gibba, Cidaris serrifera, Pentacrinus, Cretirhynchia*[*Rhynchonella*]*plicatilis, Rh. reedensis,* etc.

lata (=*gracilis*) — 89 ft

Soft white marly chalk; flint lines crowded, with abundant nodular and digitate flints; two bands of flintless marly chalk, and two pairs of yellow nodular bands.

Zone fossil, *Terebratulina lata*, large and in profusion; *Micraster corbovis, Sternotaxis* [*Holaster*] *planus, Discoidea dixoni*, and five other echinoids; *Pentacrinus, Orbirhynchia*[*Rhynchonella*] *cuvieri, Inoceramus lamarcki, Pachydiscus peramplus;* the foraminifer '*Haplophragmium*' abundant.

labiatus (=*cuvieri*) — 25½ ft

(normal facies). Very hard nodular iron-stained chalk, almost devoid of flints; locally occurs as Beer Stone; very rich in fossils; *Orbirhynchia cuvieri, Inoceramus mytiloides, Pachydiscus peramplus, Serpula avita, Gibbithyris* [*Terebratula*] *semiglobosa, Carneithyris* [*Terebratula*] *carnea; Pentacrinus;* and 10 characteristic echinoids, including *Discoidea dixoni, Conulus* [*Echinoconus*] *castanea, Cidaris hirudo, C. clavigera.*

Middle Chalk — Turonian

(*continued overleaf*)

73

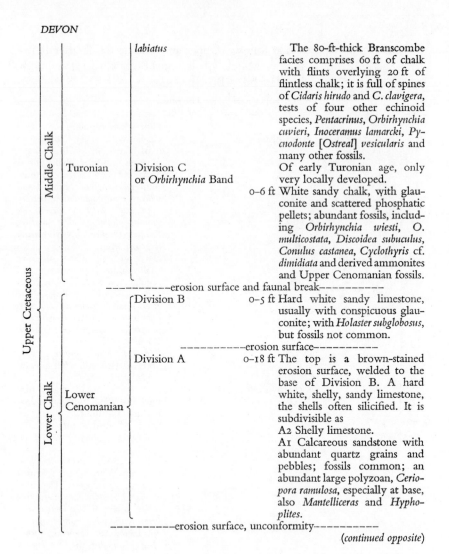

		labiatus		The 80-ft-thick **Branscombe** facies comprises 60 ft of chalk with flints overlying 20 ft of flintless chalk; it is full of spines of *Cidaris hirudo* and *C. clavigera*, tests of four other echinoid species, *Pentacrinus, Orbirhynchia cuvieri, Inoceramus lamarcki, Pycnodonte [Ostrea] vesicularis* and many other fossils.
	Middle Chalk	Turonian	Division C or *Orbirhynchia* Band	Of early Turonian age, only very locally developed.
Upper Cretaceous				0–6 ft White sandy chalk, with glauconite and scattered phosphatic pellets; abundant fossils, including *Orbirhynchia wiesti, O. multicostata, Discoidea subuculus, Conulus castanea, Cyclothyris* cf. *dimidiata* and derived ammonites and Upper Cenomanian fossils.

––––––––––erosion surface and faunal break––––––––––

		Division B		0–5 ft Hard white sandy limestone, usually with conspicuous glauconite; with *Holaster subglobosus*, but fossils not common.

––––––––––erosion surface––––––––––

	Lower Chalk	Lower Cenomanian	Division A	0–18 ft The top is a brown-stained erosion surface, welded to the base of Division B. A hard white, shelly, sandy limestone, the shells often silicified. It is subdivisible as
				A2 Shelly limestone.
				A1 Calcareous sandstone with abundant quartz grains and pebbles; fossils common; an abundant large polyzoan, *Ceriopora ramulosa*, especially at base, also *Mantelliceras* and *Hyphoplites*.

––––––––––erosion surface, unconformity––––––––––

(continued opposite)

In this most westerly exposure in England of the Chalk and associated Cretaceous strata, the most abundantly fossiliferous beds are those of the Middle Chalk, particularly the *labiatus* Zone, which has a normal lithological and fossil facies found at Beer and to the east; and an abnormal facies occurring at Branscombe that is characterised by the unusual presence of flint in the upper part and a great abundance of two species of echinoids, particularly their spines.

There are included in this Site of Special Scientific Interest nine PSD localities, six from the Beer cliffs, one at Whitecliff, one west of Dunscombe Mouth, two and a half miles east of Sidmouth, and one at Branscombe. Particularly notable throughout the Chalk here are the echinoids, of which 22 species in all were identified by Rowe.

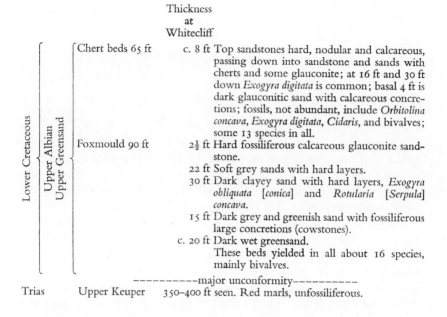

			Thickness at Whitecliff	
Lower Cretaceous	Upper Albian	Upper Greensand	Chert beds 65 ft	c. 8 ft Top sandstones hard, nodular and calcareous, passing down into sandstone and sands with cherts and some glauconite; at 16 ft and 30 ft down *Exogyra digitata* is common; basal 4 ft is dark glauconitic sand with calcareous concretions; fossils, not abundant, include *Orbitolina concava*, *Exogyra digitata*, *Cidaris*, and bivalves; some 13 species in all.
			Foxmould 90 ft	2½ ft Hard fossiliferous calcareous glauconite sandstone. 22 ft Soft grey sands with hard layers. 30 ft Dark clayey sand with hard layers, *Exogyra obliquata* [*conica*] and *Rotularia* [*Serpula*] *concava*. 15 ft Dark grey and greenish sand with fossiliferous large concretions (cowstones). c. 20 ft Dark wet greensand. These beds yielded in all about 16 species, mainly bivalves.

----------major unconformity----------

Trias	Upper Keuper	350–400 ft seen. Red marls, unfossiliferous.

From Seaton Hole as far west as Branscombe a synclinal fold of Tertiary date carries the top of the Upper Greensand down to low-water level at Beer; the beds then rise rapidly westwards, and the whole 170-foot thickness of the Upper Greensand there is exposed in the cliffs between Beer and Branscombe; beyond there is a steady rise towards Sidmouth.

At Seaton Hole a north-trending fault throwing at least 200 feet down to the west lets down Cretaceous against Trias. Because of a south-westerly dip of about 8° on the flank of the Beer syncline, the Whitecliff coast section between Seaton Hole and the east side of Beer provides the most complete and accessible succession of Cretaceous rocks in Devon. For the Upper Chalk, however, the most accessible section is at Annis' Knob, above Beer Harbour.

The Cenomanian Limestone varies greatly in thickness, from only one foot at Beer beach to 30 feet at Hooken Cliff. This is apparently related to a north-north-east-trending axis of inter-Cretaceous uplift. The Cenomanian rests with strong unconformity upon the uppermost beds of the Upper Greensand.

The inaccessible Windgate Cliff lies in the Trias marls.

1900, Jukes-Browne; 1903, Rowe and Sherborn; 1957, 1961, Smith; 1960, Tresise; 1962, Smith and Drummond; 1965, Ager and Smith.

Southacre Pit, Kingsteignton, 3 miles SE of Bovey Tracey

SX(20)854754	176	339	109NE	S, P, T
			(–)	IGC, 1967

A large working pit in white ball clay (pipeclay) interbedded with lignitic bands, exposing a section in the Oligocene Bovey Tracey Beds (Fig. 8). In these strata are found the only extensive deposits of lignite in Britain, with beds up to 10 feet thick. Both lignite and clay have been exploited since the early eighteenth century, but the lignite is no longer worked.

The Bovey Beds were laid down in an isolated lake basin formed in Devonian, Carboniferous (Culm) and Permian rocks, and at the north-west end they rest directly upon the Dartmoor granite. They now occupy an irregular lozenge-shaped area of 18 square miles, some 10 × 5 miles, elongated from north-west to south-east, and traversed by the River Teign and its tributary the River Bovey. The impression is now of a rather flat-bottomed wide open basin, heathy and wooded, with many active clay workings on a considerable scale.

The beds are found at all levels between about 30 and 442 feet O.D., indicating much erosion. Reid considered that at least 400 feet thickness of the beds had been removed by denudation, and their original total thickness may have been in excess of 1400 feet. The lake basin may have reached nearly 1600 feet deep and was presumably of tectonic origin.

In 1918 a boring east of Teingrace, at about SX855740, near the middle of the exposure, went to 667 feet below surface without reaching the base.* The ground level at the boring is given as 33 feet above O.D.; thus, the Bovey Beds descend more than 634 feet below the present sea level.

Although not mapped on the ground, the Sticklepath–Lustleigh fault zone appears to continue south-south-eastward in the direction of the axis of the Bovey basin. Possibly originating at a late Armorican date, the fault was active in Tertiary times (Blyth, 1957; Dearman, 1964).

In the British Isles today one lake alone approaches these depths, namely Loch Morar in the Scottish Highlands; 31 feet above sea level it has a maximum depth of 1017 feet. Elsewhere in the world there seem to be only about a dozen lakes over 1000 feet deep, so they are very exceptional. Whether the old Bovey lake was ever really a thousand feet deep or more may be questioned, for subsidence may have kept pace to some extent with the deposition. But these are matters for further investigation.

* Bott, Day and Masson-Smith (1958) estimated from their gravity survey that this boring may have ended only about 20 feet above the bottom of the basin.

Fig. 8. Southacre Clay Pit, Kingsteignton, near Bovey Tracey, Devon; looking north (1951). A working ball-clay pit, beds of white clay interbedded with blackish lignitic clay, of the Oligocene Bovey Tracey Beds; the dark bands show up superficial structures due to ice and formed during the Pleistocene. (W.A.M. Photo, R.21/4.)

The Bovey Beds consist of alternating fine-grained pottery clays and beds of lignite with some sands in the remaining upper part of the succession; and even sparse gravel round the margins, where it is worked commercially near Gappah in the north and at Staplehill in the south. The clay and sand indicate derivation from decayed Dartmoor Granite. Schorl (tourmaline) constitutes up to 25 per cent of some of the gravel and sand streaks. The common detrital minerals are quartz, chert, flint, limonite, zircon, rutile, tourmaline and muscovite mica, an assemblage without definite character.

The only records of a fauna are galls (Chandler) and a single partly destroyed elytron of a beetle, identified by Pengelly/Heer (1862, p. 1082) as *Buprestis Falconeri*.

However, there is a considerable flora preserved in the Bovey lignites and this has been investigated in great detail from the beds that have been accessible. It is overwhelmingly of three species, the conifer *Sequoia couttsiae* being easily the dominant element, followed by the fern *Osmunda lignitum* and the swamp-palm *Calamus daemonorops*. A modern relative of the first is *Sequoia sempervirens*, the giant Californian Redwood.

There are also found: true aquatics—*Stratiotes, Brasenia, Potamogeton;* marsh plants—*Salvinia, Caricoidea, Myrica, Microdiptera, Lysimachia;* climbers —*Rubus*, grape vines; and trees and shrubs—*Nyssa, Tilia, Hamamelis, Magnolia, Carpinus, Lauraceae* (including several species of *Cinnamomum*), oak and fig.

Chandler (1957, revised 1964) has confirmed 33 named species (and at least nine other forms) belonging to 31 families of trees and other vegetation. This flora suggests an accumulation of sub-tropical vegetation debris, mostly trunks and branches of *Sequoia*, swept down a steep valley into a lake surrounded by marshland. No trees are found in position of growth. There is a great abundance of a few forms, other species being represented by only one or a few specimens, perhaps because *Sequoia* forest and *Calamus* jungle do not provide a congenial habitat for a wide range of vegetation. A picture is given of indigenous upland vegetation and differs from that of the other Eocene and Oligocene plant beds of the south of England, which were accumulated at coast level and include the true lowland tropical flora. The presence of evergreen plants at Bovey indicates a considerable measure of warmth and humidity at all seasons, as do *Calamus* and *Mastixia*.

The age of the Bovey Beds, based upon the flora of the lignite, has long been controversial. Chandler (1957) described from these beds *Mastixia boveyana*, distinct from any species of the British Eocene, where the genus is abundant, and the genus is claimed to have disappeared from Europe after the Oligocene. The species of *Potamogeton* and *Rubus* found at Bovey do not appear until the Oligocene in other south of England fossil floras. However, the *Stratiotes* species (whose modern representative is *S. aloides*, the water soldier) are of most significance, for they have short time ranges. The Bovey species *S. websteri* is essentially Middle Oligocene.

The upper remaining part of the Bovey Beds, of which the lignites have been adequately sampled and studied palaeobotanically, appears to be definitely of Oligocene age; the possibility has not been excluded that the lowest beds might be of earlier, and the uppermost beds (now mostly eroded) of later, date. Further work on the pollen seems at present the only source from which more precise dating is likely to come.

In her 1964 revision Chandler considers that only the beds containing *Stratiotes websteri* are sufficiently firmly dated as Oligocene, simply, without mention of Middle Oligocene. 'The uppermost 197 feet at Bovey at least are probably Oligocene.'

I am indebted to Miss Chandler for kindly help and discussion.

Some superficial deposits in the Bovey basin contain *Betula nana* associated with *Pinus* and *Salix* (Pengelly/Heer, 1862, 1081) pointing to a Pleistocene date.

Problematical contorted structures affecting the upper layers of the Bovey Beds have been known for more than a century, but have only recently been studied. The undisturbed beds are almost horizontal, but in places, and separated by some 50 yards or more from the next, are found patches of disturbed strata. When excavated during the winning of ball clay, they show irregular domes or cones, roughly circular or square in plan, and some 50 feet in diameter. Only the upper strata are involved, to a depth of some 25 feet below surface, the main part of the structures being seen in about the top 10 feet, where dips of 50–60° are common, and sometimes range up to the vertical. They do not seem now to make any noticeable surface features. Some of these buried structures were truncated by erosion before the overlying gravel and alluvium were deposited. Nine have been investigated in detail (Dineley, 1963). They are explained as fossil pingos, dated to a stage of the Pleistocene when the Bovey basin was within the periglacial area of the ice sheets, with permafrost at slight depth. The clay has moved under pressure in the direction of least resistance, that is, upwards, and by its movement produced the cone-like structures clearly shown by the black carbonaceous bands.

As for the economic products, lignite appears to have been worked from early in the eighteenth century. In 1761 it was sold at the pit for half-a-crown a ton, but 'the thick heavy smoak which arises from this coal when burnt is very foetid and disagreeable'. It was then used as fuel for the neighbouring pottery, and for lime-burning near Chudleigh. In 1798 it was stated that in the winter 12 men could raise about 120 tons a week, which was perhaps a measure of the output. In 1862 there was a great open 'coal pit' a mile north of Heathfield. It was 960 feet long, 340 feet wide and nearly 100 feet at the deepest part. Tunnels opened from the bottom to mine the lignite, one of them 190 fathoms long. How it was kept from flooding is not clear.

South Devon ball clay was first used outside Devon in 1693. In 1785 more than 10 000 tons was shipped from Teignmouth to various English ports, partly for the North Staffordshire potteries, and by the end of the nineteenth century supplies were being sent abroad. In 1898 nearly 38 000 tons was dug, valued at £18 000. The Bovey clay now supports a flourishing industry, with reserves for 200 years or more at the present rate of working.

The clay is won by cutting it into seven (or eight)-inch cubes; this 'ball' is then hacked out with a special mattock called a 'tubill', and further handled with a spiked stick called a 'poge' (Sylvester-Bradley, 1939). Much is now mined, with access by sloping adits, and greater mechanisation. There are two principal types: potting or whiteware clays used in the manufacture of porcelain, and stoneware or siliceous clays used for making drain-pipes, tiles and bricks (Dewey, 1949). Different Bovey clays have a wide range of chemical and physical properties. The principal clay mineral is kaolinite, with subsidiary mica and quartz, and carbonaceous matter ranging up to some 6·5 per cent. The clays now have a number of uses other than for pottery.

1761, Milles; 1862, Pengelly; 1862, Heer; 1910, Reid; 1913, Ussher; 1923, Boswell; 1939, Sylvester-Bradley; 1949, Dewey; 1957, 1964, Chandler; 1957, Blyth; 1958, Bott, Day and Masson-Smith; 1963, Dineley; 1964, Dearman.

Tolcis Quarry, 2 miles NW of Axminster

ST(31)280010	177	326	60SW, 72NW	S, P
			(–)	1962

A large disused quarry exposing a magnificent section of Lower Lias passing down without a break into the Upper Rhaetic. Practically the complete local development of both the Hettangian and the Langport Beds is exposed. The whole sequence is fossiliferous, mainly with mollusca. The quarry had been in work for more than 150 years when it ceased working in 1956 (Fig. 9).

The succession has been measured bed by bed, and the fossils have been collected and studied by Hallam (1956). His primary object was to investigate the nature of the Lias–Rhaetic junction. At the coast this shows a sharp break marked locally by a pebble bed, whereas at Tolcis there is a gradual passage, and the most notable change observed lay in the reduction in the maximum length of *Modiolus hillanus* from 42 millimetres in the uppermost bed of the Rhaetic to 19 millimetres in the Pre-*planorbis* beds of the lowest Lias.

Separation of the zones and subzones is made from correlation with Lang's coast sections, checked by certain characteristic fossils as noted above.

Fig. 9. Tolcis Quarry, near Axminster, Devon; looking north-north-east (1934). Alternating limestone and shale bands in the Blue Lias (Lower Lias) overlying similar beds in the Rhaetic. (Crown Copyright Geological Survey photograph A.6418. Reproduced by permission of The Controller, H.M. Stationery Office.)

Hallam's section may be stated as follows.

		Zone	Subzone	Bed No.		
Lower Jurassic; Lower Lias	Lower Sinemurian	*bucklandi*	*conybeari*	156	8 in	Shale.
	Hettangian	*angulata*	*complanata*	155–142	8 ft 1 in	Seven shales interbedded with seven limestones; *Calcirhynchia calcaria* common, *Schlotheimia*. Bed 145 with *Spiriferina pinguis*.
			?	141–121	5 ft 2½ in	Ten shales with 10 limestones; *Pleuromya* cf. *crassa*, *Pholadomya ventricosa*, *Schlotheimia* sp.
			extranodosa	120–98	9 ft 2 in	Twelve shales with 11 limestones; *Pleuromya* cf. *crassa*, *Pholadomya ventricosa*.
		liasicus	*laqueus*	97–92	8 ft 8 in	Three shales with three limestones; Bed 92 with *Psilophyllites hagenowi*.
			portlocki	91–80	7 ft 4 in	Six shales with six limestones.
		planorbis	*johnstoni*	79–55	7 ft 3½ in	About 13 shales with some marl and 13 limestones.
			planorbis	54–47	4 ft 5 in	Four shales or marls with four limestones; *Pailoceras*.
		Pre-*planorbis*		46–26	10 ft 2 in	Eleven shales and 10 separately named limestones; Bed 44 with *Pleuromya tatei* var.; Bed 31 with *Meleagrinella* sp., also *Psiloceras* sp., *Placunopsis*, *Modiolus* cf. *hillanus*, *Liostrea hisingeri*, *Terquemina arietis*, *Diademopsis* spines etc. Scattered Ichthyosaur vertebrae were found in the bottom few feet.
Upper Rhaetic Langport Beds	'The Seven Beds'			25–7	6 ft 3½ in	Even-bedded cream-coloured limestones and pale-brown conchoidal marls, with *Liostrea hisingeri*, *Modiolus hillanus*, *Lima gigantes*, *Protocardia phillipsi*, *Parallelodon hettangiensis*, *Gervillia precursor*, *Atreta* [*Plicatula*] *intusstriata*, etc.
	'White Lias proper'			6–1	8 ft 5½ in	Thin-bedded, even-bedded or rubbly cream-coloured limestones with marl partings; Beds 3 and 4 are bored limestones; *Isocyprina ewaldi*, *M. hillanus*, *Lima valoniensis*, *Protocardia phillipiana*, *Chlamys* [*Pecten*] *valoniensis*.

De Luc described his visit to the quarry, near 'the hamlet of Tolshays hill', on 4 July 1806. The principal quarry, he wrote, lay at the top of the slope; it was very high and was worked in steps, both for building stone and for lime, which was burnt on the spot. He also noted many well-preserved small marine shells and that the strata dipped into the hill. Some reports like pistol shots heard from the lime kilns were said by the quarry foreman to be caused by small masses of mundic [pyrite] in the stone.

1811, De Luc (Vol. 3, p. 46); 1946, Lang; 1956, Hallam.

Torbryan Caves, 1 mile SW of Denbury, 7 miles W of Torquay

SX(20)815675	188	338	115NW	P, A, C
			(G)	1963

A group of nine small caves in Middle Devonian limestone. Six of them lie in woodland, in the face of a broken line of low cliffs forming the south-west side of the small, dry, Torbryan Valley, whose stream at some Pleistocene date was captured by the present Am Brook farther to the west. The three other caves lie on the opposite side of the Torbryan Valley.

From south to north: In the first rock cliff lies Torcourt Cave, at SX818673. In the second section of the cliff, in Dyer's Wood, are found Tornewton Cave, at 816674, fully treated below; the Old Grotto, with the remains of a mediaeval building associated with it; and an unnamed cave. In the third cliff section lie Three Holes Cave, at 813675, which has yielded a Sauveterrian (Mesolithic) industry and another which is probably late Upper Palaeolithic; and the Broken Cave and rock shelter, which have also yielded archaeological remains.

Across the valley are found Pulsford Cave, at 818678, which contained no archaeological material but gave valuable information about the history of the valley; and Levaton Cave, at 811679, which has likewise been excavated and produced Pleistocene mammalian remains (see p. 86). Also on the east side of the valley is Rectory Cave, at 817675, in the rock cliff behind Torbryan Old Rectory, which has not yet been excavated.

Tornewton Cave was first partly excavated by J. L. Widger about 1870–90 and subsequently by Sutcliffe and Zeuner from 1953 to 1960. Sutcliffe and Zeuner have deciphered a complex sequence of deposits and deduced the dating shown below. The site is further complicated because there are two sequences of deposits to be considered, one inside the cave and the other excavated in the talus at the mouth, outside the cave; and these two sequences, both significant, cannot be exactly correlated.

83

Tornewton Cave, apparently developed along a highly inclined fault fissure, has been excavated to 35 feet below the old floor, no base being found. It has accumulated the longest sequence of stratified deposits of Pleistocene age yet discovered at a British locality and its fauna has proved important for establishing a climatic chronology for the Upper Palaeolithic. The cave includes evidence of the presence of Man, and is of special interest in that it contains two horizons indicating cold climates, separated by a fully temperate horizon with remains of Hippopotamus. The succession is provisionally interpreted as representing the Riss and Würm Glaciations separated by the Eemian (Last) Interglacial.

The sequence inside Tornewton Cave is given as follows (Sutcliffe and Zeuner, 1962).

Stratum		Climate	Age
XV	Angular slabs of limestone. Black Mould. Evidence of Man	Much as today	Historic period
XIII	Stalagmite I		
XII	'Diluvium'. Evidence of Man; 6 mammal species, 29 non-marine molluscan species		Holocene
XI	Stalagmite II	Temperate	
X	Internal Reindeer stratum. Evidence of Man; 13 mammal species	Cold	⎫
IX	Dark earth. Fossils recorded by Widger included evidence of Man; ? = Elk stratum of the talus		⎬ Würm Glaciation
VIII	Shattering of Stalagmite III	Frost	⎭
VII	Stalagmite III		
VI	Hyaena stratum. c. 13 mammal species	Warm	⎫ Eemian Interglacial
V	Stalagmite IV		⎭
IV	Bear stratum. Brown Bear etc.	Less cold	
III	Glutton stratum. c. 14 mammal species; shattering of Stalagmite V	Frost	⎫
II	Stalagmite V (incorporated in Glutton stratum)		⎬ Riss Glaciation
I	Laminated Clay		⎭

The sequence in the talus excavated outside the cave is given as:

Correlation with section inside the cave		Age
XV	Soil	⎫ Post-glacial
—	Éboulis (Stratum XIV)	⎭
X	External Reindeer stratum; red earth with c. 13 mammal species	⎫
—	Grey loam, with few rodent teeth	⎬ Würm glaciation
? IX	Elk stratum. Evidence of Man; c. 13 mammal species	⎭
? VIII	Head	
	_ _	–Eemian Interglacial
I	Laminated clay	

84

Stratum I, the Laminated Clay, is an unfossiliferous, water-laid, yellow clay with layers of clastic material; the upper part was later brecciated, probably by frost.

Stratum II, Stalagmite V, was formed in some profusion. It was then all shattered and fell to the floor of the cave, becoming haphazardly mixed with mammalian remains and other material of the Glutton stratum.

Strata III and IV, the Glutton and Bear strata, have the same fauna: abundant Brown Bear; common Wolf, Fox, Cave Lion; more rare Glutton, Horse, Rhinoceros, Reindeer, small Bovid, Hare, Badger, Clawless Otter (the first British record of this animal), Rodents, Bats, Birds, Fish. The cave was a Bear den during these periods. Although the fauna of these two strata cannot be separated, the matrix was quite distinct. III contains a great deal of broken stalagmite and was quite unstratified. IV was faintly stratified and much less compact, and the few stones in it were of limestone, not stalagmite.

Stratum V, Stalagmite IV, formed a thin floor, a few inches thick, in the main chamber; the cave had apparently ceased to be a Bear den before this was deposited.

Stratum VI clearly represents a Hyaena den; some 1300 isolated Hyaena teeth, representing at least 110 animals, were recovered in the recent excavations, and some 20 000 were claimed by Widger, their remains far outnumbering all the rest of the animals together. These include Wolf, Fox, Cave Lion, and, more rare, Slender-nosed Rhinoceros, Hippopotamus, Fallow Deer, Red Deer, large Bovid, Hare, Vole and Birds. The presence of southern species, Hippopotamus and *Dicerorhinus hemitoechus* (the Slender-nosed Rhinoceros) and the absence of the northern forms, Glutton and Reindeer, point to such warmth that the stratum is identified as representing a full interglacial.

The Stalagmite floor, Stratum VII, is remarkable for containing a brown glassy material identified as the phosphate mineral collophane, no doubt derived from the abundant Hyaena coprolites.

The Head of the talus section has been identified in four facies, with rather rare and miscellaneous fossils. It is possibly the equivalent of Stratum VIII inside the cave, and indicative of frost.

The Elk stratum, also of the talus section, contains the first traces of Man, with a few flint flakes; also Wolf, Fox, Hyaena, Bear, Elephant, Horse, Woolly Rhinoceros, Red Deer, Elk, Reindeer, large Bovid, Hare and Rodents. Elk and Red Deer are woodland species, and the presence of Woolly Rhinoceros and Reindeer suggests that the amelioration of the climate was not so great as that represented by the Hyaena stratum. The Elk stratum is therefore referred to an interstadial period only. It is possibly the equivalent of Stratum IX inside the cave.

The Reindeer stratum X included one human incisor tooth, also Wolf, Fox, Hyaena, Stoat, Bear, Horse, Rhinoceros, Reindeer, small Bovid, Mole,

Hare, Rodents and a small amphibian, Frog or Toad. Reindeer and Bovid were the most abundant forms, but in curious occurrence; over 400 fragments of Reindeer antlers, nearly all of which had been shed naturally, but only 17 teeth and parts of bones. Of the Bovid nearly 100 fragments of rib, broken into pieces about four inches long. All these are interpreted as the debris of a human industry. Four flint and two bone artifacts were also found.

Reindeer is the only species of deer found, and all the shed antlers were small, probably of young or female animals. This is interpreted as marking a deterioration of climate, the animals being probably spring and summer visitors only, migrating southward to the Continent, which was still joined to this country, in the autumn because of severe winters here.

Stratum XII, which Widger termed the 'Diluvium', appears to be a cave earth, and not water-laid. In it were found some six species of animals, including Wolf, Hyaena, Bear, small Bovid, Reindeer and Rodents, and also a small fish, and 29 species of non-marine mollusca, which are still under description.

Stratum XIV. The position of the Éboulis, a deposit of angular blocks of limestone, in the talus succession is clear, but it is not certain exactly how it fits in with the succession inside the cave. It may possibly have come before Stratum XI. It contained many non-marine mollusca of species still living at this site, and a few other fossils; all of these might be later introductions. The deposit is believed to represent a cold phase.

Of Stratum XV, the Black Mould contains a variety of fossils, including flints, rounded pebbles, pottery fragments, remains of Rodents, charcoal, and a small bronze Roman coin. Both this stratum and the angular slabs of limestone are believed to date from the historic period.

Kowalski (1966) has identified the Rodent remains from Tornewton Cave as follows. Forest species: *Apodemus sylvaticus* (Common Field Mouse), *Clethrionomys glareolus* (Bank Vole). Tundra species: *Dicrostonyx* (Arctic Lemming), *Lemmus* (Norway Lemming), *Lagurus* (Steppe Lemming), *Microtus gregalis* (Narrow-Headed Vole), *Microtus nivalis* (Snow Vole).

The forest species came generally from the strata assigned above to the Holocene (Post-glacial) and to the Eemian Interglacial, and the Tundra species from the strata dated to the Riss and Würm Glaciations.

Levaton Cave was discovered in 1950 by Mr Walter A. Cheesman, who carried out some excavation; among his finds was the jaw of a very young Hyaena. Excavation was resumed in 1967 by a team under Dr Sutcliffe, as it was thought to be a favourable site on which to pursue the problems posed by Hyaena remains in cave deposits in Britain. Remains, some extensively gnawed, included Woolly Rhinoceros, Mammoth, Horse, Giant Ox or Bison, Red Deer, Fox, Badger, Rodents, a small bird and amphibia. Further evidence is required and it is planned to continue the dig (1967: *Assoc. Wm. Pengelly Cave Research Centre*, Newsl. No. 9, 38–9).

I am greatly indebted to Dr A. J. Sutcliffe for help with the Torbryan caves.

1962, Sutcliffe and Zeuner; 1966, Kowalski; 1967, Walker and Sutcliffe.

Wilmington Quarry, 3 miles E of Honiton

SY(30)208999	177	326	71NW	S, P
			(–)	PSD, 1966

A large working quarry in the arenaceous facies of the Cenomanian, the most westerly inland occurrence of the Upper Cretaceous in England. The presence of this small patch of Chalk is due to a north–south fault half a mile to the east and downthrowing west.

The section shows (after Smith, 1957b):

Recent to Tertiary Clay with Flints, which descends in pipes and fissures into the underlying strata

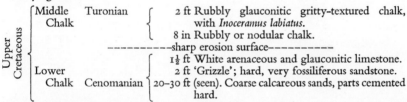

The base of the Cenomanian is not exposed here, but may be seen in Reed's Farm Pit (q.v.) half a mile to the north-east.

The Cenomanian sands and sandstone here contain a large and very abundant fauna, and the quarry is one of the finest Cretaceous fossil localities in the country. It has yielded some 300 species, including ammonites (*Mantelliceras, Calycoceras, Schloenbachia*), brachiopods (*Cyclothyris*), gastropods, bivalves (*Chlamys* [*Pecten*] *aspera, Neithea* [*Pecten*] *quinquecostata, Inoceramus etheridgei*), many echinoids (especially *Holaster*), polyzoa and corals. A curiosity of preservation shows sand grains forced into the calcareous shells of some of the fossils, notably the echinoids. This fauna is unique in Britain, but comparable with that of the Sarthe in Western France.

1898, 1903, Jukes-Browne; 1911, Woodward and Ussher; 1957b, Smith.

DORSET

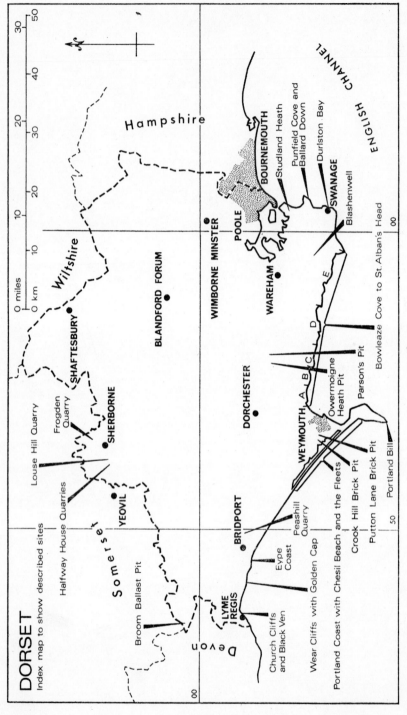

DORSET
Index map to show described sites

Halfway House Quarries

Louse Hill Quarry

Frogden Quarry

Broom Ballast Pit

Wiltshire

Somerset

Devon

YEOVIL

SHERBORNE

SHAFTESBURY

BLANDFORD FORUM

Hampshire

WIMBORNE MINSTER

POOLE

BOURNEMOUTH

Studland Heath

Punfield Cove and
Ballard Down

Durlston Bay

SWANAGE

Blashenwell

WAREHAM

DORCHESTER

Owermoigne
Heath Pit

Parson's Pit

Bowleaze Cove to St. Alban's Head

ENGLISH CHANNEL

WEYMOUTH, A, B, C, D, E

Putton Lane Brick Pit

Crook Hill Brick Pit

Portland Coast with Chesil Beach and the Fleets

Portland Bill

Peashill
Quarry

Eype
Coast

Wear Cliffs with Golden Cap

Church Cliffs
and Black Ven

LYME
REGIS

BRIDPORT

0 miles   10   20   30
0 km   20   40

Fig. 10

Black Ven and Church Cliffs, Lyme Regis

National Grid Reference	One-inch Sheet, Pop. and 7th ed. Ordnance Survey	One-inch Sheet, Geol. Survey	Six-inch Sheet, Ordnance Survey	Main Site Interests
SY(30)353930	177	326, 327	37SW (G)	P, S PSD, 1963

One and a half miles of coastal cliffs between Lyme and Charmouth (Fig. 11), exposing classic and standard sections, perhaps the best in Europe, of the greater part of the abundantly fossiliferous Lower Lias; studied and the fauna collected since the eighteenth century. Also sections of the overlying fossiliferous Cretaceous, Upper Greensand and Gault. The succession is:

THE LOWER CRETACEOUS SECTION

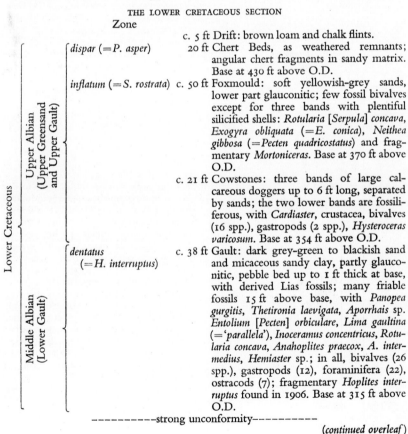

Zone

c. 5 ft Drift: brown loam and chalk flints.

dispar (=*P. asper*)   20 ft Chert Beds, as weathered remnants; angular chert fragments in sandy matrix. Base at 430 ft above O.D.

inflatum (=*S. rostrata*) c. 50 ft Foxmould: soft yellowish-grey sands, lower part glauconitic; few fossil bivalves except for three bands with plentiful silicified shells: *Rotularia* [*Serpula*] *concava, Exogyra obliquata* (=*E. conica*), *Neithea gibbosa* (=*Pecten quadricostatus*) and fragmentary *Mortoniceras*. Base at 370 ft above O.D.

c. 21 ft Cowstones: three bands of large calcareous doggers up to 6 ft long, separated by sands; the two lower bands are fossiliferous, with *Cardiaster*, crustacea, bivalves (16 spp.), gastropods (2 spp.), *Hysteroceras varicosum*. Base at 354 ft above O.D.

dentatus (=*H. interruptus*)   c. 38 ft Gault: dark grey-green to blackish sand and micaceous sandy clay, partly glauconitic, pebble bed up to 1 ft thick at base, with derived Lias fossils; many friable fossils 15 ft above base, with *Panopea gurgitis, Thetironia laevigata, Aporrhais* sp. *Entolium* [*Pecten*] *orbiculare, Lima gaultina* (='*parallela*'), *Inoceramus concentricus, Rotularia concava, Anahoplites praecox, A. intermedius, Hemiaster* sp.; in all, bivalves (26 spp.), gastropods (12), foraminifera (22), ostracods (7); fragmentary *Hoplites interruptus* found in 1906. Base at 315 ft above O.D.

Upper Albian (Upper Greensand and Upper Gault)

Middle Albian (Lower Gault)

Lower Cretaceous

----------strong unconformity----------

(*continued overleaf*)

91

		davoei (Beds 127 part–122)	25 ft Green Ammonite Beds: lower part only of thinned succession on east shoulder of Black Ven; dark marly clays and limestones; highest bed is clay 2 ft above red band (Bed 126); fossils include ammonites (c. 52 spp.), belemnites (5), gastropods (9), bivalves (11), foraminifera (55), brachiopods (4), crinoids (3). Base at 305 ft above O.D.
	Lower Pliensbachian or Carixian	*ibex* *jamesoni* (Beds 121–103b)	75 ft Belemnite Marls: the Belemnite Stone (Bed 121) at top; rather inaccessible in the highest clay precipice on Black Ven; light-grey marls with some indurated bands and some lenticular masses of lignite. Except in the Belemnite Stone, belemnites are almost the only abundant and well-preserved fossils; 26 named spp. (19 described from here as new) are recorded in five genera, two of which were described from here, *Angeloteuthis* and *Clastoteuthis;* of many ammonites, there are seven species of *Tropidoceras;* bivalves (8), gastropods (7), brachiopods (16), foraminifera (8). Base at 225 ft above O.D.
Lower Jurassic Lower Lias	Sinemurian	*raricostatum* (*oxynotum* missing) *obtusum* (Beds 103a–76b)	145 ft Black Ven Marls: the two lower precipices on Black Ven. Monotonous blue-black clays and paper shales, with occasional tabular and nodular limestones; iron pyrites abundant; non-sequence above Coinstone (Bed 89); many spp. of ammonites (12 described from here); badly preserved bivalves (12), brachiopods (8), foraminifera (21), many insects from the flatstones (Bed 83). The *Birchi* tabular at bottom (Bed 76a).
		turneri *semicostatum* (Beds 76a–54)	75 ft Shales with Beef: reefs on the foreshore at Charmouth; paper shales, marls, indurated bed and limestone nodule beds, with much fibrous calcite ('beef') showing cone-in-cone structure; much pyrite, calcite, selenite, traces of barytes; fossils mainly poor, many ammonites, badly preserved, bivalves (8 spp.), foraminifera (23); belemnites, fishes, *Ichthyosaurus*.
	Hettangian	*bucklandi* *angulata* (Beds 53–1)	85 ft (seen). Blue Lias: Church Cliffs and reefs on shore; limestones separated by thin shales; Table Ledge (Bed 53) at top; Brick Ledge (Bed 1) at bottom is the lowest bed seen under Church Cliffs; fossils include a great profusion of large ammonites, Coronocerates etc. reach more than 2 ft in diameter; also nests of *Gryphaea arcuata, Lima gigantea*, etc., *Isocrinus tuberculatus*, about 30 spp. of foraminifera; five known genera of fishes; the great marine reptiles come mainly from the Blue Lias.

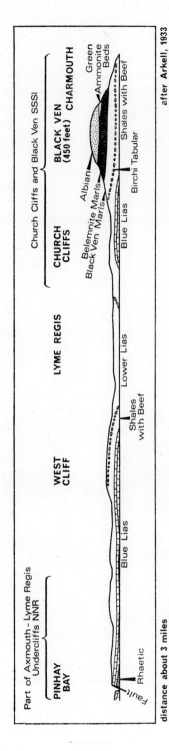

Fig. II. Pinhay Bay–Charmouth: coast section exposing Albian (Lower Cretaceous) unconformably overlying Lower Lias.

after Arkell, 1933

(the O.D. levels are given for the eastern section on Black Ven (Lang, 1904, 1907, 1923, 1924, 1926, 1928, 1936).

The lowest 18 feet of the *angulata* Zone, and the whole of the *planorbis* and pre-*planorbis* Zones down to the White Lias at the top of the Rhaetic are not exposed at the present site; they may be seen in the Axmouth–Lyme Regis National Nature Reserve to the west of Lyme (q.v.).

At Charmouth a 'submerged forest' is intermittently exposed on the shore at the mouth of the River Char. A section one foot thick was seen in 1960, driftwood, branches and twigs lying horizontally, resting upon small blackish cherts with white rind. Red Deer remains have been found in the lowest level only.

Exposures of the Cretaceous beds are to be seen in landslipped ground along, above, and below the old abandoned coast road between Lyme and Charmouth at about SY355932. The basal pebbly layer is porous and springs are thrown out where it overlies the Lias Clay.

This Upper Greensand near Lyme Regis has been dated by the radio-metric potassium–argon method on the glauconite as 96 million years old (Evernden *et al.*, 1961: *Geochim. Cosmochim. Acta* **23,** 78).

It is impossible to separate the fossil records for the Lower Lias of this site alone; thus the fauna outlined above and below refers to that of the relevant beds from the whole of the Dorset coast sections. Many Lias fossils are recorded only from the 'Dorset Coast' and may have come from anywhere between four miles west and eight miles east of Lyme. The exact horizons of many are also unknown, since a score of good specimens are picked up loose or from displaced blocks for every one collected *in situ*.

The name 'Green Ammonite Beds' derives from ammonites embedded in limestone nodules in these particular beds, their chambers filled with green calcite. Specimens were cut down the middle, polished and sold as curios. However, these and other fossils are no longer sold in the shops at Lyme.

EARLY RECORDS

John Woodward (1728) seems to record no fossils from the present site in his collection, but he states: 'Mr Hutchinson in the searches he made by my direction in the year 1706 observed incredible numbers of these shells [ammonites] thus flatted and extremely tender in shivery stone about *Pyrton* Passage, *Lime* and *Watchet*'.

The earliest general description found, that of Maton (1797), proves that scientific interest in the sections was well established by that date. He wrote that the *ludus Helmontii* [septarian structure] is common here and it is difficult to presuade the vulgar that it is not a fossil turtle. There is a good deal of pyrite and bituminous matter in the soil which has often taken fire after heavy rains and produces an appearance of flames at a distance. A remarkable instance of this occurred in 1751.*

De Luc gave a brief description of the shore section from Charmouth to Lyme in 1805, and Townsend (1813) gave another early account.

The earliest detailed geological studies of both Lias and Cretaceous here were published by De la Beche in 1822 and 1826, and the richness and variety of the fossil fauna have resulted in a very large literature, culminating in the extremely detailed and meticulous work of Dr W. D. Lang and his collaborators, described in a dozen important memoirs and many other communications, spanning some 60 years.

THE LIAS OF THE DORSET COAST

'The Lias is cut by the Dorset coast in such a way that a complete section is laid bare across the extension of the trough of deposition that stretched from the Mendip Hills and South Wales in the north to the massif of the Cotentin and Normandy in the south. One after another its zones rise from the beach to build the line of magnificent cliffs extending from Bridport in Dorset westwards past Charmouth and Lyme Regis to Seaton in Devon' (Arkell, 1933).

In addition to the general slight easterly tilt of the Mesozoic strata hereabouts, the Lias between Lyme and Charmouth was thrown into a series of gentle folds, and the sections are broken by a number of small faults. The net result has been to impart a slight easterly dip to these strata at an early date.

After this came marine planation, and the subsequent Cretaceous transgression overstepped lower and lower strata from east to west. In the east, at Thorncombe Beacon, the Gault rests upon Inferior Oolite and the underlying Upper Lias; at Golden Cap it rests upon Middle Lias, and at Black Ven upon the eroded Lower Lias. Six miles farther west, near Seaton, it rests directly upon the Trias.

Three Sites of Special Scientific Interest on the Dorset coast between them illustrate the whole Lias succession from the top of the Upper Lias in the Eype coast site, through the Middle and the upper part of the Lower Lias in the Wear Cliffs and Golden Cap site, and the majority of the Lower Lias in the Black Ven and Church Cliffs site. The succession downward to the base of the Lower Lias, and the underlying Rhaetic down to the Trias is exposed in the adjoining Axmouth–Lyme Regis National Nature Reserve along the coast of Devon.

* There was a similar fire on the cliffs between Charmouth and Lyme in 1908, and iron pyrites was formerly collected from the Black Ven Marls for manufacturing sulphuric acid.

In no other part of the country can the Lias as a whole be seen so well. In these coast sections the Upper Lias, of sands and sandy clay, is about 200 feet thick; the Middle Lias, of sandy clays, about 345 feet thick; and the Lower Lias, of marls and limestones, and the outstandingly fossiliferous part of the whole Lias succession, is about 485 feet thick.

In former times limestones of the Blue Lias on the shore were wrought for cement-making and for road stone and paving slabs, an industry which seriously increased coast erosion. Roberts recorded that 90 feet breadth of Church Cliffs was lost in 31 years between 1803 and 1834. Such working has now been stopped. One result was that each worked limestone bed (but this excluded many in the lowest part of the section) was given a name by the quarrymen to identify it; these names have been usefully embodied by authors, and notably by Lang, in their geological descriptions of the section.

It has been suggested since Day in 1863, and finally by W. A. Richardson (in Lang *et al.*, 1923) that the Lower Lias limestones here are not original but secondary, the result of rhythmic precipitation of a concretionary limestone suite after deposition of an original calcareous clay. This view was generally accepted until Scott Simpson in 1957 and Hallam (1956, 1960, 1961) concluded otherwise. Simpson claimed that 'the bed-junction preservation of *Chondrites* (the borings of problematic marine worms) in the Blue Lias limestones of Lyme Regis and the Belemnite Marls of Charmouth proves that the limestone bands are original depositional features and eliminates the possibility of chemical segregation on the lines suggested by Richardson'. Hallam investigated the deposition of these beds in great detail, and added other grounds, reaching the same conclusion as to the primary nature of most of the limestone beds. This conclusion, however, was again queried by Kent (1957: *Geol. Mag.* 429).

CONDITIONS OF DEPOSITION OF THE LOWER LIAS OF THE DORSET COAST

General marine conditions are indicated by the abundant fauna of ammonites, belemnites, foraminifera and other groups. The rarity of echinoids (except in the basal beds) and of corals may be due to the generally muddy facies.

Other, but sparse, elements of the fauna, such as 'Pterodactyl' and Dinosaur, lignite and plants, and insects at one notable horizon, suggest proximity of the land, and the possibility of some admixture of fresh water, as in an estuary.

A general lack of oxygen and the presence of much sulphuretted hydrogen in the highly calcareous water are indicated by the abundance of pyrite, which, finely disseminated, accounts for the general dark or slate-grey colour of the Lower Lias, here as elsewhere. In some of the beds it is also found

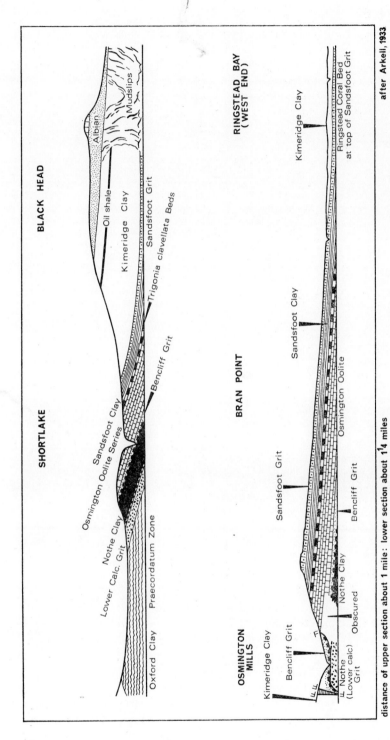

SHORTLAKE   BLACK HEAD

Albian
Oil shale
Mudslips
Kimeridge Clay
Sandsfoot Grit
Trigonia clavellata Beds
Bencliff Grit
Sandsfoot Clay
Osmington Oolite Series
Nothe Clay
Lower Calc. Grit
Praecordatum Zone
Oxford Clay

RINGSTEAD BAY
(WEST END)

Kimeridge Clay
Ringstead Coral Bed
at top of Sandsfoot Grit

BRAN POINT

Sandsfoot Grit
Sandsfoot Clay
Bencliff Grit
Osmington Oolite
Nothe Clay
Obscured

OSMINGTON
MILLS

Kimeridge Clay
Bencliff Grit
Nothe
(Lower calc.) Grit

distance of upper section about 1 mile: lower section about 1¼ miles

after Arkell, 1933

Fig. 13. Shortlake–Ringstead Bay: coast section showing Albian (Lower Cretaceous) unconformably overlying Upper Jurassic.

in the clay, forming ponds and marshes, and mud slides which creep down the cliffs.

At Osmington Mills, SY735818, there is complex trough faulting. Beyond, the cliff shows a nearly complete succession of the Corallian.

CORALLIAN SUCCESSION MEASURED BETWEEN OSMINGTON MILLS AND RINGSTEAD
(Arkell, 1947a)

(Kimeridge Clay above)

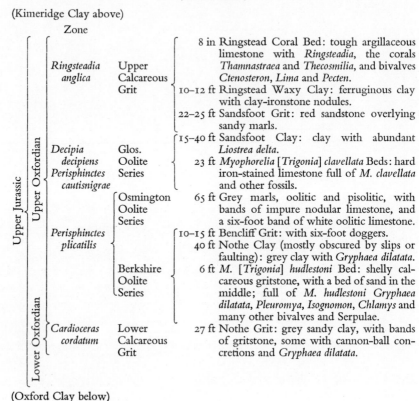

(Oxford Clay below)

MICROFAUNA

From the Corallian succession here Gordon (1965) has described and well figured 59 species of foraminifera, four new. In numbers two species, *Ammobaculites coprolithiformis* and *Lenticulina münsteri*, make up more than half the fauna at almost every horizon. The family Lagenidae is by far the most conspicuous, being recorded in 15 genera and 35 species.

From the Osmington Mills section at SY736817 Sarjeant (1960) described two new Hystrichospheres: *Baltisphaeridium lumectum* from the Ringstead Waxy Clay, and *Polystephanosphaera valensii* from the Osmington Oolite.

SECTOR B. RINGSTEAD BAY

From near Ringstead, and eastwards of SY748816, lies Kimeridge Clay. The shore follows the strike, with the dip to the north. The low cliff in the *Pictonia* and *Rasenia* Zones is capped by flint gravel. The cliff increases in height near the middle of Ringstead Bay, SY760814, where the *Aulacostephanus* Zone comes in, and, still higher, the Oil Shale horizon. Sections are poor, however, and in the last half mile, nearly to White Nothe, the Kimeridge Clay is concealed by slipped Cretaceous material.

Burning Cliff, SY7682, west of Holworth House, takes its name from a fire in 1826, when rapid oxidation of pyrite ignited the oil-shales of the Kimeridge Clay. These smouldered below the surface for about four years, baking the clay to the condition of red tiles or even to a scoriaceous slag.

Two north-trending faults with downthrow to the east bring forward the outcrop of Portland Sand and Stone to the brow of the cliff, which is capped by Lower Purbeck Limestones with ostracods and *Neomiodon*. On the downthrow side of the fault the sequence shows (Arkell, 1947a):

Upper Jurassic	Lower Purbeck {		Broken Beds.
			Gastropod Bed with *Hydrobia* and *Valvata*.
	Portland Stone {	21 ft	Freestone Series, with a 'roach' at the top.
		34 ft	Cherty Series, with a bed full of *Glomerula* [*Serpula*] *gordialis* just above the base.
	Portland Sand	72 ft	A bed full of *Exogyra nana* 20 ft below top.
	Kimeridge Clay	185 ft	Paper shales and clays with fauna of the *Pavloviae* and *pectinatus* Zones down to the White Band. Two phosphate nodule beds 82 ft and 90 ft above the White Band yield small Pavlovids and bivalves.

Below Holworth House, SY764816, Cretaceous beds dipping gently eastwards progressively overstep the upturned edges of the Kimeridge Clay, Portland Sand, Portland Stone and Purbeck Beds, giving a striking example of angular discordance. Here is the finest section in the county of the upper part of the Upper Greensand and Gault; the rest of the stage is not well exposed (Fig. 14).

SECTOR C. WHITE NOTHE TO SWYRE HEAD (Fig. 14)

From White Nothe eastwards is found the Chalk, and the mile of magnificent Chalk cliffs from Bat's Head to Durdle Door presents an unrivalled section of the vertical and sometimes overturned beds.

The top of White Nothe lies in the *coranguinum* Zone; to the east, at first under a gravel cap, the *testudinarius* Zone appears; and beyond, on both sides of West Bottom and onward to Bat's Head, SY795804, *quadrata* Chalk forms the ground surface. East of Bat's Head the Chalk is much disturbed by folds and thrusts, the bedding being partly vertical at Swyre Head; but it

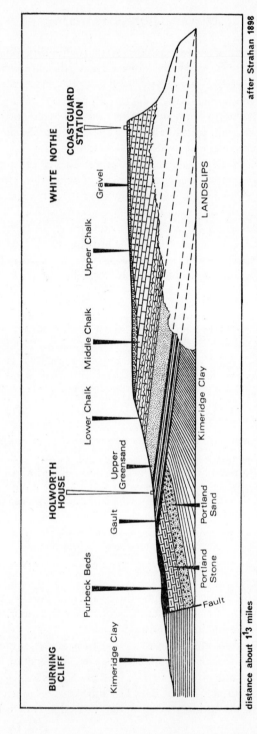

after Strahan 1898

distance about 1⅓ miles

Fig. 14. Burning Cliff–White Nothe: coast section showing Chalk and Lower Cretaceous unconformably overlying Upper Jurassic.

continues to Hambury Tout, 454 ft above O.D. at SY816803, where *quadrata* Chalk again occupies the summit. The bedding there is vertical, or in places inverted.

At White Nothe there are two famous fossil localities. The first is the Basement Bed of the Cenomanian, where fossils are very abundant, mostly ammonites, echinoids and brachiopods; fallen blocks of this bed furnish the best collecting from this horizon in Dorset. The second is the *dispar* Zone of the Upper Greensand, with nearly 50 species of ammonites, including *Engonoceras*, one of the two British species known of the Pseudoceratitids; and very numerous other fossils of at least 16 species, including bivalves, gastropods, echinoids and brachiopods.

THE CRETACEOUS SUCCESSION AT WHITE NOTHE, SY775809
(Arkell, 1947a)

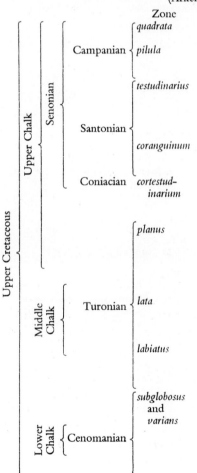

			Zone	
Upper Cretaceous	Upper Chalk	Senonian	Campanian	*quadrata* — c. 60 ft Fairly hard white chalk with flints, many marl seams, few fossils.
			pilula — 136 ft Fairly hard white chalk with flints, many marl seams, abundant *Offaster pilula, Terebratulina*, starfish ossicles.
			Santonian — *testudinarius* — 111 ft Fairly hard white chalk, with lines of flints; *Uintacrinus* band at base, *Marsupites testudinarius* in a higher band; *Echinocorys, Bourgeticrinus* and Terebratulids.
			coranguinum — 170 ft Fairly hard chalk, flints generally in regular lines; few echinoids; *Micraster, Echinocorys, Conulus, Cidaris*.
		Coniacian	*cortestud-inarium* — 113 ft Hard discoloured chalk, with bands of yellow nodular chalk, and much flint in irregular layers; poorly fossiliferous, few *Micraster, Echinocorys*.
	Middle Chalk	Turonian	*planus* — 51 ft Soft marly grey chalk with many irregular flint bands; rich in fossils: *Micraster, Echinocorys, Cidaris, Tylocidaris, Isocrinus*, the calcareous sponge *Pharetrospongia*, and a band of the polyzoan *Bicavea* near the base.
			lata — 58 ft Hard marly chalk, with bands of yellow and green nodules 12–16 ft from the top; flints rare; *Terebratulina lata* common, few echinoids.
			labiatus — 76 ft Hard white nodular chalk, with flint band at the top; *Inoceramus labiatus* abundant, also *Tylocidaris, Discoidea, Hemiaster, Cardiaster*.
	Lower Chalk	Cenomanian	*subglobosus* and *varians* — 45 ft Blocky grey marly chalk with few fossils; at base a 5-ft sandy glauconitic bed, richly fossiliferous, with *Schloenbachia, Mantelliceras, Calycoceras, Turrilites, Scaphites, Cyrtocheilus, Holaster, Hemiaster, Glyptocyphus, Conulus*, and numerous Terebratulids.

(continued overleaf)

117

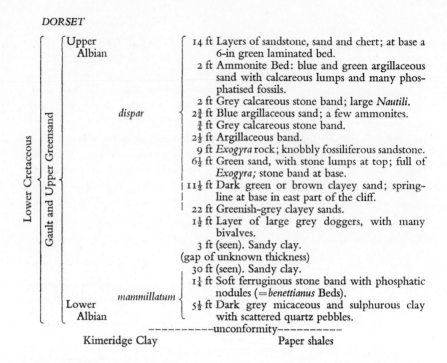

	Upper Albian		14 ft Layers of sandstone, sand and chert; at base a 6-in green laminated bed. 2 ft Ammonite Bed: blue and green argillaceous sand with calcareous lumps and many phosphatised fossils.

Lower Cretaceous — Gault and Upper Greensand

Upper Albian — *dispar*

14 ft Layers of sandstone, sand and chert; at base a 6-in green laminated bed.

2 ft Ammonite Bed: blue and green argillaceous sand with calcareous lumps and many phosphatised fossils.

2 ft Grey calcareous stone band; large *Nautili*.

2¾ ft Blue argillaceous sand; a few ammonites.

¾ ft Grey calcareous stone band.

2½ ft Argillaceous band.

9 ft *Exogyra* rock; knobbly fossiliferous sandstone.

6½ ft Green sand, with stone lumps at top; full of *Exogyra;* stone band at base.

11½ ft Dark green or brown clayey sand; spring-line at base in east part of the cliff.

22 ft Greenish-grey clayey sands.

1½ ft Layer of large grey doggers, with many bivalves.

3 ft (seen). Sandy clay.

(gap of unknown thickness)

30 ft (seen). Sandy clay.

Lower Albian — *mammillatum*

1¼ ft Soft ferruginous stone band with phosphatic nodules (=*benettianus* Beds).

5½ ft Dark grey micaceous and sulphurous clay with scattered quartz pebbles.

----------unconformity----------

Kimeridge Clay   Paper shales

C. V. and E. W. Wright (1949) recorded six or more species of *Discohoplites* (five of them new) and one new variety of *Hyphoplites* from the Upper Albian of White Nothe.

SECTOR D. DURDLE DOOR TO WORBARROW TOUT

The name of the little promontory of Durdle Door, SY805801, is derived from an Anglo-Saxon (Old English) word *thirl* meaning to pierce. A spectacular monument of unexpected permanence, its name shows that it has stood in the breakers for at least 800 years, and probably over 1000 years.

Durdle Door bears witness to tectonic movements of Tertiary date culminating in the Miocene, for it forms part of the northern limb of the Purbeck Monocline. This is one of related folds *en echelon* which can be traced along the South Coast from the Weymouth Anticline in the west to the Isle of Wight in the east. Of all the folds thrown up across the Chalk plains of north-west Europe during the Alpine mountain-building epoch, the most interesting and the most perfectly exposed is that of Purbeck. In the cliffs of the Dorset coast the sea has laid bare all parts of the structure. In the Durdle promontory are to be found some of the consequences of the folding. The Albian, Wealden and Purbeck rocks have been attenuated in three ways. It is estimated that about 130 feet of Upper Wealden beds was removed by denudation before the deposition of the Gault. Strike-faulting has cut out about half the Lower Wealden, the whole of the Upper Purbeck and part of the Middle Purbeck; and finally the incompetent Wealden shales remaining have been squeezed to a mere 215 feet thickness. At Worbarrow, four miles to the east, they are measured as 1393 feet thick. In Durdle Cove are to be seen the low-angled thrust planes which form a striking feature of this coast (Arkell, 1938).

Phillips (1964) has re-studied the structures in the rocks along the coast between White Nothe and Mupe Bay, and arrived at somewhat different conclusions. He suggests that they resulted from '. . . the accommodation of the blanket of largely unconsolidated sedimentary beds above a major thrust fault in the basement. At first a broad flexure developed and this became most pronounced in the highest division, the Chalk, because of the buckling and northward sliding of these beds off the rising land to the south. As the magnitude of the fault increased, successively higher beds fractured and eventually the fault extended to the surface and brought about the displacement and modification of the structures formed during the earlier phase of folding'.

The hinterland is of Chalk, involved in the Purbeck folding; south of the axis the Chalk stands vertically or slightly inverted, traversed by thrusts, and ending against a vertical sequence towards the south of Upper Greensand and Gault, Wealden, Purbeck, and a selvage of Portland Stone at the coast. From Durdle Door this sequence continues for four miles eastward to Worbarrow Tout at SY870797. The barrier of Portland and Purbeck beds has been breached by the sea, which has exposed Cretaceous rocks in the embayments at St Oswald's Bay, Lulworth Cove, and from Mupe Bay to Worbarrow Bay. However, stretches of the barrier remain at Durdle Door itself, from Dungy Head to Lulworth Cove, from Lulworth Cove to Mupe Bay, and from Worbarrow Tout a little to the east, before the pattern changes. Finally, the Chalk forms the coast for a mile between Mupe and Worbarrow Bays.

The beauty spot of Lulworth Cove offers a great variety of interesting tectonics, the different strata recording the results of the severe stresses they suffered during the folding. The Lulworth Crumple is a superbly exposed accomodation fold in the Purbeck Beds, best seen nearby to the west at Stair Hole. Lulworth Cove itself is explained by Burton (1937) as being due to marine erosion exploiting a small stream channel cut through the barrier of the hard Purbeck and Portland limestones, while the low ground behind, in the soft strata of the Wealden, already part-eroded by short streams draining from both east and west, was easily washed away by the sea through the gap. A little east of Lulworth Cove is found the best strike section of the Purbeck Broken Beds, and underlying them the famous Lulworth Fossil Forest, at

SY831797, first described by Buckland and De la Beche in 1836. The section, which dips northward, shows:

Lower Purbeck	25 ft Broken Beds.
	(gap)
	2–4 ft Soft Cap: irregular bed full of giant calcareous tufa 'burrs' enclosing the boles and prostrate trunks of large silicified trees, Cycads and Conifers.
	½–1 ft Dirt Bed: black gravelly earth full of white stones; with the above trees rooted in it.
	15 ft Hard Cap.

The algae of some of these beds are described by Pugh (1968). The beds rest upon Portland Stone, which forms the vertical cliff descending below sea level.

Mupe Bay and Bacon Hole, SY842798, afford the most complete section from the Wealden (750 feet thick) through the Purbeck (250 feet thick) down to the Portland Stone, exposing the critical junction between Wealden and Purbeck. Dips range from 40 to 75°; and at least 16 feet of the Wealden Sands with shaly layers is impregnated with oil.

The Upper Cretaceous succession at Mupe Bay, Arish Mell, Worbarrow Bay and Worbarrow Tout is comparable with that given above at White Nothe, although higher beds are found, and the thicknesses differ considerably at the two sites. In view of the general similarity only an outline of the beds here is given below (Arkell, 1947a).

					Zone	Thickness, in feet
Upper Cretaceous		Upper Chalk	Senonian	Campanian	*mucronata*	30
					quadrata	180 at least
					pilula	120
				Santonian	*testudinarius*	39
					Uintacrinus	30
					coranguinum	200
				Coniacian	*cortestu-dinarium*	60
		Middle Chalk	Turonian		*planus*	(?)
					lata	(?)
					labiatus	49
					plenus subzone	6
		Lower Chalk	Cenomanian		*subglobosus* and *varians*	c. 90
Lower Cretaceous			Albian and Aptian		Gault and Upper Greensand	c. 190
					Lower Greensand	115
			Neocomian		Wealden	1393
					Purbeck	290
Upper Jurassic			Portlandian		Portland Stone Portland Sand	

120

At Worbarrow Bay the basal beds of the Middle Chalk show the most remarkable development of fracture cleavage to be seen on this coast. The close-set cleavage planes, about vertical, have been etched out to perfection on the sea- and spray-washed cliff at the farthest accessible point of the bay. In the cliffs below Flowers Barrow in Worbarrow Bay the 190 feet of Gault and Upper Greensand is the greatest thickness of these beds recorded on the coast. They contain *Exogyra*, ammonites, etc. The variegated Wealden marls and sandstone give to the beautiful Worbarrow Bay its brilliant red, orange, yellow and purple cliffs, with the 20-feet-thick Coarse Quartz Grit in the middle.

Complete sections of the Purbeck are found on the west and north sides of Worbarrow Tout; the beds have a steep northerly dip, the limestones standing up in wall-like slabs. The Lower Purbeck contains large masses of gypsum. The best sections of the basal Purbeck Beds, from the Cypris Freestones downwards, and dipping 35°N, are exposed on the east side of the Tout. The Broken Beds here show unusual features. The Purbeck Beds rest on the Shrimp Bed of the Portland Stone.

The view both east and west from Worbarrow Tout gives a good impression of the structure of Purbeck; the lowland core of Kimeridge Clay with its black cliffs, and the escarpment of the Portland Stone which forms a nearly level plateau at St Alban's Head and Swyre Head and then turns over to plunge down northwards at 30–40° all along the ridge of hills which ends in the splendid precipice of Gad Cliff.

SECTOR E. WORBARROW TOUT TO ST ALBAN'S HEAD
(Figs. 15 and 16)

From a quarter of a mile east of Worbarrow Tout the six and a half miles of coast to St Alban's Head is formed of Kimeridge Clay, including its type section. In places the outcrop is concealed by landslipping on a large scale. Portland Sand (with its type section at Emmit Hill), Portland Stone and Purbeck Beds occupy the higher ground behind. The Kimeridge Clay consists of a thick succession of marine black shales, often bituminous, and clays, broken only by occasional bands of cementstone rarely more than one foot thick. At Kimeridge the exposed part of the formation is about 1056 feet thick, measured by Arkell along three and a half miles of cliffs, with the bottom three zones not exposed. In 1959 a boring at Kimeridge Bay proved 581 feet of the clay below the lowest bed exposed on the surface in Hobarrow Bay. In the Ringstead district, eight miles to the west, there is found not much more than half this thickness. The type section of the Kimeridge Clay on the Kimeridge coast is given by Arkell (1933, 1947a) as on p. 124. (The thicknesses are those found east of Broad Bench, SY894796.)

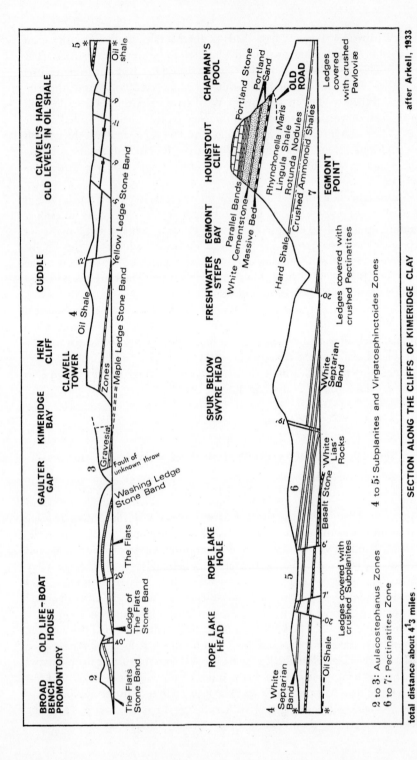

Fig. 15. Broad Bench–Chapman's Pool: coast section exposing the type section of the Kimeridge Clay, Upper Jurassic.

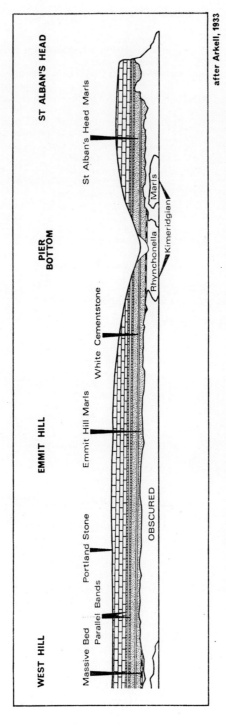

Fig. 16. West Hill-St Alban's Head: coast section showing Portlandian overlying Kimeridge Clay, Upper Jurassic.

KIMERIDGE CLAY TYPE SECTION

	Zone	Thickness

Upper Jurassic

Upper Kimeridgian

Pavlovia pallasioides and *Pavlovia rotunda* 339 ft

50 ft Hounstout Marl: blue sandy marl with thin sandy cementstone bands.

20 ft Hounstout Clay: dark unfossiliferous clay.

160 ft {
Rhynchonella Marls: grey marls with *Rh.* cf. *subvariabilis, Oxytoma*, small belemnites, and *Cidaris* spines.
Lingula Shales: dark shales with *Lingula*.
Rotunda Clays and Nodules: calcareous mudstones with abundant well-preserved ammonites (*Pavlovia rotunda*).
}

109 ft Crushed Ammonoid Shales, with many crushed *P. rotunda, Lucina minuscula, Orbiculoidea latissima.*

Pectinatites pectinatus 127 ft

2 ft Hard shale band.
25 ft Clay.
18 ft Clays with several horizons of large nodules.
} *Paravirgatites Clays* of Spath 45 ft

2 ft Hard shale band.
20 ft Shales.
} *Pectinatites Clays* of Spath 22 ft

1½ ft Freshwater Steps Stone Band.
28 ft Shales.
2 ft Middle White Stone Band.
24 ft Dicey clays and cementstones.
1½ ft Paper shale.
3 ft The White Band; with Saurian vertebrae.
} The Three White Stone Bands 60 ft

Middle Kimeridgian

Subplanites wheatleyensis 115 ft

33 ft Shales and clays.
23 ft Dicey clays.
3¾ ft The Basalt Stone: hard black cementstone with cuboidal fracture.
55 ft Dicey clay (estimated thickness).
} Dicey Clays 115 ft

Subplanites spp. 170 ft

4 ft Shale.
½ ft Cementstone.
8 ft Shale and clay.
1½ ft Rope Lake Head Stone Band.
5½ ft Shales, mainly hard.
3 ft Hard shale with *Blackstone* at bottom.
} Bituminous shales 22½ ft

27 ft Shale with hard shale bands.
14 ft Dicey clay, with abundant *Subplanites* (forms base of Clavell's Hard promontory).
5½ ft Double hard shale band.
20 ft Clay with hard band, *Subplanites* common.

Upper Jurassic	Middle Kimeridgian	*subplanites* spp.	2½ ft Cementstone (forms Grey Ledge). 16 ft Dicey clay. 17 ft Laminated clays and hard shaly bands. 1½ ft Cattle Ledge Stone Band. — Cattle Ledge Shales 37 ft 42 ft Bituminous shales. 1½ ft Yellow Ledge Stone Band.
		Gravesia gigas and *G. gravesiana* 70 ft	70 ft Shales with crushed *Lithacoceras* and the giant Peresphinctid *Gravesia*. — Hen Cliff Shales 70 ft
	Lower Kimeridgian	*Aulacostephanus pseudomutabilis* 235 ft (seen)	¾ ft Laminated shaly cementstone. 68 ft Shales with *Lithacoceras* and *Aulacostephanus*. 1 ft Maple Ledge Stone Band. — Maple Ledge Shales 70 ft
			c. 20 ft Shales. 46 ft Shales with abundant *Aulacostephanus*. 24 ft Shales west of Gaulter's Gap, with abundant *Aulacostephanus*. — Gaulter's Gap Shales 90 ft
			1¼ ft Washing Ledge Stone Band: conspicuous double band with shale parting; large *Aulacostephanus* common in Brandy Bay. 40 ft Washing Ledge Shales, with abundant *Aulacostephanus*. 1½ ft The Flats Stone Band. — Washing Ledge Shales 43 ft 32 ft (seen). Shales with *Aulacostephanus*, *Aspidoceras*, and *Amoeboceras*.

LOWEST ZONES EXPOSED AT BLACK HEAD

Upper Jurassic	Lower Kimeridgian	*Aulacostephanus pseudomutabilis*, c. 140 ft	Shales and clays; 10 feet below the top of the zone are found abundant crushed *Amoeboceras krausei*, *A. anglica*, *Protocardia lotharingica*, sometimes with lenticles of cementstone with very many well preserved small ammonites, the gastropod *Dicroloma* and small bivalves. At the base the shales are crowded with *A. pseudomutabilis*, *A. eudoxus*, etc.
		Aulacostephanoides mutabilis, c. 120 ft	Black clays full of iridescent shells, e.g. fine-ribbed *Rasenia* of *mutabilis* style; the three-feet-thick *Astarte supracorallina* Bed lies about 82 feet above the base of the zone. There are lines of large septaria on at least two horizons.
		Rasenia cymodoce, c. 15 ft	Shaly blue clay with layers of *Liostrea delta* and crushed *Rasenia* of *uralensis* style.
		Pictonia baylei, 3 ft	1 ft. *Exogyra nana* Bed, a mass of small oysters locally hardened to a dark limestone; *Pictonia* and *Prorasenia*. 2 ft. *Rhactorhynchia inconstans* Bed; purplish-grey clay, with some rolled and bored phosphatic pebbles, large Serpulae, *Pictonia densicostata*, *Rh. inconstans*, *Exogyra nana*, *E. praevirgula*, and many casts of other mollusca.
	Oxfordian (Corallian)		Ringstead Coral Bed

The three lowest zones of *Aulacostephanoides mutabilis*, *Rasenia cymodoce* and *Pictonia baylei* are not exposed near Kimeridge, but may be studied at Black Head, 11 miles to the west. I have found no detailed descriptions of the *pseudomutabilis* and *mutabilis* Zones there, and there are about 220 feet of clays above, extending into the Upper Kimeridgian. The approximate succession of the lower zones is on p. 125. (Section at east side of Black Head, Osmington Mills, about SY735820, according to Arkell (1947a) in Sector A of this Site of Special Scientific Interest; the basal strata are generally exposed close above the beach.)

FOSSILS IN THE KIMERIDGE CLAY OF THE DORSET COAST

FISH

Fish scales are extremely abundant in the Upper Kimeridgian.

REPTILES

The Dorset coast is outstanding for the richness and variety of its saurian remains, found especially in the Purbeck, Kimeridgian, and Lower Lias.

A very large number of Ichthyosaur bones have been found in the Kimeridge Clay, but a single fragment of bone may represent the sole relic of other forms. The following 32 named forms are recorded by Delair (1958–60) from the Kimeridge Clay of the Dorset Coast, from three general localities—Kimeridge Bay, Weymouth and the Isle of Portland. Arkell (1947a) states that they are specially characteristic of the Lower Kimeridge and of the *Aulacostephanus* Zones in particular.

   Chelonia (Turtles and Tortoises):
     Plesiochelys sp.
     Pelobatochelys blakei (more than one foot long)
     Tropidemys langi
   Crocodilia:
     Teleosaurus megarhinus (about five feet long) ⎫ both of these are marine
     Steneosaurus sp. ⎭   forms
     Machimosaurus mosae (one of the largest known Jurassic crocodiles with jaws some 52 inches long)
     Dakosaurus maximus
     Metriorhynchus superciliosus (a marine form)

Pterosauria (Flying reptiles):
  Rhamphorhynchus sp. (a long-tailed form)
  Pterodactylus manseli, P. pleydelli, P. suprajurensis
Sauropterygia:
  Plesiosaurs (Carnivores):
    Muraenosaurus truncatus (many vertebrae found)
    Colymbosaurus manseli, C. trochanterius, C. brachystospondylus (all three
      species were very large animals)
    Cryptocleidus aff. *richardsoni*
    Cimoliosaurus brevior
  Pliosaurs (Carnivores):
    Pliosaurus brachydeirus (teeth and vertebrae found)
    P. brachyspondylus, P. macromerus (the largest described species)
    P. ferox
Sauropoda (Dinosaurs, giant, herbivorous, amphibious):
  Ornithopsis humerocristatus (humerus about 63 inches long)
  O. manseli, O. ? leedsi
Ichthyopterygia (Ichthyosaurs):
  Macropterygius trigonis (up to 25 feet long)
  M. dilatatus, ? M. ovalis, ? M. thyreospondylus
  Ophthalmosaurus pleydelli (a form with relatively enormous eyes)
  Brachypterygius extremus
  Nannopterygius enthekiodon

INVERTEBRATES

Fossil invertebrates are often numerous but usually poorly preserved, being crushed flat. Ammonites are abundant, and the Upper Kimeridgian is characterised by a great abundance of Perisphinctids. Occasionally ammonites are well preserved in cementstones. Bivalves abound in the Lower Kimeridgian, especially *Protocardia, Astarte, Lucina* and *Exogyra;* and abundant *Liostrea delta* in the lowest zone. The brachiopods *Lingula ovalis* and *Discina latissima* are more common in the Upper Kimeridgian.

MICROFAUNA

The *Subplanites* spp. Zone of the Middle Kimeridgian near the horizon of the Blackstone (oil shale) has yielded pyritised radial plates of the pelagic crinoid *Saccocoma* which has proved a useful indicator of this horizon in several parts of England.

Lloyd investigated the foraminifera of the whole Kimeridge Clay type section, including the lowest zones at Black Head. He found the Lagenidae to be dominant, but as yet he has published no details of them. In 1959 he described and figured 25 forms of agglutinating foraminifera in eight genera, three species as new; and in 1962 some 23 calcareous forms, four species as new, the commonest being forms of *Eoguttulina*.

Microplankton

Mainly from Kimeridge, Downie (1956) has recorded, from the *Subplanites* Zone, six named species of Dinoflagellates and four of Hystrichospheres, and from the *pectinatus* Zone eight of Dinoflagellates and three of Hystrichospheres; there are also many unidentified microfossils.

In the *Subplanites* Zone coccoliths are numerous, occasionally forming thin argillaceous or bituminous limestones, such as the Rope Lake Head Stone Band. In the clays both groups of microfossils are abundant and consist mostly of three genera, *Palaeoperidinium*, *Gonyaulax*, and the form called *Hystrichosphaeridium pattei*, which was renamed *Baltisphaeridium downiei* by Sarjeant (1960).

In the *pectinatus* Zone the Three White Stone Bands are laminated chalky limestones largely composed of coccoliths. Each band contains up to three per cent of brown organic matter, largely Dinoflagellates and Hystrichospheres, with spores, pollen and small plant fragments.

OIL-SHALE

'Kimeridge Coal', found in the Middle Kimeridgian, consists of one or more bands totalling about two feet ten inches thick with its shale partings, of a dark brown bituminous oil-shale with marine fossils, with the local name of 'Blackstone'. It rises from the beach 200 yards east of Clavell's Hard headland (SY926775) and was mined in adits and in surface workings. It was used as a fuel from very early times, and until recently in local cottages, but it burns with an offensive smell, leaving copious ash. In the middle of the eighteenth century it could be bought for sixpence a ton, but by 1879 it was fetching twelve shillings, and by 1885 a sovereign. From this Blackstone was made what was once called 'Kimeridge Coal-Money', discs from one to three inches in diameter and a quarter of an inch thick, with two or four shallow holes on one side. These were found on the top of the cliff and in fields round Kimeridge. They appear, however, to be the waste centres of armlets; when smoothed and polished with beeswax, the material looks very like jet.

Armlets and also buttons, rings and even beads, were made by fashioning the material on a type of lathe using flint tools, in Neolithic times, and throughout the Bronze Age, the Iron Age and Romano-British times.

abundantly as small segregations. This points significantly to some form of limited and partially enclosed and stagnant basin rather than open sea as the site of deposition. However, it must in some way have been freely open and attractive to the marine fauna which is found in such profusion.

The fine-grained nature of the sediments, calcareous muds and limestones (some of these of considerable purity) and the lack of coarser material, other than scattered sand grains, indicates deposition off a mature coast of low elevation, the calcareous matter presumably being derived from denudation of Carboniferous Limestone.

After studying the insect fauna of the *obtusum* Zone, Zeuner (1961) has suggested that the conditions then may be compared with those obtaining today off the Indian coast 'north of Bombay', with many low islands which, towards the land, fuse into strips of land interrupted by water inlets of varying width. These last link up with rivers, so that 20 miles inland the water is fresh.

If it be true that the limestones are mainly of primary origin, then the frequent alternation of marls and limestones must reflect some change in the conditions of deposition or in the material deposited. Variation of sea depth has been suggested, caused by often-repeated movements of the land, but the problem remains speculative.

THE LOWER LIAS FAUNA AND FLORA

Fossils were collected from the sections near Lyme Regis certainly as early as 1790 for sale to visitors, who included geologists. The most celebrated collector was Mary Anning (1799–1847). A fine collection is preserved in the British Museum (Natural History) and for important specimens mentioned below this resting place is indicated by (BM).

An attempted conspectus, derived from the literature, is given below. Although this no doubt contains errors, the attempt may be thought worth while in relation to the ecology reflected by the deposits, and because for some groups no summary is available.

REPTILES

Skeletons of giant reptiles, largely marine, are a conspicuous feature of the Lower Lias of Lyme and Charmouth. The exact horizons of the earlier specimens are unknown, but many seem to have come from the Saurian Shales at the top of the Blue Lias, and from the overlying Shales with Beef.

In 1811 Mary Anning, then about 12 years of age, found the first associated skeleton of an Ichthyosaur from Lyme to be described scientifically; the specimen is now stated to be lost. In 1823 she discovered the first known associated bones, which formed an almost complete skeleton nearly 10 feet long, of a Plesiosaur, which, sold to the Duke of Buckingham for £100, was described by Conybeare as *P. dolichodeirus* (BM). In 1828 she found a 'Ptero-dactyl' new to England, described by Buckland, and now known as *Dimorphodon macronyx* (BM).

Twenty-one named species of reptiles and three other forms named only generically (two doubtfully) are now recognised from this site (Delair). They are marine except for the 'Pterodactyl' and the Dinosaurs. Many are represented by fairly complete skeletons as well as by loose bones. They are as follows. Plesiosaurs: *Plesiosaurus dolichodeirus, P. hawkinsi, P. macrocephalus, P. costatus, P. macromus, P. platydeirus, P. rostratus, P. eleutheraxon, P. conybeari, P. sp., ? Eretmosaurus sp., Eurycleidus arcuatus.*

Ichthyosaurs: *Eurypterygius communis, E. communis* var. *hyperdactyla, E. intermedius, E. breviceps, E. conybeari, Leptopterygius tenuirostris, L. platyodon, L. lonchiodon, L. latifrons.*

Dinosaurs: ? *Zanclodon* sp., *Scelidosaurus harrisoni.*

Pterosaurs (flying reptiles, 'Pterodactyls'): *Dimorphodon macronyx.*

Plesiosaurs

The Plesiosaurs resembled 'a snake threaded through the shell of a turtle', a description ascribed to Buckland. They generally possessed a long slender neck, a small head with short snout, and jaws with strong, pointed, conical teeth. Propulsion was effected by rowing over the surface of the sea with their two pairs of paddles, the short tail being used for steering. Fishes and cuttle-fish were included in their food and the females laid their eggs upon the shore. 'Stomach stones' found in some individuals have been held to have served to triturate the harder parts of the food; but they might perhaps have been swallowed for ballast, as in modern crocodiles (see p. 201). The type species of Plesiosaurus, *P. dolichodeirus*, from Lyme, was about 10 feet long, but others from here reached nearly 20 feet.

Ichthyosaurs

The present site is one of the chief repositories for the Ichthyosaurs, marine reptiles, rapid swimmers, completely adapted to life in the open sea. They were shaped like the modern porpoise, and had a smooth-skinned brownish body devoid of scales. They were essentially surface swimmers, breathing by lungs. The large head was produced into a slender, pointed snout, and the eyes were enormous. The neck was very short. There was a dorsal fin, and propulsion was effected by a large, vertical, fan-shaped tail. The two pairs of paddles were used for balancing and steering. The jaws and teeth, which were long and sharply pointed, must have formed a very effective fish trap and they swallowed their prey whole. They must have had enormous stomachs and lungs. The fish *Pholidophorus* was a principal food. Ichthyosaurs were ovo-viviparous, and skeletons have been found elsewhere containing up to seven unborn young. Flattened coprolites have been found within Ichthyosaur skeletons, but the abundant spirally marked coprolites assigned by Buckland to these animals are now considered to have come from the large sharks. The Lyme and Charmouth Ichthyosaurs range in size from *Eurypterygius communis* and *E. intermedius*, not more than 10 or 12 feet long, to the great *Leptopterygius platyodon*, which reached about 30 feet.

Dinosaurs

The Dinosaurs were essentially land animals. Of the two represented at this site, that doubtfully identified as *Zanclodon* belonged to the Theropods, which include the largest and most fearsome land carnivores the world has ever known. For running and walking they probably used the hind limbs only. The only Lias relic of these animals to be found in Dorset, ? *Zanclodon*, is represented by a single tooth. The second Dinosaur is *Scelidosaurus harrisoni*, represented by a nearly complete skeleton (BM) found at Charmouth in the early 1860s and some further bones. Generally regarded as a primitive Stegosaur, *Scelidosaurus* was a quadruped animal and comparatively low on the ground. It was protected by armour, a series of longitudinal rows of bony scutes and low spines, and, on neck and tail, series of low vertical plates. The oldest of these armoured forms, *Scelidosaurus* was about 12 feet long, and was a plant-eater. A graphic idea of the living animal is given on the title page of *Dorset: A Shell Guide* (new ed., 1966).

Pterosaur

Dimorphodon macronyx (BM) from the Lower Lias of Lyme is stated by Swinton to be the earliest flying reptile known; a graphic restoration of it in flight is given as the frontispiece of his *Fossil Amphibians and Reptiles* (1958). The size of a raven, it had a wing span of four feet and a tail 21 inches long. The skeleton was very light but composed of hard compact bone, the limb bones being hollow. The head was very large but lightly constructed, the jaws having large teeth in sockets in front and small teeth behind. The body was relatively small. The wings were disproportionately large, consisting of a membrane supported by the greatly elongated fourth finger, but without any other support in the membrane itself. The flying structure was

99

therefore unlike that of either bird or bat. The muscular power for flapping the wings was not so well developed as in birds, and *Dimorphodon* must have floated on air currents rather than have flown by strong movements of the wings.

FISHES

In 1841 Egerton listed 54 species of fossil fishes from the Lower Lias of Lyme Regis, and in 1879 (in Wright 1878–86) he gave a revised list of 86 species in 35 genera. Most of the specimens were obtained from workmen who kept their sites secret, but the chief repositories of the Lyme fishes appear to be the Saurian Shales (Lang's Beds 50–52) at the top of the Blue Lias, and the overlying Shales with Beef. A mass of the Saurian Shales at the 'Cockpits', a little east of Lyme, has since been removed by coast erosion, and now fish remains are seldom found at Lyme and belong to only a few species.

The following conspectus of the 43 Lower Lias fishes now recognised from Lyme Regis has been kindly revised by Dr E. I. White and Dr C. Patterson with the group and generic names in modern usage. They are all marine forms.

Chondrichthyes (skeletons cartilaginous). Selachii (sharks and rays): *Acrodus*, 2 spp.; *Hybodus*, 4; *Palaeospinax*, 1. Holocephali (chimaeroids— deep sea forms): *Squaloraja*, 2 spp.; *Myriacanthus*, 2.

Osteichthyes (skeletons largely bony). Crossopterygii: *Holophagus gulo* (a Coelacanth).

Actinopterygii (bony fishes). Chondrostei: *Chondrosteus*, 2 spp.; *Cosmolepis*, 1; *Centrolepis*, 1; *Coccolepis*, 1; *Platysiagum*, 1; *Ptycholepis*, 3; *Saurorhynchus*, 2. Holostei: *Dapedium*, 5 spp.; *Furo*, 5; *Caturus*, 1; *Pholidophorus*, 5; *Osteorhachis*, 2; *Heterolepidotus*, 1; *Leptolepis*, 1.

Of the Selachians, the large sharks *Hybodus* and *Acrodus*, which had teeth adapted to crush hard food, are represented by some nearly complete skeletons. Their remains are as abundant as those of the marine reptiles. The very common spirally marked coprolites (in early days known as 'Bezoar stones') are considered to have come from these fishes, whence they would be normal, rather than from the Ichthyosaurs (as Buckland believed) whence they would be abnormal.

The Crossopterygians, with paddles for fins, are more closely related to the four-legged land animals than any other fishes. They are represented at Lyme by the coelacanth *Holophagus* (previously *Undina*) *gulo*, a fine specimen being preserved (BM). Until 1938 the coelacanths were thought to have become extinct by the end of the Cretaceous, but in that year a representative was discovered living in deep water off the East African coast, and other specimens have now been caught.

The Chondrostei are the forerunners of the modern sturgeons.

Some of the many Holostei (with thick, rhombic, enamel scales of the type known as ganoid), and particularly *Pholidophorus*, are represented by beautiful specimens (BM).

The Jurassic fishes with a future, represented by *Leptolepis*, were, in Lower Lias times, 'little more than food for the fashionable races of the time' (Smith Woodward).

INSECTS

Since about 1954 Mr J. F. Jackson of Charmouth has collected many specimens of insects, some of them well-preserved, the first known from the Lower Lias of Dorset. Nearly all came from the Flatstones, Lang's Bed 83, of the Black Ven Marls *obtusum* Zone, but a few from lower in the same marls. Some are associated with plant debris, as well as with the general marine fauna, especially ammonites. This fauna, represented by 434 specimens which have been acquired by the British Museum (Natural History) is in course of description, and the late Professor Zeuner has given a preliminary summary of it as: beetles, including primitive wood-borers, 38 per cent; grasshoppers, 20 per cent; bugs, 6 per cent; cockroaches, 3 per cent; dragonflies, 3 per cent; flies, 3 per cent; sawflies, 1 per cent; indeterminable, 26 per cent.

Zeuner (1961) has given a detailed description of all the remains of the dragonflies (which are among the largest known, some of the wings being over 80 millimetres in length), an orthopteron of special interest, and three species of beetles, as follows. Dragonflies: *Petrophlebia anglicanopsis* (described as new), *Liassophlebia magnifica*, *L. jacksoni* (described as new), *L. gigantea* (described as new), *L.* sp. *Diastatommites liasina*. Grasshoppers: *Protohagia langi* (described as new genus, new species). Beetles: *Liassocupes parvus* (described as new), *Holcoëlytrum giebeli*, *H. schlotheimi*.

Zeuner considered that the insects had undergone both transport and decay before they were embedded in the sediment. This and the composition of the fauna suggested that they may be interpreted as flotsam which was finally deposited in a calm sea.

CRUSTACEA

The only record I have found from the present site is of a chela doubtfully assigned to the crab *Palaeopagurus*, from Bed 127 of the Green Ammonite Beds (Lang, 1936).

Crustaceans, however, are not uncommon in the lower beds of the Blue Lias exposed to the west of Lyme.

BELEMNITES

Belemnites form an important element in the fauna of the Green Ammonite Beds (two genera with five species) and particularly of the Belemnite Marls (five genera with 26 species). In the underlying beds they are represented only sparingly. Belemnites from this site have been described since 1826, and those of the Belemnite Marls and Green Ammonite Beds have been studied in detail by Lang (1928), who has proved them of value as zonal markers. He described two new genera, *Clastoteuthis* and *Angeloteuthis*, and 19 new species. Preservation here has in some cases been so good that Buckland (1826) obtained specimens with remnants of their ink and had a drawing prepared with that material; he later described *Belemnosepia* (*Geoteuthis*). Remains of *Xiphoteuthis* have been recognised, and Huxley also obtained fossil ink-bags. According to Hallam (1960), the important influx of new belemnites at the base of the Pliensbachian (Belemnite Marls) appears to be Europe-wide.

Belemnites were among the first fossils noticed by man. As the hard part of an extinct kind of cuttle-fish with no modern creature just like them, they provided a difficult problem of identification. The word 'Belemnite', derived from a Greek word for a dart, was first used for the fossil by Agricola in Germany in 1546, and he gave the earliest reasonable observations on them. Examples were probably first figured by Gesner in 1565. In Great Britain they were recorded by Merret in 1667, and by Plot, Lister, Grew, Sibbald and Llwyd all before 1700. There have been many vernacular names, such as Devil's Fingers and Spectre's Candles; later they were ascribed to the action of lightning and called *Lapis Fulmineus*, and Thunderbolts. In the seventeenth century they were held to be merely a *lusus naturae*, truly mineral, either stalactites or crystals.

In 1724 Erhart gave his celebrated description of belemnites, regarding them as marine shells near to *Nautilus* and *Spirula;* but it was another century before their more detailed study began with Miller's paper of 1826, in which he described and figured 11 species from England, including three from Lyme. This was quickly followed by de Blainville's great monograph of 1827.

AMMONITES

Ammonites, considered to have lived as crawlers on the sea-bed, but also able to swim well and quickly, are found here generally flattened in the clays, but well preserved in limestone nodules. In size they range up to some two feet in diameter. At Lyme they have aroused interest at least since 1706, when

Mr Hutchinson (recorded in Woodward, 1728) observed there 'incredible numbers of these Shells'. From 1812 to 1846 the Sowerbys figured some 17 ammonite species from the Lyme and Charmouth cliffs.

The value of ammonites as zonal indices was recognised at an early date by, for example, Buckland and by Hunton (on the Yorkshire coast) in 1836; but the earliest serious attempt to apply this knowledge in detail was made by Oppel (1856–58), who first classified the beds of this district in a table of zones. The soundness of his method has been demonstrated after a century of research and controversy. The method was furthered by T. Wright (1878–86) and very greatly extended by S. S. Buckman in the early years of this century. Finally, by the work of Spath and Lang the latter was able to establish a remarkable sequence of some 50 ammonite zones covering the whole of the Lower Lias of the Dorset coast, and later shown to be of much wider application. As an example of the volume of the original data may be cited Lang's record of 52 species of ammonites in the Green Ammonite Beds alone.

The importance of the ammonites for stratigraphy has resulted in a vast literature, much of it extremely specialised. A summary of the results as applied to the Lias of the north-west European province is given by Hallam (1961). That part referring to the ammonite families of the Lower Lias he gives as follows. 'The Hettangian is characterised by Psiloceratids and Schlotheimiids with the latter occurring spasmodically in the Sinemurian. The Lower Sinemurian is marked by the incoming at the base of a very important group, the Arietitids. These became extinct at the top of the Lower Sinemurian. The Upper Sinemurian is characterised by the Oxynoticeratids and Eoderoceratids, the latter having first appeared in the *turneri* Zone. The former have completely, and the latter almost completely, disappeared by the beginning of Pliensbachian times. No individual genus appears to cross the Sinemurian-Pliensbachian boundary. The Carixian (Lower Pliensbachian) is marked by the incoming of two new families, the Polymorphitidae and the Liparoceratidae. Both families became extinct at the end of the Carixian.'

There are a few isolated records of *Nautilus striatus* or *N*. sp. from the Black Ven Marls, Shales with Beef, and Blue Lias.

GASTROPODS

Gastropods are extremely rare at all horizons below the uppermost beds of the Belemnite Marls, where Cox (1936) recorded seven species. He recorded nine species from the Green Ammonite Beds, five described as new. Small forms are abundant a few feet above the Belemnite Stone. The commonest are *Coelodiscus wrighteanus*, *C. aratus*, *Buvignieria carixensis*, and *B. biornata*. These forms are all small, ranging from one and a half millimetres high to seven millimetres in diameter.

BIVALVES

Like the gastropods, many of the species are small, but they are distinctly more abundant. The five named stages have each a bivalve fauna ranging from about seven to 14 species, only a few being common in each stage. In the Blue Lias most are oysters and *Plagiostoma*, and the incoming of *Gryphaea arcuata* marks the Hettangian–Sinemurian boundary. Thirteen species in the Shales with Beef are recorded by Lang but not described. In the *obtusum* shales in the Black Ven Marls *Plagiostoma giganteum* and *Oxytoma inequivalvis* are fairly abundant but very badly preserved, and there are four other species, one described as new. In the Belemnite Marls eight species are recorded, *Inoceramus ventricosus* alone being abundant in some beds, and *Chlamys rollei* is not uncommon. Three species were described by Cox as new. In the Green Ammonite Beds 11 species were recorded, three new, including the new genus and species *Anningia carixensis*. Common are *Nucula ungulella, Palaeoneilo [Nuculana] galatea* and *Parallelodon trapezium* (Cox, 1936).

BRACHIOPODS

Brachiopods are generally rather rare, although at a few horizons they may be abundant. They are usually small, poorly preserved, distorted and crushed. Huddles of Rhynchonellids are found in the Blue Lias, and in Table Ledge (Bed 53) at the base of the Shales with Beef, but they are not further identified. In the Black Ven Marls some eight species include *Spiriferina walcotti* (Bed 82) and other forms occur abundantly in Bed 87. The Belemnite Marls contain scattered brachiopods, but they are abundant only at two horizons: Beds 111 and 112, where species of *Cincta* form a well-marked band; and Beds 118 and 119, where crushed specimens of *Tropiorhynchia thalia* predominate. In the Green Ammonite Beds are four species, including one that was described as new; they are generally rare, but many specimens of *Scalpellirhynchia scalpellum* have been found (Muir-Wood, 1936).

POLYZOA

A single occurrence of *Stomatopora antiqua* is recorded by Lang from the Blue Lias.

ECHINOIDS

Apart from abundant remains of four species from near the base of the Blue Lias of Pinhay Bay, and so not included in the present sections, echinoid remains are very rare. They have not been recorded from the Belemnite Marls. From the Green Ammonite Beds Lang (1936) recorded *Eodiadema* cf. *minuta* from a number of horizons between Beds 122 and 128a, inclusive. From the Black Ven Marls Wright (1860) recorded *Acrosalenia minuta* and *Cidaris* spines and, from the Shales with Beef *Cidaris edwardsi;* Lang recorded a Diademoid spine from Bed 70c; and from the *bucklandi* beds of the Blue Lias Wright recorded spines of *Pseudodiadema* and other forms.

CRINOIDS

Crinoids are recorded in each group of beds except the Belemnite Marls. In the Green Ammonite Beds Lang has found *Isocrinus basaltiformis* (Beds 122b and 123); *Balanocrinus* cf. *laevis* (124 to 130); and cf. *Pentacrinites subangularis* (129). In the Black Ven Marls (*obtusum* Zone) and only developed on Black Ven, is the one-inch-thick Pentacrinite Bed (84b) much noted by early authors. It yields portions of the fossil now known as *Isocrinus tuberculatus* and also *Pentacrinites fossilis* [=*Extracrinus briareus*], fine specimens of which have been collected in the past. In both the Shales with Beef and the Blue Lias (*bucklandi* Zone) *Isocrinus tuberculatus* and *I.* cf. *tuberculatus* are recorded.

? WORMS (TRACE FOSSILS)

The casts of branching burrows, of both from a half to one millimetre and from two to three millimetres in diameter, known as *Chondrites*, are present here (Simpson, 1957) in practically every bed of the Blue Lias and of the Belemnite Marls. A band of 'mottled Marl' from two to four inches thick made up of them typically occurs beneath each limestone band, and often on top of a limestone. Although the organism concerned remains problematical, its works are important, since they seem to prove that the limestones must be original deposits and not secondary, as had been suggested.

Cylindrites is a larger trace fossil, with a diameter of from five to 20 millimetres, that is sometimes associated with *Chondrites* here.

U-shaped burrows with a septum, known as *Corophioides*, are also recorded by Simpson from Bed 70f of the *turneri* Zone of the Shales with Beef. The tube has a diameter of from seven to eight millimetres, and the U descends to a depth of 10 centimetres. Similar burrows, which he attributes to a Polychaet worm, are described by Coysh (1931) under the name *Arenicolites lymensis*, but he gives no precise horizon. From the Black Ven Marls Lang records 'cf. the supposed worm *Tissoa siphonalis*' (Bed 95).

CORALS

Corals are extremely rare; only two specimens seem to be known. There is a record of *Oppelismilia* [*Montlivaltia*] *victoriae* apparently from the Green Ammonite Beds; and a worn specimen of *Isastrea* sp. was found on the beach below Black Ven, and probably came from the Blue Lias.

MICROFOSSILS

Foraminifera

Although known from here for a century and although a few specimens were preserved in two old collections, foraminifera have been described from the Lower Lias of Lyme and Charmouth only since 1941, when Macfadyen recorded 55 species in 20 genera from the Green Ammonite Beds. These included one genus and species described as new, *Carixia langi*, and another new species, *Lagena davoei*. Barnard (1950) studied samples covering the whole of the Lower Lias, finding that foraminifera abound at many horizons, and he recorded 47 species, eight described as new. Although his paper did not mention the fact, this represents only about one-third of the foraminiferal fauna that he identified, and the majority have not yet been published. The Lagenidae account for some 80 per cent of the species, marking the explosive outburst of that family in the Lower Lias.

Ostracods

Macfadyen (1941) recorded only two specimens of a smooth form likened to *Bairdia liassica* from the Green Ammonite Beds, but Barnard (1950) found smooth, unornamented forms common in most of his samples. The ostracod fauna has yet to be described.

Fish teeth and otoliths

From the Green Ammonite Beds two minute fish-teeth have been recorded by Macfadyen (1941) and identified by Dr E. I. White as *Palaeospinax priscus* and *P.* ? sp., respectively. A few fish otoliths were also found, which could then not be further identified. From Bed 120, part of the Crumbly Bed of the Belemnite Marls, Frost (1926) described three new species of the form genus '*Otolithus*'. Barnard (1950) recorded numerous small fish-teeth which were not further identified.

106

Holothurians

Two wheels of *Chiridota* and 20 holothurian spicules (perforated plates) were recorded by Macfadyen (1941) from the Green Ammonite Beds.

LAND PLANTS (CONIFERS, CYCADS, FERNS)

Seward (1904) stated that from the Lower Lias near Lyme Regis had been obtained the majority of British Liassic plants then known. It may be added, however, that from there he recorded only eight species, of which three were single specimens and three others were represented by a total of eight specimens. The most abundant forms were the fern *Thinnfeldia rhomboidalis* (10 specimens) and the conifer *Pagiophyllum peregrinum* (11 specimens). One specimen of *P. peregrinum* consisted of a trunk nearly 13 feet long, and another, a small branch, was found in the Flatstones (Bed 83) on Black Ven in 1959. Also recorded were an indeterminable cone and four specimens of coniferous wood. Although a considerable amount of driftwood occurs, it is practically valueless botanically because the tissues have not been petrified. A specimen of lignite from Lyme Regis has recently been found to be comparatively rich in certain trace elements, of which germanium is one (Hallam and Payne, 1958). Plant spores and hystrichosphaeroids have been recorded as of rare occurrence in the Blue Lias (Hallam, 1960).

1728–29, Woodward; 1797, Maton; 1811, de Luc; 1813, Townsend; 1822, 1826, De la Beche; 1841, 1879, Egerton; 1863, Day; 1860, 1878–86, Wright; 1904, Seward; 1904, 1907, 1914, 1923, 1924, 1926, 1928, 1932, 1957, Lang; 1936, Cox; 1936, Muir-Wood; 1936, Lang and Thomas; 1933, Arkell; 1941, Macfadyen; 1950, Barnard; 1957, Simpson; 1956, 1960, 1961, Hallam; 1958, Hallam and Payne; 1958, Swinton; 1958, Wilson *et al.* ; 1958–60, 1966, Delair; 1965, Ager and Smith.

Blashenwell, 5 miles WNW of Swanage

| SY(30)952805 | 179 | 343 | 56NW | P, S, A |
| | | | (G) | PSD, 1963 |

A long-studied deposit of calcareous tufa, remarkable for its inclusion of Mesolithic flint artifacts and bones of mammals, shells and leaves of trees. The section is given as follows (modified after Reid, 1896).

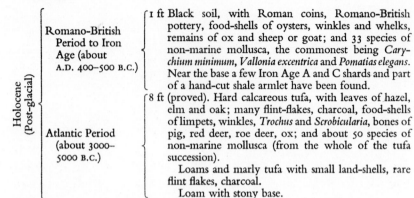

Holocene (Post-glacial)

Romano-British Period to Iron Age (about A.D. 400–500 B.C.)

1 ft Black soil, with Roman coins, Romano-British pottery, food-shells of oysters, winkles and whelks, remains of ox and sheep or goat; and 33 species of non-marine mollusca, the commonest being *Carychium minimum*, *Vallonia excentrica* and *Pomatias elegans*. Near the base a few Iron Age A and C shards and part of a hand-cut shale armlet have been found.

Atlantic Period (about 3000–5000 B.C.)

8 ft (proved). Hard calcareous tufa, with leaves of hazel, elm and oak; many flint-flakes, charcoal, food-shells of limpets, winkles, *Trochus* and *Scrobicularia*, bones of pig, red deer, roe deer, ox; and about 50 species of non-marine mollusca (from the whole of the tufa succession).

Loams and marly tufa with small land-shells, rare flint flakes, charcoal.

Loam with stony base.

Most of this succession rests directly upon Lower Cretaceous Wealden Beds, but the southern tip of the deposit rests upon the Upper Purbeck.

The tufa is mapped over an area of some 23 acres, measuring some 590 by 220 yards. It was apparently first described by Mansel in 1857. A recent carbon-14 dating of a bone from the middle of the deposit gave an age of 6450 ± 150 years B.P. (say 4490 B.C.) (Rankine, 1962); this confirms the Atlantic age which had been deduced from the artifacts.

The main pit was being exploited for marl in 1938, and was later soiled over and had become an arable field by 1952. In 1963 a good section of the tufa was seen in a cutting on the east side of the road at SY952805, alongside the position of the old main pit; it was some five feet thick (exposed minimum) and 25 yards long. Another exposure was seen in the stream bed east of the road at that point.

A collection of the flint artifacts now in the Dorchester Museum consists entirely of cores and flakes of a microlithic (pygmy) industry unmistakably of the later Mesolithic period. Although most bear a white patina, their uniformly fresh and sharp condition shows them to be contemporaneous with the deposit. From the species of marine food-shells found in the tufa, and the absence of oysters, cockles and whelks, Reid showed that they were probably collected on the rocky coast near Chapman's Pool, two miles due south, rather than from elsewhere. The Mesolithic people seem to have lived on a diet consisting mainly of wild pig, deer and limpets.

The fossil snail fauna from the tufa indicates a habitat of damp ground and not permanent pools. Nevertheless, Bury had 47 specimens of *Limnaea truncatula* and also recorded *Pisidium cinereum*; and later Dance recorded *P. subtruncatum* and *P. personatum*. Bury wrote: 'My specimens are definite proof of water, but only perhaps towards the end of the tufa-making period.'

The tufa derives from the calcium bicarbonate held in solution in the water which now issues as two springs, the one large but intermittent, the other very small but permanently flowing. Certainly the main spring is the overflow of a natural trough, a small sharp syncline (an accommodation fold, according to Arkell, 1947) in Upper Purbeck strata lying within an outcrop of Middle Purbeck.

The production of calcareous tufa has often been given a physicochemical explanation, the escape of carbon dioxide from the emergent spring water leading to the simple precipitation of calcium carbonate. Much tufa, however, is formed by various specialised plants known as tufa formers, including particular species of blue-green algae, mosses and liverworts (Sernander). In some manner as yet imperfectly understood they abstract calcium in solution in spring waters. Unfortunately, it may be very difficult to detect traces of these plants in the tufa, and often, as at Blashenwell, they have not yet been sought. Mansel (1857) wrote of the top layer of the tufa as 'a narrow stratified band of indurated concretions resembling limestone'. That suggests the possibility of an algal origin, in part, for ball-like algal concretions are found at the present day in running water, and range in size from peas to footballs or larger.

The tufa was deposited in the Atlantic period, ceased before the Iron Age to Romano-British soil layer formed, and has not been produced since. This could be correlated with the behaviour of the spring. A heavier rainfall in Atlantic times may have caused a perennial flow in the main spring, which allowed the growth of the specialised vegetation, whatever it may have been, and the resulting tufa deposition. A reduced rainfall in the succeeding Sub-Boreal period, about 3000–500 B.C., resulting in the main spring flowing only intermittently, as at present, could have allowed the specialised vegetation to dry out seasonally, and so to die and halt the tufa deposition.

In the absence of an available analysis two water samples were collected on 1 November 1963, and were analysed by the Government Chemist; the results may be re-stated as follows:

	Main (intermittent) Blashenwell spring at SY 951803, flow roughly estimated at 100 000 gallons per day or more	'Luke's Well' permanent spring at SY 953805, flow roughly estimated at 10 000 gallons per day
	parts per million	
Chloride (as Cl)	40	44
Sulphate (as SO_4)	30	29
Nitrate (as NO_3)	9	31
Bicarbonate (as HCO_3)	197	197
Calcium (as Ca)	132	141
Magnesium (as Mg)	8·5	7·5
Sodium (as Na)	24	27
Potassium (as K)	1·7	4·7
Total dissolved solids (dried at 180°C)	444	489
pH (after storage in glass bottles for two months)	7·2	7·1

This water is comparable with that of some other springs which at present are depositing calcareous tufa through the agency of certain kinds of vegetation. But the water has about double the total salinity, with twice as much calcium bicarbonate in solution, as have two such springs known to the present writer, at Binn in Switzerland, and at Southstone Rock in Worcestershire. Since the pH here is below eight, the alkalinity will be wholly as bicarbonate.

The Blashenwell water would appear to be eminently capable of depositing tufa now if the requisite vegetation and other conditions were present.

1857, 1886, Mansel (later Mansel-Pleydell); 1857a, Austen; 1896, Reid; 1901, Kennard and Woodward; 1915–16, Sernander; 1938, Clark; 1947a, Arkell; 1950, Bury; 1955, Carreck and Davies; 1962, Rankine; 1966, Brown.

Bowleaze Cove to St Alban's Head, East of Weymouth

SY(30)702820	178,	342,	53NE, 54NW, NE,	S, P, T, U
to 962754	179	343	55NW, SW, SE,	PSD, IGC
			56SW, 59NW	1966
			(G)	

A 17-mile stretch of coastal cliffs, half of that coast of which Arkell (1947a) wrote that it is so richly endowed as a training ground and museum of geology that few tracts of equal size could raise so many claims, scientific, aesthetic and literary, for preservation as a national park. It is impossible here to do justice to it.

For the description below this *embarras de richesse* has been divided into five sectors, with stratigraphic elements of various Jurassic and Cretaceous ages, exposed in order from west to east. Although most of the coast land is privately owned, nearly five miles in Sectors D and E, from the east side of Lulworth Cove at SY828796 to Kimeridge Bay at SY903792, including Mupe Rocks and Bay, Worbarrow Bay, Worbarrow Tout, Gad Cliff, Brandy Bay and Hobarrow Bay, is in War Department ownership, and is closed to the public unless special permission has been first obtained.

Sector A, 2·7 miles: Bowleaze Cove–Redcliff Point–Black Head–Osmington Mills–Ringstead; after a Holocene site, Corallian and Oxford Clay, with Kimeridge Clay behind.

Sector B, 1·5 miles: Ringstead Bay; Kimeridge Clay, with Portland and Lower Purbeck beds behind.

Sector C, 2·5 miles: White Nothe–Bat's Head–Swyre Head; Cretaceous: Lower Chalk, with some Upper Greensand and Gault.

Sector D, 4·0 miles: Durdle Door–Hambury Tout–Dungy Head–Lulworth Cove–Mupe Rocks and Bay–Arish Mell–Worbarrow Tout; cliffs of Purbeck and Portland beds, with erosion inbreaks exposing Wealden, Upper Greensand and Chalk behind.

Sector E, 6·5 miles: Worbarrow Tout–Gad Cliff–Brandy Bay–Hobarrow Bay–Broad Bench–Kimeridge Bay–Kimeridge Ledges–Chapman's Pool–Emmit Hill–St Alban's Head; Kimeridge Clay coast, the type section, with Portland and Purbeck beds forming the high ground behind.

SECTOR A. BOWLEAZE COVE–RINGSTEAD (Figs. 12 and 13)

At Bowleaze Cove, SY703820, excavations in the banks of the River Jordon exposed a series of Holocene alluvial loams, one with a Romano-British occupation layer. The loams included some 42 species of non-marine mollusca, beetles and remains of voles etc. (Carreck and Davis, 1955).

East of the River Jordon nearly flat-bedded Corallian rocks form the cliffs as far as Redcliff Point, with the Nothe Grit at the base. Among fallen blocks on the shore are oval, flattened doggers up to six feet across, from the Bencliff Grit; and cuboidal blocks from the *Myophorella* [*Trigonia*] *hudlestoni* Beds.

At the tip of Redcliff Point, SY712816, an east–west fault brings up Oxford Clay capped by Nothe Grit. Some 80 feet thickness of Oxford Clay, *cordatum* Zone, is present, and the lowest 52 feet has yielded 23 species (four new) of foraminifera; the two commonest forms are *Ammobaculites suprajurensis* and *Lenticulina münsteri* (Barnard, 1953). It has also yielded holothurian spicules (Hodson, Harris and Lawson, 1956).

To the east a minor anticline brings up Oxford Clay again, the most easterly exposure of these beds on the South Coast. The upper part is pale blue-grey clay with many large *Gryphaea dilatata*, and lines of septaria, some containing *Modiolus* and *Astarte*. Lower beds with small pyritised ammonites are exposed along the anticlinal axis. The northern limb of the anticline dips about 50° to the north, and the Corallian Nothe Grit and *M. hudlestoni* Beds rapidly descend from the cliff top to sea level. The Nothe Clay reaches the shore at the Short Lake ravine, and the higher beds of the Corallian form the cliffs beneath the great slips of Kimeridge Clay at Black Head, SY730818.

The basal part of the Kimeridge Clay, with *Rhactorhynchia* [*Rhynchonella*] *inconstans* and *Liostrea delta*, is seen near the shore below Black Head,* and the section extends upwards to above the Oil Shale horizon. The dip is still about 50° to the north, and the Upper Greensand, dipping gently east, truncates successive zones of the Kimeridgian and rests upon the *Rasenia* Zone near Osmington Mills. Water seeping from the Upper Greensand causes landslips

---

* For a description of this section see p. 125.

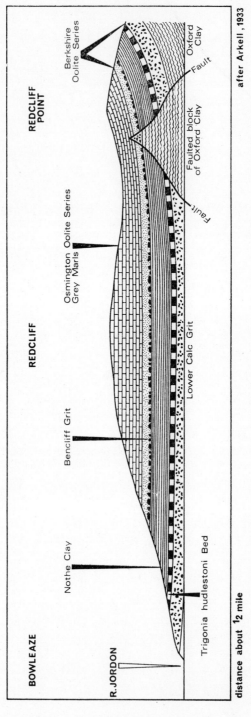

BOWLEAZE — REDCLIFF — REDCLIFF POINT

R.JORDON

Nothe Clay — Bencliff Grit — Osmington Oolite Series Grey Marls — Berkshire Oolite Series

Trigonia hudlestoni Bed — Lower Calc Grit — Fault — Faulted block of Oxford Clay — Oxford Clay

distance about ½ mile   after Arkell, 1933

Fig. 12. Bowleaze–Redcliffe Point: coast section showing the lower part of the Corallian succession and Oxford Clay.

Under Roman direction the industry attained its zenith and armlets were made on a large scale. The Saxons who followed did not appreciate these products, and the industry died after flourishing for upwards of 800 years, from about 400 B.C. to A.D. 400 (Calkin, 1955).

Oil

Prospecting for oil was carried out by British Petroleum Ltd., and Dr G. M. Lees found oil-impregnated sands at Worbarrow Bay, Mupe Bay, Lulworth Cove and Dungy Head. The first shallow test was drilled to 943 feet near Broad Bench in 1938.

Kimeridge Well No. 1, on the edge of the bay at SY908792, was completed to 1816 feet in 1959, proving the following strata (House, 1963b), including 581 feet thickness of Kimeridge Clay below the lowest horizon exposed in Hobarrow Bay.

		Thickness
Kimeridge Clay	0–798 feet	798 feet
Corallian	–1115	317
Oxford Clay	–1703	588
Kellaways Beds	–1763	60
Cornbrash	–1816	53 (drilled)

Oil was found in the Cornbrash, and in a 50-day test flowed 40 tons per day of 44° API gravity. The yield in 1961 totalled 8738 tons, and increased until in 1967 it was 16 700 tons. It is still the paramount English oil well and by 1968 had produced 100 000 tons (28 million gallons), nearly half of the total British oilfield production over the last 30 years. The oil was taken by road tanker to Wareham and thence by rail to Pumpherston refinery in Scotland until that closed in January 1965, and afterwards to Ellesmere Port refinery near Liverpool.

Three other wells were drilled on land nearby, and one in the sea on Lulworth Banks, 2½ miles to the south-west, the first off-shore well to be drilled in British waters. None of them, however, were successful producers.

THE PORTLAND BEDS OF THE DORSET MAINLAND

Although portions of these beds have been referred to in the literature since Fitton in 1836, the greater part remained relatively unknown until Arkell's description in 1935. The outcrop is 27 miles long, stretching from Durlston Head in the east to Portisham near Abbotsbury in the west. Except at St

PORTLANDIAN OF DORSET MAINLAND

Freestone Series (50 ft)

	Bed		
	V	10 ft	Shrimp Bed: white sublithographic limestone, with fragmentary ? *Callianassa* (a small shrimp-like decapod crustacean); abundant *Trigonia, Chlamys, Protocardia, Isognomon, Pleuromya, Isocyprina* and coarsely ribbed ammonites. This is the *Paracraspedites* horizon.
	U	10 ft	*Titanites* Bed (otherwise *Perna* Bed, Blue Bed or Spangle): hard grey shelly limestone with giant ammonites, *Trigonia, Ostrea expansa,* etc.
	T	5 ft	Pond Freestone: oolitic freestone, comminuted fossils.
	S	5 ft	Chert Vein (otherwise Flint Vein): limestone with nodular chert.
	R	1 ft	Listy Bed: soft grey limestone (or freestone) with ready vertical and horizontal fracture.
	Q	8½ ft	House Cap: hard grey shelly limestone; *Ostrea expansa, Isognomon, Chlamys,* giant *Titanites* and a chert band.
	P	3 ft	Underpicking Cap: hard freestone.
	O	8 ft	Under or Bottom Freestone: fine oolite, excellent-quality freestone, comminuted fossils.

Upper Jurassic: Portlandian — Portland Stone — *Titanites giganteus* Zone

Cherty Series (65 ft)

	Bed		
	N	11 ft	Cherty limestones.
	M	3–5 ft	Cherty and serpulitic limestones: *Isognomon, Chlamys, Exogyra, Trigonia, Serpula gordialis.*
	L	4½–6 ft	Highly serpulitic cherty limestone; *Behemoth,** *Isognomon, Chlamys,* etc.
	K	1½–2¼ ft	Persistent cherty limestone band; *Behemoth.*
	J'	1½ ft	Prickle Bed or Puffin Ledge: soft chertless bed, conspicuous all along the cliffs, with a ropy structure, and a mass of ramifying ? algae; *Behemoth.*
	J	4–5½ ft	Sparsely cherty limestone, cemented to J'; with J' it is the most noticeable block in every section.
	H	3–4 ft	Limestone, little chert; serpulitic, *Behemoth.*
	G	2½–4 ft	Limestones with bands of chert nodules; *Behemoth, Lima, Isognomon, Ostrea.*
	E and F	17 ft	The Sea Ledges: limestone with massive bands of confluent nodular chert; usually weathering into two blocks; at sea level; *Behemoth.*
	D	3½ ft	Cherty limestone; much chert in large nodules.
	C	7 ft	Cherty limestone.
	B	2 ft	Soft sandstone with bands of chert.

* Buckman's generic name *Behemoth* is in this section used by Arkell for large ammonites at least some of which are now considered to belong in *Titanites* Buckman; but further study is needed.

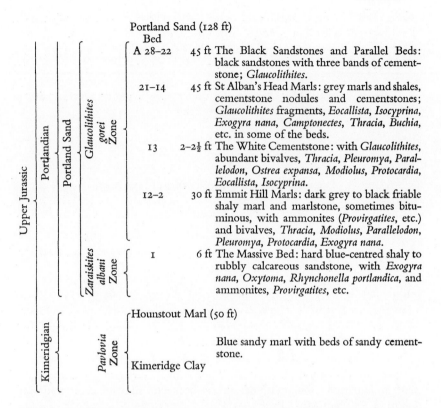

Portland Sand (128 ft)

Bed

A 28–22   45 ft The Black Sandstones and Parallel Beds: black sandstones with three bands of cementstone; *Glaucolithites*.

21–14   45 ft St Alban's Head Marls: grey marls and shales, cementstone nodules and cementstones; *Glaucolithites* fragments, *Eocallista, Isocyprina, Exogyra nana, Camptonectes, Thracia, Buchia,* etc. in some of the beds.

13   2–2½ ft The White Cementstone: with *Glaucolithites*, abundant bivalves, *Thracia, Pleuromya, Parallelodon, Ostrea expansa, Modiolus, Protocardia, Eocallista, Isocyprina.*

12–2   30 ft Emmit Hill Marls: dark grey to black friable shaly marl and marlstone, sometimes bituminous, with ammonites (*Provirgatites*, etc.) and bivalves, *Thracia, Modiolus, Parallelodon, Pleuromya, Protocardia, Exogyra nana.*

1   6 ft The Massive Bed: hard blue-centred shaly to rubbly calcareous sandstone, with *Exogyra nana, Oxytoma, Rhynchonella portlandica,* and ammonites, *Provirgatites,* etc.

Hounstout Marl (50 ft)

Blue sandy marl with beds of sandy cementstone.

Kimeridge Clay

(vertical labels: Upper Jurassic; Portlandian — Portland Sand — Glaucolithites gorei Zone, Zaraiskites albani Zone; Kimeridgian — Pavlovia Zone)

Alban's Head, where the strata are nearly horizontal, the beds mostly dip steeply in the northern limb of the Weymouth–Purbeck Anticline. The outcrop is intricately dissected by the sea coast and provides a series of unrivalled sections, although not all are accessible, some of them being sea cliffs rising from deep water. The beds can be studied, in the present Site of Special Scientific Interest, at St Alban's Head, Emmit Hill and Hounstout. The type section of the Portland Sand, named by Fitton, is at Emmit Hill.

Important changes take place as the Portland Beds are traced from east to west: the Freestone Series thins by 70 per cent, the Cherty Series by 60 per cent, and the Portland Sands by 40 per cent. In the Portland Stone the commercially valuable freestones deteriorate and become cherty. The succession above is chiefly after Arkell (1935); the Freestone Series measured at Seacombe, and the Cherty Series and Portland Sand at Seacombe, Winspit, Worth and St Alban's Head, all near the eastern end of the exposure.

1836, Buckland and De la Beche; 1836, Fitton; 1901, Rowe; 1933, 1935, 1938, 1947a, Arkell; 1949, C. W. and E. V. Wright; 1953, Barnard; 1955, Calkin; 1955, Carreck and Davis; 1956, Hodson, Harris and Lawson;

1956, Downie; 1958–60, Delair; 1958, 1963b, House; 1960, Sarjeant; 1964, Phillips; 1965, Gordon. See also 1967, Casey, *Proc. geol. Soc. Lond.* No. 1640, 128; 1969, West, Shearman and Pugh, *Proc. Geol. Ass.* **80**, 331; 1969, Wilson, *ibid.* 341; 1969, Torrens (A).

Warning. Beware of adders in the undergrowth under Emmit Hill, St Alban's Head, and Gad Cliff (Arkell, 1935).

Broom Ballast Pits, 3 miles NE of Axminster

ST(31)328024	177	326	27NE	A, S
		(–)		1965

Extensive disused ballast pits in Pleistocene valley gravels of the River Axe, on the left bank, just north of the junction with the tributary River Blackwater which forms the county boundary with Devon. The pits are located in one of a dozen patches of the gravels mapped in the neighbourhood. Their main interest lies in the contained palaeoliths.

The section now exposes a face of about 25 feet of gravels, which are believed to total some 50 feet thick. They consist chiefly of greensand chert, flint, and quartz pebbles, with occasional dark grit and black chert, and some schorl-rock. Reid Moir describes the section as follows.

Pleistocene
{
1 ft Surface soil, with implements said to be Neolithic.
25 ft Tumbled gravel with partings of clay; derived implements.
8 ft Stratified gravel with old land surfaces; 'Clacton III and Late Acheulian floors'.
17 ft Stratified gravel and sand; ? derived implements.
}

The gravels appear to extend in a flooded condition below the level of the present river. The base of the old pit lay at about 150 feet above O.D. and that of the present one lies some 10 feet higher; and the top of the gravels lies at about 200 feet. From the one-inch Geological Survey sheet they rest upon Lower Lias shales.

Neither vertebrate remains nor non-marine mollusca have been found in these pits, but they have yielded artifacts representing a West Country palaeolithic culture, now assigned to the Middle Acheulian. They are mainly finely wrought hand-axes, both ovates and slender pointed implements, very variable in size, ranging from nine and a half to two and a half inches in length. Most are made from Greensand chert, some of a distinctive honey colour, easily recognised, and others from chert of other tints. A few are made from Chalk flint.

Some of the implements are fresh and sharp; others are heavily rolled. Although several hundreds have been found, nearly all have been collected by workmen, and their exact levels in the section are not precisely known. Scientists seeking them *in situ* have been singularly unsuccessful. Two artifact horizons have been supposed, both near the base of the section; but Reid Moir claims that the sharp implements come from old land surfaces which he identifies in the middle eight feet of gravel.

The old pits, partly in Devon south of the River Blackwater, were worked for ballast for the London and South Western Railway from about 1863 until some date between 1911 and the 1940s. In 1877 D'Urban, Curator of the Exeter Museum, obtained from the workmen a fine series of specimens, apparently the first known. Other collections are preserved in Salisbury and Brighton Museums, and there are 205 specimens in the Sturge Collection in the British Museum. The slopes of the steep hills bordering the Axe valley here are strewn with debris from the Greensand scarps, so that there was a plentiful source of material nearby for making the implements.

The gravels are held to be at least in part equivalent to the 100-foot or 'Boyn Hill' Terrace of the Lower Thames, of the Hoxnian (Great) Interglacial. However, a wider spread of age has been claimed, for Reid Moir thinks they represent two glacial periods and the intervening Hoxnian Interglacial. Hawkes is of similar opinion, but regards the Interglacial as the Eemian.

1897, Evans; 1936, Moir; 1944, Hawkes; 1947, Green; 1948, Dewey.

Crook Hill Brick Pit, Chickerell, 2 miles NW of Weymouth

SY(30)644797	178	341	53SW	S, P
			(G)	IGC, PSD
				1963

A large working brick pit alongside the main road B3157, exposing a nearly vertical face some 70 feet high, as on p. 134 (Arkell, 1947a).

The top of the *athleta* Zone is not exposed in this pit and, despite some earlier statements, Arkell later found that no *jason* Zone is represented at the base. Nevertheless R. T. Smith (in Torrens (A), 1969) includes part of the *jason* Zone in his very detailed section (p. A41).

From the top 25 feet, 13 species of ammonites are listed (Arkell, 1939) and the fauna is particularly interesting on account of the Reineckias, which are very rare in this country. A dibranchiate cephalopod, *Geopeltis brevipennis* (Owen), and a fragmentary teleostean fish, *Leptolepis macrophthalmus* Egerton, were found in this pit in 1960, both new to the Oxford Clay of this

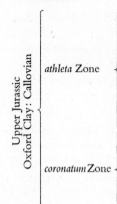

	15 ft Yellowish-brown weathered clays, with yellow-coated septaria; ammonites with well-preserved body chambers but more rare inner whorls. Large *Peltoceras athleta* abundant, and *Kosmoceras* spp.
athleta Zone	10 ft Blue-grey clay, said to have the properties of fuller's earth, with grey septaria; similarly preserved ammonites, but with more inner whorls; *Pseudopeltoceras, Kosmoceras, Reineckia, Hecticoceras.*
	15 ft Greenish-brown shales, bituminous; surfaces covered with crushed *Meleagrinella*, and *Kosmoceras* spp.
coronatum Zone	30 ft Greenish and chocolate-brown shales, bituminous, with strong odour; lines of large flat septaria ('turtle stones') sometimes containing *Erymnoceras coronatum* and *E.* spp. In the shales are abundant but flat-crushed ammonites chiefly *Kosmoceras* (3 spp.), and abundant crushed bivalves, *Protocardia, Nucula, Procerithium, Meleagrinella*, and uncrushed tubes of *Serpula vertebralis* and fossil wood.

district; and Sarjeant (1960) described a new Hystrichosphere, *Polystephano-sphaera paracalathus*, from the *jason* Zone here.

The Oxford Clay of the Dorset coast covers a small area completely separated from the main outcrop to the north by the overstepping Chalk of the Downs. Owing to its softness, natural outcrops are poor, but the Crook Hill and Putton Lane (q.v.) brick pits afford magnificent sections of several zones. They have long formed the principal sources of bricks and tiles in the area. One of the pits here was at work in 1895.

1897, Woodward; 1899, Lydekker; 1939, 1947a, Arkell; 1960, Carreck; 1960, Sarjeant; 1969, Torrens (A).

Durlston Bay, Southern outskirts of Swanage

SZ(40)034780	179	343	57SW (G)	S, P PSD, IGC, 1963

A one-mile stretch of coastal cliffs rising from sea level at Peveril Point to 243 feet above O.D. at Durlston Head which exposes the type and finest section of the Purbeck Beds in England, famous for their very diverse fauna and flora, particularly the unique mammals (Fig. 17). The bed numbers and thicknesses given below are those of Bristow's section (1857); the uppermost 11 feet of beds is added from Arkell to complete the tally of the Purbeck Beds known to exist here.

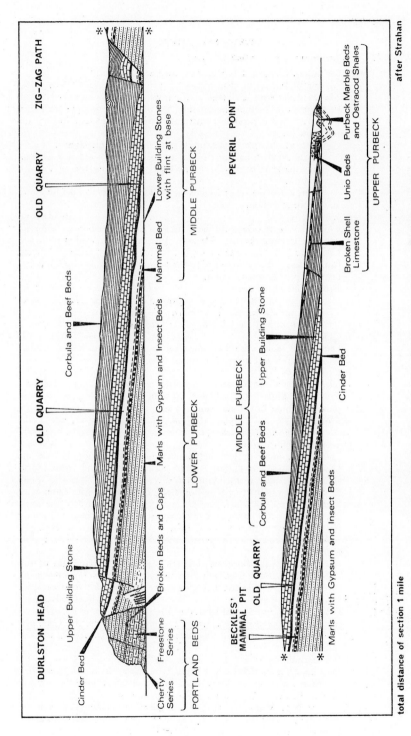

total distance of section 1 mile

Fig. 17. Durlston Bay, Swanage: type coast section of the Purbeck Beds.

after Strahan

DURLSTON BAY SECTION

Wealden   Junction not exposed at Swanage

11 ft (or more). *Viviparus* Clays (not exposed at Swanage).

46 ft Marble Beds and Ostracod Shales   (Beds 93–84)
Mainly shales with little sand and sandy limestone, with two 4-ft beds
of compact shell-limestone (marble) 11 ft apart, the upper dark green,
the lower red. *Viviparus cariniferus* and ostracods abundant, also *Unio* and
fish remains.
  Anderson's Marine Bands, marked by marine ostracods are:
    Battle M.B. at base of Bed 93, 4 ft below top of exposed section.
    Tyneham M.B. at base of Bed 90, 20 ft below top of exposed
    section.
    Durdle M.B. at 1½ ft above base of Bed 88, 28 ft 8 in below top of
    exposed section.

5 ft Unio Beds   (Beds 83–79)
Sandy clays with occasional 'beef' (fibrous calcium carbonate) and with
greenish limestone bands; *Neomiodon*, *Viviparus*, *Unio*; Bed 81 is the 6 in
Crocodile Bed, limestone with remains of vegetation, shells, coprolites,
fish, turtles, and crocodiles; with the Mupes M.B. at its base, 49 ft 1 in
below top of exposed section.

10 ft Broken Shell Limestone   (Bed 78)
Hard massive limestone with comminuted shells of *Neomiodon*, *Limnaea*,
Viviparus, remains of fish and turtles.
    Lulworth M.B. at base, 60 ft 7 in below top of exposed section.

30 ft Chief 'Beef' Beds   (Beds 77–71)
Dark sandy shales with thin limestone bands, layers of 'beef'; selenite;
with ostracods and bands of perished *Neomiodon* shells.

34 ft *Corbula* Beds   (Beds 70–59)
Shelly limestone and shales, with a little 'beef' and selenite; includes the
Toad's Eye Limestone, Bed 68, 2 ft 4 in below the top, with shales and
pink and blue limestone bands; marine mollusca, *Corbula* (4 spp.),
Modiolus, *Ostrea*, '*Pecten*', *Isognomon*, *Thracia*, *Protocardia*; also fresh-
water *Melanopsis*, *Neomiodon*; turtle and fish remains, insects, ostracods.

50½ ft Upper Building Stones   (Beds 58–45)
Limestones, more or less shelly, with subsidiary shales; include the
Leaning Vein (Beds 57, 56) the Freestone (Bed 50) and Downs Vein
(Beds 49–45) once worked commercially; the highest 4½ ft, the White
Roach (Bed 58; Bristow's 'Scallop Bed') contains a marine fauna with
'*Pecten*', *Ostrea*, *Gervillia*, *Corbula* spp., *Protocardia*. Nearly the same
assemblage is met at several horizons, separated by beds with only
fresh-water or estuarine forms, such as *Viviparus*, *Limnaea*, *Hydrobia*,
Valvata, etc. Bed 50 in the Roach includes the Pink Bed with reptile
footprints.
    The basal 3½ ft, *Tombstone*, etc. (Bed 45), contains remains of fish
(*Microdon*, *Hybodus*, *Lepidotus*), fresh-water tortoises (*Pleurosternum*),
Pterodactyl, crocodile, etc. Beds 57–45 were Bristow's 'Intermarine Beds'.

8½ ft Cinder Bed   (Beds 44, 43)
Dark limestone composed largely of the shells of *Liostrea distorta*; with
occasional *Hemicidaris purbeckensis*, *Laevitrigonia gibbosa*, the *Corbicula*-
like *Myrene*, a thin-shelled *Protocardia*; also *Isognomon*, *Serpula* and fish
remains.

33 ft Lower Building Stones (The 'Turtle Beds' of Austen) (Beds 42–23)
Beds 42–25 were Bristow's 'Cherty Fresh-water Beds' which overlay
his 'Marly Fresh-water Beds' Beds 24, 23. They include the 2 ft 2 in
Feather Bed (Bed 40) 1 ft 4 in below the base of the Cinder Bed; the
4 ft 1 in New Vein, Bed 34, 9 ft 3 in below the Cinder Bed; and the
3 ft 0 in Flint Bed, Bed 31, 15 ft 0 in below the Cinder Bed. Beds 42–36
were collectively known as the Feather Quarry. Limestones and shales,

Upper Purbeck:
Zone of *Cypridea setina*
72 feet

Middle Purbeck:
Zone of *Cypridea granulosa*
157 feet

often bituminous. The Flint Bed is a white bituminous marly limestone with black chert nodules and fresh-water gastropods, *Viviparus, Physa, Planorbis*; generally are found ostracods, *Neomiodon, Corbula*, insects, fish and turtle remains. From the Feather Bed is recorded the Dinosaur *Nuthetes destructor*.

1 ft Mammal Bed   (Bed 22)

Dark-grey and brown carbonaceous shale, filling inequalities in the bed below; with selenite, remains of vegetation, ostracods, fresh-water gastropods, remains of the lizard *Macellodus brodiei* and of dwarf crocodiles, and 19 species of mammals.

137 ft Marls with Gypsum and Insect Beds   (Beds 21–7)

90½ ft Soft Cockle Beds, Beds 21–13: marls with two partings of brecciated limestone, some large botryoidal masses of gypsum near the base; pseudomorphs of rock-salt crystals; plant and insect remains, ostracods, *Archaeoniscus brodiei*, fish remains and gastropods.

10 ft Hard Cockle Beds, Bed 12: pale blue marly limestone and shales, with insects, etc.

36½ ft *Cypris* Freestone, Beds 11–7: marly shales and marly limestone, with many ostracods, *Archaeoniscus* near base, and marine bivalves, *Protocardia purbeckensis*, and *Corbula*, occasional fresh-water *Neomiodon, Planorbis*, etc.

15 ft Broken Beds   (Bed 6)

Shattered limestone, thin slaty beds of bituminous sandy limestone, of controversial origin (see below).

19 ft Caps and Dirt Bed   (Beds 5–1)

Bituminous limestone and shale; the upper 7½ ft is the Soft Cap, Beds 5, 4, the beds being fissured, with bands of chert. The next 11 ft is the Hard Cap, Beds 3, 2. The stone composing the caps is a peculiar porous, tufaceous and botyroidal limestone, occasionally enclosing such mixed fossils as fish remains, ostracods and *Archaeoniscus*. At the base is a three-inch Dirt Bed, or ancient surface soil, Bed 1.

Portlandian   Portland Stone; Shrimp Bed.

(margin) Lower Purbeck: Zone of '*Cypris' purbeckensis* 171 feet

In the Broken Beds of the Lower Purbeck the brittle limestone is broken into small fragments and blocks of all sizes up to some three by five feet or larger. They are considered by Arkell (1947) to be small-scale accommodation effects due to adjustment between the competent Portland Stone and the incompetent Purbeck Beds during the Tertiary folding. Other hypotheses have been put forward, however, and none is certain.

At Peveril Point the Upper Purbeck Beds from the Broken Shell Limestone upward, including the Marble Beds, are well exposed in contortions within a sharp synclinal fold trending west-north-west obliquely across the main strike. Two small faults are mapped within 300 yards of the Point, but the section is unfaulted for the next 500 yards, as far as the Zigzag Path at SZ035780. There, two faults have a net downthrow of about 130 feet to the south. Beyond, in the southern half of Durlston Bay, which exposes a dip section dipping 8° north, the sequence is repeated, unfaulted again up to Durlston Head, where there is some complex faulting with net downthrow to the north, exposing the underlying Portland Stone.

HISTORY

John Middleton (1812) gave a brief account with what purported to be a measured geological section to show the levels of four beds, the Leaning Vein, Freestone, Downs Vein and New Vein, which were then worked for flag-paving, and shipped to London from Durlston Bay or Swanage. Further but incomplete descriptions were given by Webster (1816, 1826), Forbes (1851), Austen (1852) and Fisher (1856). Bristow (1857) of the Geological Survey made the first one-inch geological map of the district, and published his vertical section, now complete for the first time, dividing it into 93 beds. This records the fossils found in each bed and is accepted as the standard section. This was usefully re-published by Damon but, at least in the later edition of his book (1884), there are a few inaccuracies, two of them of some importance. More up-to-date accounts are given by Strahan (1898) and Arkell (1933, 1947) and authors in Torrens (A) (1969, pp. 57–62).

Durlston, sometimes spelt Durdlestone or Durlstone, is a name of Saxon origin, with the same root as Durdle Door, indicating a 'pierced stone'. A natural arch was presumably present in Saxon times but has since been removed by marine erosion, leaving only the name.

Because of the attractive green, and less frequently red, mottled colour, and the property of taking a high polish, Purbeck 'Marble', obtained from the Upper Purbeck Beds in the district, has been used for decorative purposes from time immemorial. A freshwater limestone made up of myriads of *Viviparus* shells, it gave rise to one of the oldest stone-quarrying industries in the country. Its earliest known examples are found in Roman Silchester and Verulamium, Chichester, Cirencester, Colchester, London and Chester. Used for church decoration in the Middle Ages, it was taken to Scotland, Ireland and the Continent. Extensive use in Salisbury Cathedral testifies to active quarrying in the year 1258. Excellent examples may also be seen in Westminster Abbey and the Temple Church, and in the cathedrals of Canterbury, Lincoln, Winchester and Worcester. Its last considerable use, according to Arkell, was in the Eldon memorial church at Kingston in Purbeck, opened in 1880. The industry is now dead, killed by the import of more varied and durable true marbles from Italy.

The Portlandian shallow sea gave place abruptly, but without any sign of unconformity, to an oscillating area on which the Purbeck Beds were laid down, about 140 million years ago. Sometimes it was a land surface, some-times covered by swamps and fresh-water lakes, or occupied by mud flats. From time to time the sea invaded it. A varied series of deposits was accumu-lated, particularly well developed in the Isle of Purbeck, where, at Durlston Bay they reach their maximum thickness of about 400 feet; they thin fairly rapidly to the west. Comparison of the Purbeck Beds has been made with the English Rhaetic.

Lithologically there is much thinly laminated limestone, mudstone and shale, with layers of fibrous calcium carbonate ('beef'); sometimes the limestone has mammilated surfaces and arborescent markings in the interior, reminiscent of the Rhaetic Cotham Marble. Lenses of gypsum and pseudo-morphs of rock salt indicate land-locked lagoons, and lenticles of celestine have been found in the Caps and Broken Beds near the faulting at Durlston Head (West, 1960).

'The Purbeck Beds although highly fossiliferous require much patient search to yield results' (Arkell). They include beds of ostracods, pond-snails, mussels, oysters, insects, isopods, fish, reptiles and small mammals. Dirt Beds, or fossil soils with tree-trunks of conifers and cycads, often with stumps and roots still in position, mark former land surfaces. Remains of such forests are seen at Lulworth and elsewhere, but are absent at Durlston Bay. Dinosaurs, small mammals and abundant insects lived on the land, on the vegetation and on one another, and crocodiles, some of them 'dwarf', inhabited the marshes, with water tortoises, fish, ostracods, shells and *Chara*. Occasional inroads of the sea left deposits with marine shells, such as *Trigonia* and *Isognomon*, and the most widespread marine episode is marked by the Cinder Bed, made up largely of the shells of *Liostrea distorta*.

Early geologists grouped the Purbeck Beds with the Wealden by reason of their largely fresh-water fauna. Forbes, however, influenced by his dis-covery of *Hemicidaris* in the Cinder Bed and certain other fossils, concluded that their affinities lay rather with the Jurassic, and that view has prevailed until very recently. They were considered the equivalent of the marine Tithonian Stage of Southern Europe. Nevertheless, they pass up into the Cretaceous Wealden Series without a break, forming a perfectly continuous and conformable series (Arkell, 1933).

Anderson (1962) has recognised eight Marine Bands in the Upper Purbeck Beds, five of which have been so far identified in the Durlston Bay section (see above) and he correlates them with beds in the German Wealden. He names them, in descending order, as the Hastings, Battle, Tyneham, Durdle, Tisbury, Brede, Mupes and Lulworth Marine Bands. The Hastings Marine Band is not found here, as it would lie four feet above Bed 93, the top of the exposed section. Anderson proposed to take this Hastings Marine Band as marking the base of the Wealden in England. Casey (1963), however, claims that from palaeontological evidence the Dorset Purbeck Beds straddle the Jurassic–Cretaceous boundary. He bases his new grouping upon correlation between the South of England Purbeck and the marine Middle and Lower Spilsby Sandstone of Lincolnshire. In those beds he has discovered am-monites which, newly identified, link the beds with the succession in Poland and the Moscow Basin. The Cinder Bed makes a firm datum line along a front of 150 miles from Dorset to Kent, indicating a salt-water invasion, which he claims as marking the incoming of the Cretaceous. Although some of the Cinder Bed fossils, as *Hemicidaris* and *Laevitrigonia* also occur in normal marine deposits, others such as *Liostrea*, *Myrene* and *Protocardia* have modern analogues which thrive near river mouths. The absence of cephalopods, corals and brachiopods supports the conclusion that there was dilution of the sea-water by fresh water here.

Casey's dating is as follows.

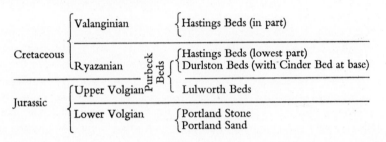

Casey would thus divide the Purbeck into two parts only, newly named as the Durlston Beds (lowest Cretaceous) and the Lulworth Beds (uppermost Jurassic).

THE FOSSIL FAUNA AND FLORA

The fauna of the Dorset Purbeck, that of Durlston Bay being aggregated with that of the Swanage neighbourhood, is of exceptional interest, but the many bivalves, of marine, brackish and fresh-water habitat, await detailed description, and no attempt is made to include them here.

Certain of the fossils, such as the reptiles, especially the tortoises, and the fish, came largely from the underground mining of the building stones; since working has now ceased, that source of supply is no longer available, and opportunities for securing specimens are much less favourable than formerly.

MAMMALS

The Purbeck mammals of Durlston Bay are world-famous and of great importance to palaeontology. Remains of *Spalacotherium tricuspidens* were first discovered by Brodie in 1854 from the Dirt Bed now known as the Mammal Bed, taken as the base of the Middle Purbeck, or, in Casey's revision, near the top of the Lulworth Beds of Upper Volgian, uppermost Jurassic, age. Brodie obtained further specimens in 1855–56. In 1857 S. H. Beckles exposed an area of the same bed some 40 feet by 10 feet (*fide* Austen, 1857) in which he was so fortunate as to find nearly all of what are now recognised as 19 species, belonging to 13 genera in four Orders. The remains are nearly all of jaws and teeth, with one or two crushed skulls. His specimens were obtained from a fossiliferous 'pocket' in the thin Dirt Bed in what is known as the Beckles Pit, near the top of the cliff, sited a few yards north of the Belle Vue Restaurant, and at about SZ035782. The Beckles Collection, monographed by Owen, was bought by the British Museum (Natural History) in 1876. Extensive subsequent search by the Willetts in 1881 yielded only one more mammal specimen, of *Triconodon mordax*, together with crocodilian and other vertebrate remains.

Except for the few mammalian remains found in the Stonesfield Slate and the Rhaetic beds, these Purbeck mammals are unique in Europe although they are paralleled in the Morrison Formation of about the same age in Wyoming in the United States. They have been found only at Durlston Bay,* and the specimens were all obtained, according to Austen (1857) from an area less than 500 square yards in extent, and from a single stratum a few inches thick. They have been re-described by Simpson (1928), who classified them as follows.

Multituberculata (up to the size of a fox; herbivorous, ? arboreal): *Plagiaulax becklesi* (5 specimens); *Ctenacodon minor* (1), *C. falconeri* (1); *Bolodon crassidens* (1), *B. osborni* (1), *B. elongatus* (1).

Triconodonta (the size of a cat; carniverous): *Triconodon mordax* (15 specimens), *Trioracodon ferox* (18), *T. oweni* (1), *T. major* (1).

Symmetrodonta (the size of a mole or shrew; ? carniverous): *Spalacotherium tricuspidens* (6 specimens); *Peralestes longirostris* (1).

Pantotheria (the size of a rat; ? insectiverous, ? arboreal): *Peramus tenuirostris* (7 specimens); *Peraiocynodon inexpectatus* (1); *Amblotherium pusillum* (14), *A. nanum* (6); *Kurtodon pusillus* (2); *Peraspalax talpoides* (1); *Phascolestes mustelula* (3).

In 1949 Heap (1958) found in the outcrop of the Mammal Bed under the beach shingle near the Zigzag Path 'four mammal jawbones with some teeth intact'. In 1960 he presented them to the British Museum (Natural History), where two of them have been identified as *Trioracodon* but not published. Three other specimens were described by Clemens (1963). Thus, there are now known some 93 specimens in all.

Simpson considered the four Orders to have been derived independently from Cynodont reptiles, which were dog-shaped animals with dog-like dentition. The Symmetrodonts and the Pantotheres may, he thinks, have diverged from the same stock as all the living mammals but they branched off long before the marsupials and placentals became differentiated. In the nineteenth century these mammals were thought to have been marsupials, on the authority of Owen's opinion. All these animals were small; the jaws

---

* However, Charles Moore (1874) states that he got very fragmentary remains of some of the same mammals and other vertebrates from Dirt Beds in the Purbeck of the Town Gardens Quarry at Swindon, Wiltshire. These specimens seem to be lost, for it has not been possible to trace them in recent years.

range mostly from 14 to 20 millimetres and a few from 27 to 38 millimetres, and the largest is 80 millimetres in length.

The Multituberculata were the largest, and they appear to have been mainly or entirely herbivorous, probably living on the fruit and bark of the cycads and conifers. They were probably arboreal, and were ecologically analogous to the smaller rodents of the Tertiary faunas. The plagiaulacoid Multituberculates appear probably for the first time, although they must have existed since the Trias. They continued through the Cretaceous and Palaeocene into the base of the Eocene before vanishing without issue.

The Triconodonts were, from the evidence of their teeth, definitely carnivores. They make their last appearance in the Purbeck, where they differ little from the Middle Jurassic forms.

The Symmetrodonts, also probably carnivorous, are known only from the Purbeck and the Morrison Formations. They seem to represent a very early and distinctive offshoot from the Pantotherian or pre-Pantotherian stock. They do not appear again and have left no recognisable descendants.

The Pantotheres were possibly arboreal and insectivorous. They are the most important and interesting group, for they are not so aberrant as the other three Orders and show early stages in the development of the marsupial and placental dentitions. The teeth seem to be adapted to a diet chiefly of invertebrates, but in part omnivorous, somewhat analogous to that of the smaller opossums and many insectivores. Although still very primitive, they have lost the generalised nature of the genus *Amphitherium* from which they could have been derived. Just 100 named species of insects, half of them beetles, and another 34 unnamed forms, are recognised by Handlirech from the combined Lower and Middle Purbeck of Durlston Bay. These must be minimum figures probably much exceeded in fact, and they may have furnished a large part of the diet of the Pantotheres.

Simpson concluded that the main mammalian stem is not represented here; these animals were marginal, probably swamp-dwelling groups. They represent the main stock in a structural way, however, and may give an idea of the stage of evolution reached by the ancestral Theria of the Upper Jurassic.

REPTILES

Thirty-three forms, mostly named specifically, are now recognised (Delair, 1958-60) as follows. Those recorded actually from Durlston Bay are starred.

> *Chelonia* (Turtles and Tortoises):
> > Amphichelydia:
> > > *Pleurosternum bullocki, P. obovata, *P.* sp.; Tretosternum punctatum,* *T. bakewelli*; **Platychelys* (?) *anglica*; *Glyptops ruetimeyeri*; *Plesiochelys* sp.

Pleurodira:
 Hylaeochelys latiscutata, H. emarginata, H. belli, H. sollasi.
Rhynchocephalia:
 *Homoeosaurus sp.
Squamata:
 Lacertilia (Lizards):
  *Macellodus brodiei, and four or five obscure and unnamed genera.
Crocodilia:
 *Goniopholis crassidens, G. simus, *G. tenuidens, *G. sp.; *Petrosuchus laevidens; *Nannosuchus gracilidens; *Theriosuchus pusillus; *Oweniasuchus major, *O. minor; Pholidosaurus decipiens, P. laevis.*
Pterosauria (flying reptiles):
 *Pterodactylus sp.; *Doratorhynchus validus; Ornithocheirus validus.*
Saurischia:
 Theropoda (carnivorous Dinosaurs):
  *Nuthetes destructor.
Ornithischia (herbivorous Dinosaurs):
 Ornithopoda:
  Iguanodon hoggi, I. sp.
 Stegosauria:
  Stegosaurus sp. ; **Echinodon becklesi;* *? Stegosaurian indet.
Reptilian footprints (Ichnites):
 **Three-toed: assigned to the Dinosaur *Iguanodon* or *Megalosaurus.*
 Five-toed: possibly crocodilian.

In addition to the above there are indeterminate remains assignable to the *Chelonia, Lacertilia and *Crocodilia.

Of the above Chelonians, found in the *Unio* Beds and Broken Shell Limestone of the Upper Purbeck, and in the Upper and Lower Building Stones of the Middle Purbeck, the genera *Pleurosternum, Tretosternum* and *Platychelys* were fresh-water tortoises with flattened shells, primitive forms in which the head was not retracted into the shell. *Pleurosternum* is the best-known genus, and *P. bullocki* the most commonly met Purbeck Chelonian; it reached about 20 inches long. *P. obovata* was of nearly circular contour, 10 inches long by 9 inches wide. *Tretosternum* was a somewhat similar genus, but had a very characteristic tuberculated carapace. *T. punctatum* was about 13 inches long, and *T. bakewelli* was 'of moderate size'. *Platychelys (?) anglica* was a small form, and *Glyptops ruetimeyeri* was 11 inches long by 10 inches wide. *Plesiochelys* sp. was a thick-shelled form represented by specimens lacking specific characters. *Hylaeochelys* was a genus of turtles in which the head was retracted by a sideways movement of the neck; *H. latiscutata* was 15 inches long and *H. emarginata* some 20 inches; *H. sollasi* was about 18 inches long but 19 inches wide.

The Rhynchocephalian *Homoeosaurus* was found at Durlston Bay only in 1890. The Order is represented today by a single genus *Sphenodon*, the Tuatara, a little, slow-moving, burrow-dwelling, lizard-like animal, living only on small islands off the north coast of New Zealand. All members of the Order are characterised by an overhanging beak on the upper jaw.

Lizards are represented by some of the earliest known types. *Macellodus brodiei*, with a skull one inch long, and reaching a total length of about nine inches, is known from more than 40 specimens. There are also the remains of four or five obscure and unnamed genera which cannot be assigned to any of the recognised lacertilian families.

Recent crocodiles are nearly all found in tropical or sub-tropical regions; they are the most specialised group of reptiles. All have a dermal armour of scutes characterised by a network of irregular pits over the surface, pits up to some 10 millimetres across and one or two millimetres deep, but many are smaller.

The Family Goniopholidae was typical of the Purbeck and Wealden. The genera had stout, rounded, broad-faced skulls, with moderately long snouts of the modern crocodilian shape, as contrasted with the long slender snout of the Gavial. There were two groups, one of moderately large crocodiles and the other of small forms. All seem to have been marsh-living, and the majority of these genera are peculiar to Dorset. *Goniopholis crassidens*, found here in 1837, and called by Mantell the 'Swanage Crocodile', was of large size, perhaps 20 feet long, with a two-foot-long skull, and teeth remarkably thick and stunted. *G. simus* was smaller, about 7 feet long, with more slender teeth and less tapering head, the skull some 16 inches long and 9 inches wide. *G. tenuidens*, described from a fragmentary mandible, was also small, with slender teeth. *Petrosuchus laevidens* was of moderately large size, with slender teeth and a skull probably some 15 inches long and tapering almost as abruptly as that of a Gavial. The mandible was perhaps 16 inches long.

Of the remarkable group of small rather than 'dwarf' Crocodiles, *Nannosuchus* was like a miniature *Goniopholis*, with a skull not more than five inches long, but with teeth that were long, slender, curved and sharp, well suited for catching fish. Joffe (1967), however, considers it merely a juvenile of *G. simus*. Some of these 'dwarf' crocodiles may have preyed upon the contemporary small mammals. Of *Theriosuchus pusillus* a nearly complete skeleton, 18 inches long, was obtained by Beckles from the Mammal Bed, and Owen states that its scattered teeth, scutes, vertebrae and limb bones are very numerous, and a few skulls (about three and a half inches long), mandibles and considerable portions of naturally connected skeletons have been found. It had teeth varied in shape and more specialised than those of any other crocodile. *Oweniasuchus major* and *O. minor* were both of small size, described from lower jaws. *Pholidosaurus decipiens*, of the Family Pholidosauridae, was about 12 feet long, with a 14-inch skull; and *P. laevis* was of similar size.

Of the flying reptiles, *Pterodactylus* sp. was a small indeterminate species. *Doratorhynchus validus* was of very large size, and quite unlike any other Protosaurian; but both it and the moderate-sized *Ornithocheirus validus* are imperfectly known.

Of the Dinosaurs, the carnivorous *Nuthetes destructor* from the Feather Quarry of Durlston Bay is known from a left mandibular ramus with teeth; the animal appears to have been about five feet long. The herbivore *Iguanodon hoggi*, known from a single imperfect mandible with teeth (BM), is estimated to have been somewhere about eight feet long, small for an *Iguanodon*.

The Stegosaur *Echinodon becklesi*, known likewise from teeth and jaw fragments, may have been only some six or eight inches in total length. If this was so, the animal would be the smallest known Stegosaur. The indeterminate ? Stegosaurian is known only from 'Granicones', conical dermal ossicles up to 14 millimetres in height and eight millimetres across the base, although mostly smaller. Described by Owen in 1879, they were ascribed to *Nuthetes* because they were found mixed with *Nuthetes* fragments. However, they are now considered to be best regarded as belonging to an unknown reptile tentatively referred to the Stegosauria.

Three-toed footprints (some, in the Corbula Beds, 12 inches long with the middle toe seven by five inches) are preserved in thin-bedded limestones of the Lower, Middle and Upper Purbeck. They have been attributed to *Iguanodon*. Of late years, however, three-toed footprints in the Pink Bed (Bristow's Bed 50) of the Upper Building Stones in a quarry at Herston, near Swanage, are thought just as likely to be those of a large carnivorous bipedal Dinosaur such as *Megalosaurus*, which was probably present in Purbeck times, although its bones have so far not been recognised in that formation (Charig and Newman, 1962). The Herston animal had a stride of about 46 inches, making a single row of imprints 23 inches apart.

FISHES

The rich fish fauna monographed by A. Smith Woodward (1916–19) comes almost entirely from the Upper and Lower Building Stones of the Middle Purbeck, although fish remains are recorded from many beds throughout the succession.

Of 41 species of fishes now recognised from the British Purbeck, 32 species are known from the Swanage district, as follows.

Chondrichthyes (skeletons cartilaginous):
  Selachii (Sharks and Rays):
    Hybodus, 2 sp.; *Asteracanthus*, 2 (definitely marine forms).

Osteichthyes (skeletons largely bony):
  Crossopterygii:
    Holophagus purbeckensis Smith Woodward (a Coelacanth, definitely a
    marine form).
  Actinopterygii (skeletons bony):
    Holostei (grouped in their different families):
      Lepidotes, 2 spp. (a genus of fresh-water fishes with large grinding
      teeth).
      Macromesodon, 2 spp.; *Eomesodon*, 2; *Microdon*, 1; *Coelodus*, 2; *Gyrodus*,
      1 (these five genera, with large grinding teeth, belong to the
      Pycnodonts, and are the last survivors of this Mesozoic family).
      Caturus, 2 spp.
      Amiopsis, 2 spp.
      Ophiopsis, 2 spp.; *Histionotus*, 1.
    Halecostomi (essentially Mesozoic ancestors of the Teleosts, they are
    grouped in their different families):
      Aspidorhynchus, 1 sp.
      Pholidophorus, 1 sp.; *Pholidophoristion*, 1.
      Ichthyokentema, 1 sp. (apparently willing to live in hypersaline con-
      ditions).
      Pleuropholis, 1 sp.
      ? *Oenoscopus*, 1 sp.
    Teleosti (the common Recent type of fishes, such as herring):
      Thrissops, 1 sp.; *Pachythrissops*, 1 (these are among the earliest types of
      Teleosts).

From the fish itself it is not possible to distinguish between those normally living in fresh water and those living in the sea. From their known habitats, both living and fossil, however, it is in some cases possible to be reasonably certain, and where this is so, notes have been added above.

I am indebted to Dr C. Patterson of the British Museum (Natural History) for an up-to-date list of the British Purbeck fishes from which the above list is taken, and for other information.

INSECTS

Insects were apparently first recorded from Durlston Bay by Austen in 1851 (1852) and Brodie noted that they were remarkable for their beautiful state of preservation; as many as 60 or 70 elytra and several wings and bodies have been counted on a single small slab. They were claimed to suggest a temperate or warm-temperate climate.

Insects are recorded on Bristow's section in the Middle Purbeck *Corbula* Beds, Bed 68 (Toad's Eye Limestone); in the Lower Building Stone Beds 39,

27 and 26; and in the Lower Purbeck, Beds 19 and 14 (Soft Cockle—the Insect Beds of Austen) and Bed 12, Hard Cockle.

Woodward (1895) lists 99 species of insects from the combined Middle and Lower Purbeck of Dorset; this is increased to 176 species from the same beds if the Wiltshire records are included.

Handlirsch (1906–8) revised earlier work and recognised just 100 named species in a dozen Orders, nearly all from the Lower, but a few from the Middle Purbeck, of Durlston Bay, as tabulated, with another 34 unnamed species. The 'lengths' below give some inkling, however imperfect, of the size of the insect remains that are found.

Order	Common names	Number of named species	Lengths of the remains (in mm)
Orthoptera	Locusts, grasshoppers, crickets	6	10–32
Phasmoidea	Stick insects	3	–30
Blattoidea	Cockroaches	13	6–12
Coleoptera	Beetles	52	2–29
Hymenoptera	Saw-flies	2	–25
Odonata	Dragonflies	4	–75
Neuroptera	Ant-lions, Alder and Lacewing flies	3	–22
Phryganoidea	Caddis-flies	2	
Lepidoptera	Moths, Butterflies	2	
Diptera	Two-winged flies (midges, gnats, etc.)	4	3–8
Hemiptera			
(Heteroptera)	Plant-bugs	4	–13
Homoptera	Cicads, Aphids	5	–8

One new genus of cockroaches, *Durdlestonia*, was described by Handlirsch from the Lower Purbeck of Durlston Bay.

Insects are thought to have furnished a large part of the diet of the Panto-there mammals whose remains are found in the Mammal Bed. The many beetles, with remains up to an inch and more long, would seem to have offered some of the 'best buys' for the shopping lists of these insectivorous mammals, whose jaws were of comparable length.

OSTRACODS

Ostracods are the most abundant of all the Purbeck fossils; minute bivalved Crustaceans mostly around one millimetre in length, they swarmed in shallow water, whether fresh, brackish or marine. Ostracods were apparently first recorded from the Purbeck of Dorset by Edward Forbes in 1851, and in the absence of ammonites or other more suitable groups he used them as zone fossils. Sylvester-Bradley (1941) states that Forbes divided the Dorset Purbeck into three divisions mainly on the ground that the ostracod *Cypridea fasciculata* characterised the Middle Purbeck. Since then this species or the earlier form *C. granulosa* has been used to recognise the Middle Purbeck elsewhere.

The three full zones of the Purbeck are now given as:

Cypridea setina   = Upper Purbeck
C. granulosa   = Middle Purbeck
'*Cypris*' *purbeckensis* = Lower Purbeck

Anderson (1962) has recently used the ostracods to define eight Marine (or quasi-marine) Bands in the Upper Purbeck, and of them five have so far been recognised at Durlston Bay, as noted above in the detailed section, and the presence of two more, the Brede and Tisbury Marine Bands, may both be sought there in Bed 84. These Marine Bands are indicated by the incoming of the marine or marine-brackish forms *Macrocypris horatiana*, *Gomphocythere striata*, *Limnocythere fragilis*, *Rhinocypris jurassica* and *Fabanella boloniensis*. *F. boloniensis* is the commonest of all the Purbeck marine species and is probably tolerant of hypersaline conditions, since it is very abundant in gypsiferous strata of the Lower Purbeck. These same species occur in all the marine bands, so that each band has to be identified either by superposition or by changes in the fresh-water ostracod faunas of the many forms of the genus *Cypridea* which separate them.

Darwinulinids are often very abundant and tend to occur throughout the Purbeck; they show no apparent preference for a particular degree of salinity.

ISOPODS

The land Arthropod *Archaeoniscus* (some 10–20 millimetres in length) was a creature resembling and related to the woodlouse. It has been recorded since the 1850s mainly as *A. brodiei* Edwards in the Middle Purbeck, Bed 50 of the Upper Building Stones (Bristow); in the Cinder Bed (Austin); and in a thin mudstone 'a few feet below the Feather Bed' in the Lower Building Stones (Arkell). In the Lower Purbeck it is found in Bed 19 of the Soft Cockle Beds (Bristow); and in Bed 7 of the *Cypris* Freestone (Bristow).

GASTROPODS

Arkell (1941) calls the Purbeck gastropod fauna 'the earliest assemblage of a dozen genera of unequivocally fresh-water molluscs', which indicates the great palaeontological interest. He records and figures from the whole English Purbeck 16 genera with 26 species of gastropods, some of them minute. From Durlston Bay he records 13 genera with 18 species, of which *Valvata helicoides* is the only one found in all three divisions.

Of the three Purbeck species of *Viviparus* (formerly called *Paludina*) recognised by Arkell, *V. cariniferus* is the chief constituent of the Purbeck Marble; *V. inflatus*, more highly inflated with shorter whorls, occurs with *V. cariniferus* in the Upper and Middle Purbeck; and *V. subangulatus*, a keeled form, is restricted to the Cherty Freshwater Beds of the Lower Building Stones, Middle Purbeck.

GASTROPODS RECORDED FROM THE PURBECK OF DURLSTON BAY
(Arkell, 1941)

	Lower	Middle below Cinder Bed	Middle above Cinder Bed	Upper	
Theodoxus fisheri	—	×	×	×	A
Viviparus cariniferus	—	×	×	×	
V. inflatus	—	×	×	×	
V. subangulatus	—	×	—	—	
Valvata helicoides	×	×	—	×	
Hydrobia chopardiana	×	×	×	—	
H. forbesi	×	×	—	—	A
Pachychilus manselli	—	—	×	—	
P. attenuatus	—	—	×	—	
Peverilia perisphincta	—	—	×	—	AAD
Paraglauconia strombiformis var. *purbeckensis*	—	—	×	—	D
'Turritella' *minuta*	—	×	—	—	
Promathilda microbinaria	—	—	×	—	AD
Physa bristovii	—	×	—	—	
Planorbis fisheri	—	×	—	—	A ?D
Ellobium durlstonense	—	×	×	—	AD
Ptychostylus harpaeformis	—	×	×	—	
P. cf. philippii	—	×	×	—	

A = species; AA = genus and species described by Arkell, 1941, as new; D = holotype from Durlston Bay section.

ECHINOIDS

Hemicidaris purbeckensis (Forbes), a regular echinoid two and a half inches in diameter, was found by Forbes about 1850 in the Cinder Bed of the Middle Purbeck of Durlston Bay. Although later found in the Portlandian of Boulogne in France, it was not again recorded in England until in 1924 Professor H. L. Hawkins, at the original locality, collected '38 tests (mostly crushed but otherwise complete) and innumerable detached plates and radioles'. Characteristically a Jurassic genus, it carried weight with Forbes in dating the beds. It is the unique echinoid species known in the Purbeck here.

ANNELIDS

The worm *Serpula coacervata* Blumenbach, which builds the Serpulite of North-west Germany, occurs in the Cinder Bed of the Middle Purbeck, where it forms small agglomerations of its calcareous tubes, which measure about one millimetre in diameter.

SPONGES

Some chert nodules in the Dorset Purbeck are largely composed of the fresh-water *Spongilla purbeckensis* Young, the only known British Jurassic Mon-actinellid sponge.

FLORA

The following have been recorded from the Lower Purbeck Beds of Dorset, but the conifers and cycads are not well represented at Durlston Bay. There the Dirt Beds contain none of the massive stools, or trunks, or the cycads (known locally as 'Crows Nests') which are numerous and striking at some places, such as Lulworth (where the tree trunks are probably conifers) and Portland.

Coniferae

Araucarites. The present-day Araucarieae are tropical conifers.

Cycadeae

The present-day Cycads are woody cone-bearing plants confined to the tropics and sub-tropics.

Bennettites portlandicus Carruthers, *Cycadeoidea gigantea* Seward, *Mantellia intermedia* Carruthers, *M. microphylla* Buckland, *M. nidiformis* Brogniart. *Bennettites* and *Cycadeoidea* combine the characters of Angiosperms, Gymnosperms and Ferns.

Characeae

Charophytes (Stoneworts) were discovered in the Middle Purbeck of Bacon Hole by Forbes about 1850, but the fossils were described for the first time by

Harris in 1939. Abundant Charophytes are found at Durlston Bay and elsewhere only within a range of the Cherty Limestone, Lower Building Stones, of the Middle Purbeck, from about two feet to 13 feet below the Cinder Bed. But the same species are found scattered in certain beds of the Upper Purbeck, and also in lower beds of the Middle Purbeck. The Lower Purbeck beds have not yet been adequately searched. Harris recognised six species, of which five were new, with two new genera *Perimneste* and *Charaxis* as follows. Their relative abundance may be judged from the number of specimens that he had.

	No. of specimens
Clavator reidi Groves	238
C. grovesi Harris	131
C. bradleyi Harris	31
Perimneste horrida Harris	111
Charaxis durlstonense Harris	1 (unique specimen)
Algacites clavatoris Harris, a filamentous alga enclosed in calcareous tubes 0·1 mm in diameter; no gyrogonites.	

The diameter of the gyrogonites (the calcified inner layer of the walls of the spiral cells which invest the egg) range from 0·23 to 0·85 millimetres in the different species. In some of the Middle Purbeck material these fossils have been silicified, the matrix remaining unaltered.

These Charophytes occur widespread in the Middle Purbeck from Dorset to Wiltshire and Buckinghamshire. Their great abundance in the Cherty Limestone suggests a very large lake, shallow enough (some to 3–30 feet deep) to allow these gregarious plants to grow over large areas.

SOME RECENT WORK

Two recent publications (Brown, 1963; Walker, 1964) claim that the depositional environment of the basal beds and of two limestones in the Middle and Upper Purbeck was essentially marine and not fresh-water. Brown finds that algae of the basal beds include the marine genera *Ortonella*, *Girvanella* and *Solenopora*. Walker cites geochemical evidence, the proportion of boron in illite in sedimentary rocks, which is believed to be an indicator of palaeo-salinity. He finds this to be similar both in the marine Cinder Bed and in two limestones in the Middle and Upper Purbeck, respectively, and hitherto considered to be of fresh-water facies. Thus, he concludes that all three are marine. This new evidence reinforces that of Anderson's Marine Bands, based on ostracods, in the Upper Purbeck, to suggest more marine conditions than was hitherto supposed. However, the other abundant fossils clearly indicating land, fresh-water and brackish environments must not be over-looked in any reappraisal of the complex facies represented.

1812, Middleton; 1816, 1826, Webster; 1851, Forbes; 1852, 1857b, Austen; 1857, Bristow and Fisher; 1884, Damon; 1885, Jones; 1895, Wood-ward; 1886, 1916–19, Smith Woodward; 1898, Strahan; 1906–8, Handlirsch; 1928, 1933, Simpson; 1933, 1934, 1938, 1941, 1947a, Arkell; 1939, Harris; 1941, 1949, Sylvester-Bradley; 1958–60, Delair; 1958, Heap; 1960, West; 1939, 1962, Anderson; 1963, Casey; 1963, Clemens; 1963, Brown; 1964, Walker; 1965, Salter and West; 1966, Cosgrove and Hearn; 1967, Joffe; 1969, Torrens (A).

Eype Coast, East from Thorncombe Beacon, 1½ miles SW of Bridport

SY(30)448910	177	327	38SW, 37SE (G)	S, P PSD, IGC, 1963

One and a half miles of coastal cliffs stretching east from Thorncombe Beacon (508 ft above O.D.) past Eype Mouth to West Bay, including the finest Dorset coast sections of the Upper Lias, unfortunately rather inaccessible (Fig. 18).

Because of faulting, three very different sections are exposed. The first extends from Thorncombe Beacon for 1600 yards to the east, as far as the Eype Mouth Fault at 'Fault Corner', and shows the undisturbed and nearly flat-bedded succession of Cretaceous unconformable upon Upper and Middle Lias. The second, at West Cliff, known to geologists as Watton Cliff, extends for 880 yards farther east up to the West Cliff Fault, and exposes a down-thrown section of the Great Oolite Series, also nearly flat-bedded. The third extends some 320 yards farther east still, to the West Bay Fault, and is much disturbed by the faulting; it exposes Lower fuller's Earth Clay and the under-lying Bridport Sands. The cliffs (and this site) end at West Bay with a sliver of the strikingly layered Bridport Sands on the south side of the West Bay Fault.

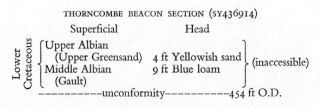

THORNCOMBE BEACON SECTION (SY436914)

	Superficial	Head	
Lower Cretaceous	Upper Albian (Upper Greensand)	4 ft Yellowish sand	(inaccessible)
	Middle Albian (Gault)	9 ft Blue loam	

----------unconformity----------454 ft O.D.

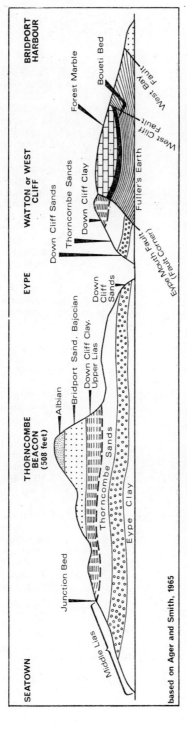

Fig. 18. Seatown–Thorncombe Beacon–Bridport Harbour: coast section; at Thorncombe Beacon a cap of Albian (Lower Cretaceous) unconformably overlies a succession of Bajocian and Upper Middle Lias; at West (or Watton) Cliff is exposed a faulted section of Middle Jurassic, Forest Marble and Fuller's Earth.

THORNCOMBE BEACON SECTION
(*continued*, after Howarth, 1957)

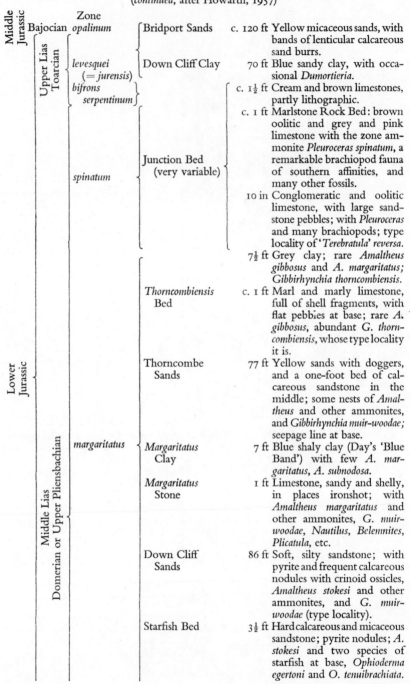

	Zone			
Middle Jurassic — Bajocian	*opalinum*	Bridport Sands	c. 120 ft	Yellow micaceous sands, with bands of lenticular calcareous sand burrs.
Upper Lias — Toarcian	*levesquei* (= *jurensis*)	Down Cliff Clay	70 ft	Blue sandy clay, with occasional *Dumortieria*.
	bifrons *serpentinum*		c. 1½ ft	Cream and brown limestones, partly lithographic.
	spinatum	Junction Bed (very variable)	c. 1 ft	Marlstone Rock Bed: brown oolitic and grey and pink limestone with the zone ammonite *Pleuroceras spinatum*, a remarkable brachiopod fauna of southern affinities, and many other fossils.
			10 in	Conglomeratic and oolitic limestone, with large sandstone pebbles; with *Pleuroceras* and many brachiopods; type locality of '*Terebratula*' *reversa*.
Lower Jurassic — Middle Lias — Domerian or Upper Pliensbachian	*margaritatus*	*Thorncombiensis* Bed	7½ ft	Grey clay; rare *Amaltheus gibbosus* and *A. margaritatus*; *Gibbirhynchia thorncombensis*.
			c. 1 ft	Marl and marly limestone, full of shell fragments, with flat pebbles at base; rare *A. gibbosus*, abundant *G. thorncombiensis*, whose type locality it is.
		Thorncombe Sands	77 ft	Yellow sands with doggers, and a one-foot bed of calcareous sandstone in the middle; some nests of *Amaltheus* and other ammonites, and *Gibbirhynchia muir-woodae*; seepage line at base.
		Margaritatus Clay	7 ft	Blue shaly clay (Day's 'Blue Band') with few *A. margaritatus*, *A. subnodosa*.
		Margaritatus Stone	1 ft	Limestone, sandy and shelly, in places ironshot; with *Amaltheus margaritatus* and other ammonites, *G. muir-woodae*, *Nautilus*, *Belemnites*, *Plicatula*, etc.
		Down Cliff Sands	86 ft	Soft, silty sandstone; with pyrite and frequent calcareous nodules with crinoid ossicles, *Amaltheus stokesi* and other ammonites, and *G. muir-woodae* (type locality).
		Starfish Bed	3½ ft	Hard calcareous and micaceous sandstone; pyrite nodules; *A. stokesi* and two species of starfish at base, *Ophioderma egertoni* and *O. tenuibrachiata*.

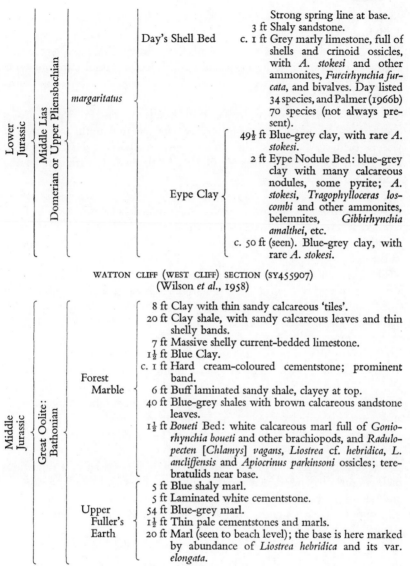

Lower Jurassic	Middle Lias	Domerian or Upper Pliensbachian	*margaritatus*	Day's Shell Bed

Day's Shell Bed — Strong spring line at base.
3 ft Shaly sandstone.
c. 1 ft Grey marly limestone, full of shells and crinoid ossicles, with *A. stokesi* and other ammonites, *Furcirhynchia furcata*, and bivalves. Day listed 34 species, and Palmer (1966b) 70 species (not always present).

Eype Clay —
49½ ft Blue-grey clay, with rare *A. stokesi*.
2 ft Eype Nodule Bed: blue-grey clay with many calcareous nodules, some pyrite; *A. stokesi, Tragophylloceras loscombi* and other ammonites, belemnites, *Gibbirhynchia amalthei*, etc.
c. 50 ft (seen). Blue-grey clay, with rare *A. stokesi*.

WATTON CLIFF (WEST CLIFF) SECTION (SY455907)
(Wilson *et al.*, 1958)

Middle Jurassic — Great Oolite: Bathonian

Forest Marble —
8 ft Clay with thin sandy calcareous 'tiles'.
20 ft Clay shale, with sandy calcareous leaves and thin shelly bands.
7 ft Massive shelly current-bedded limestone.
1½ ft Blue Clay.
c. 1 ft Hard cream-coloured cementstone; prominent band.
6 ft Buff laminated sandy shale, clayey at top.
40 ft Blue-grey shales with brown calcareous sandstone leaves.
1½ ft *Boueti* Bed: white calcareous marl full of *Goniorhynchia boueti* and other brachiopods, and *Radulopecten* [*Chlamys*] *vagans, Liostrea* cf. *hebridica, L. ancliffensis* and *Apiocrinus parkinsoni* ossicles; terebratulids near base.

Upper Fuller's Earth —
5 ft Blue shaly marl.
5 ft Laminated white cementstone.
54 ft Blue-grey marl.
1½ ft Thin pale cementstones and marls.
20 ft Marl (seen to beach level); the base is here marked by abundance of *Liostrea hebridica* and its var. *elongata*.

THE FAULTING

'Fault Corner' (at SY449909) lies at the western visible end of the Eype Mouth Fault, where this intersects the coastal cliff, about 160 yards east of Eype Mouth stream. (Eype is pronounced 'Eep'.)

The Eype Mouth Fault, probably the largest pre-Albian dislocation in the district, trends due east, and downthrows 600–700 feet to the south, exposing the best section of the Great Oolite Series on the Dorset Coast, as far east as the West Cliff Fault. Near the east end of West Cliff, at about SY457906, the West Cliff Fault trends north-east, and downthrows about 120 feet to the north-west. It appears to be a tear fault, which has produced a 30-foot-wide shatter belt, and exposes steeply dipping Lower Fuller's Earth on the south-east.

Near the same spot the West Bay Fault trends due east, and downthrows about 150 feet north, exposing in the cliff to the south a section of yellow Bridport Sands on the western side of Bridport Harbour (West Bay); these contrast with the grey Fuller's Earth on the north side of the fault.

The Thorncombe Beacon sequence is continued eastwards as far as Fault Corner, where, at the western end of the 200-feet-high Watton Cliff, west of the fault, the beds from the Down Cliff Clay to the Eype Clay are exposed in the cliff.

The Junction Bed, between Upper and Middle Lias, seen in these cliffs is a deposit of exceptional interest, investigated in much detail by Buckman (1922) and Jackson (1922, 1926). Mainly inaccessible *in situ*, it must be studied in fallen blocks. It varies markedly in development, character and fauna within very short distances west and east of Eypemouth. A much condensed stratum, only about one to three feet thick, it represents some 200–300 feet of strata developed elsewhere in Britain, including the *jurensis* (part), *bifrons* and *falciferum* Zones of the Upper Lias, and the *spinatum* Zone of the Middle Lias. It consists mainly of various limestones, rubbly, lithographic, conglomeratic and oolitic, coloured white, greenish, pink or grey, with a little clay; at the base there is usually found the brown conglomeratic Marlstone of the *spinatum* Zone.

The rock contains many fossils; Jackson gives a list of partial identifications including three forms of belemnites, 40 of ammonites, and 80 of brachiopods, and he also records four genera of bivalves, seven of gastropods (small gastropods sometimes occur in crowds) *Serpula*, *Isocrinus*, ? *Cidaris* spines, ? *Montlivaltia* and lignite.

At the western end of Watton Cliff the Junction Bed is abnormal, and was named by Buckman 'The Watton Bed'. Some five to eight feet thick, fragments of the *thorncombiensis* Bed are incorporated; part of the fauna is distinctive; and the lowest part, the *spinatum* Zone Marlstone, is missing.

Watton Cliff is unscalable from above and below, but near the western end there is a deep recess known in geological literature as 'Fault Corner', where a path leads to the top. The Fuller's Earth is hidden by talus, but the path gives access to the *Boueti* Bed and the Forest Marble above. These are bent up steeply against the Middle Lias in the corner. There was formerly a tiny patch of Cornbrash on the summit of Watton Cliff, since removed by coast erosion, good evidence that the whole thickness of the Forest Marble is present. It may be noted that Fuller's Earth is a purely stratigraphical term, for none of it has the properties of commercial fuller's earth.

THE *WATTONENSIS* BEDS, LOWER FULLER'S EARTH

A series of about 20 beds of black limestone separated by clays, in all estimated at about 25 feet thick, is found in the upper part of the Lower Fuller's Earth. These beds are nowhere well exposed at this site, but a broken and disturbed section may generally be seen beneath the beach shingle at Fault Corner, where the beds dip about 75° south-south-east. They contain a rich brachiopod fauna with *Wattonithyris*, *Rugitela*, etc., and foraminifera (see below). A second and similar exposure beneath the beach shingle is occasionally to be seen farther to the east.

East of the West Cliff Fault are bands of argillaceous limestone twisted and broken and in places vertical. They are shown by their fossils to be *wattonensis* beds. They may occur in large blocks in the 30-feet-wide fault conglomerate associated with that fault.

The Forest Marble limestones are largely made up of comminuted shells, especially of *Liostrea hebridica*. The extremely shallow water nature of the deposit is shown by ripple marks, pittings, and worm tracks, etc. The clay divisions of the Forest Marble are dark grey-green and weather brown, in contrast to the grey colour of the Fuller's Earth.

The Inferior Oolite normally found between the Fuller's Earth and the Bridport Sands has been cut out of the Watton Cliff sections by the faulting. It appears, however, in the Black Rocks, uncovered at low tide, at about SY455905.

MICROFAUNA

The following microfossils have been noted at this site.

FORAMINIFERA

Cifelli (1959) has recorded foraminifera from Watton Cliff as follows:

From nine samples of Forest Marble at Fault Corner, 36 species.
From one sample of Upper Fuller's Earth, 21 species.
From eight samples of *wattonensis* beds, Lower Fuller's Earth, from the exposure below the beach shingle at Fault Corner, 51 species.

As in many Jurassic deposits, species of the Lagenidae greatly predominate, although here forms identified as *Cornuspira*, *Spirillina* and *Spirophthalmidium* are widespread and often common.

HOLOTHURIA

From the Watton Cliff Section Hampton (1957) figured some holothurian spicules: from the Upper Fuller's Earth assigned to *Achistrum* and *Etheridgella*; and from the Forest Marble to *Rhabdotites*.

POLYZOA

Six specimens from that part of the Junction Bed now known to belong to the Middle Lias *spinatum* Zone, at Thorncombe Beacon, were identified by A. G. Davies (in Jackson, 1926, 510) as *Berenicea compressa*, *B.* sp. indet., *Macroecia lamellosa* and *Stomatopora dichotoma*.

1922, Buckman; 1922, 1926, Jackson; 1957, Hampton; 1957, Howarth; 1958, Wilson *et al.*; 1959, Cifelli; 1965, Ager and Smith; 1966a, 1966b, Palmer; 1967, Martin; 1969, Torrens (A).

Frogden Quarry, Oborne, 1 mile NE of Sherborne

ST(31)649183	178	OS18	6SW	P, S.
			(–)	PSD. 1963

A small disused quarry, largely overgrown, in a pasture field, Quarry No. 36 of Richardson (1930). Roughly the uppermost 15 feet of the section remained exposed in 1952. The whole succession that was once visible is described on p. 159 (after Richardson, 1930).

The old *niortensis* and *blagdeni* Zones used to be quarried for road metal, and they yielded abundant mollusca, including a large series of ammonites.

Frogden quarry was first described by Hudleston in 1885. He claimed that his 'Cadomensis Bed' was almost the only place in England where the wavy colour bands of *Pseudomelania* are preserved to any great extent, as they are in the Normandy beds. There was also a most curious assemblage of ammonites, many of them (at that time) 'neither named, described nor figured'. Hudleston also claimed that the *humphriesianum* (=old *blagdeni*) Zone in this quarry was the best example in England of the beds of Bayeux, the *'oolithe ferrugineuse'*, d'Orbigny's typical Bajocian.

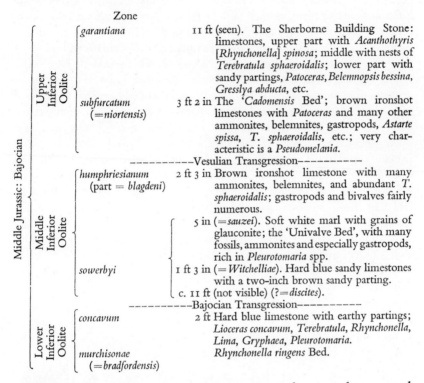

		Zone	
	Upper Inferior Oolite	garantiana	11 ft (seen). The Sherborne Building Stone: limestones, upper part with *Acanthothyris* [*Rhynchonella*] *spinosa*; middle with nests of *Terebratula sphaeroidalis*; lower part with sandy partings, *Patoceras, Belemnopsis bessina, Gresslya abducta*, etc.
		subfurcatum (=niortensis)	3 ft 2 in The 'Cadomensis Bed'; brown ironshot limestones with *Patoceras* and many other ammonites, belemnites, gastropods, *Astarte spissa, T. sphaeroidalis*, etc.; very characteristic is a *Pseudomelania*.
		----------Vesulian Transgression----------	
Middle Jurassic: Bajocian	Middle Inferior Oolite	humphriesianum (part = blagdeni)	2 ft 3 in Brown ironshot limestone with many ammonites, belemnites, and abundant *T. sphaeroidalis*; gastropods and bivalves fairly numerous.
		sowerbyi	5 in (=*sauzei*). Soft white marl with grains of glauconite; the 'Univalve Bed', with many fossils, ammonites and especially gastropods, rich in *Pleurotomaria* spp.
			1 ft 3 in (=*Witchelliae*). Hard blue sandy limestones with a two-inch brown sandy parting.
			c. 11 ft (not visible) (?=*discites*).
		----------Bajocian Transgression----------	
	Lower Inferior Oolite	concavum	2 ft Hard blue limestone with earthy partings; *Lioceras concavum, Terebratula, Rhynchonella, Lima, Gryphaea, Pleurotomaria*.
		murchisonae (=bradfordensis)	*Rhynchonella ringens* Bed.

This is stated to be the only remaining section in the area where a good sequence can still be made out.

CONSPECTUS OF THE DEVELOPMENT OF THE INFERIOR OOLITE IN DORSET

In Dorset and Somerset the Lower and Middle divisions of the Inferior Oolite are much condensed. Four famous Inferior Oolite sites in Dorset are described in this Handbook, and in three of them, Frogden, Halfway House, and Louse Hill Quarries, the whole thickness of these two divisions amounts to only some 17, five and eight feet, respectively, as compared with some 200 feet at Leckhampton in Gloucestershire (q.v.). At the fourth site, Peashill Quarry, 15 miles to the south, all but the lowest Zone of the Lower division, the whole of the Middle division, and the *garantiana* and *subfurcatum* Zones at the base of the Upper division, are missing.

In the Sherborne district the Lower and Middle divisions comprise the fossil beds which have yielded perhaps the richest store of well-preserved mollusca of all kinds obtained from any part of the Jurassic system in England. They indicate a shallow sea. Buckman's paper of 1893 made this classic ground. In the many quarries then open the detailed zonal succession of the Inferior Oolite was first worked out, the crowded ammonites being carefully collected inch by inch from their proper beds. There is an unparalleled richness of the ammonite fauna (largely figured by Buckman) and many of the gastropods in a superb state of preservation were figured by Hudleston.

The Middle Inferior Oolite closed with the most important episode of orogenic (i.e. mountain-building) movements that had occurred in England since the beginning of Jurassic time. After long periods of comparative quiescence the Armorican (east–west) anticlines revived their activity. Epeirogenic (i.e. continent-wide) subsidence followed, lowering both anticlines and synclines beneath the sea, and normal sedimentation was resumed. The crests of the anticlines had been planed off, and the new Upper Inferior Oolite strata were deposited evenly across the basset edges. It can be said that the foundations of modern stratigraphy were laid down by the studies of the intricacies of the Inferior Oolite.

The most interesting deposit of all is that of the *subfurcatum* (=*niortensis*) Zone. This is only known at a few localities in the district, including Frogden and Louse Hill, and hardly anywhere else in the British Isles. It seems best to consider it as having been laid down in a greatly restricted sea during a period of regression while the Bajocian uplifts and denudation were actively in progress, and as marking the resumption of sedimentation prior to the great transgression of the *garantiana* Zone.

1887, Hudleston; 1893, Buckman; 1930, Richardson; 1958, Wilson *et al.*; 1969, Torrens (A).

Halfway House Quarries, Nether Compton, 2½ miles W of Sherborne

---

ST(31)602164	178	312	5SE	P, S
			(G)	PSD, 1965

Three neighbouring disused quarries astride the main road A30: Rock Cottage Quarry, Section IV of Buckman (1893) and No. 17 of Richardson (1930), the western quarry, south of the road; Chapel Quarry, Section V of Buckman (1893) and No. 17a of Richardson (1930), north of the road; and Limekiln Quarry, No. 18 of Richardson (1930), the eastern quarry, south of the road. In 1952 all three quarries were disused and overgrown, the Chapel and Limekiln Quarries with bushes, and these two were also used as local rubbish tips in a small way, but some six feet or more of strata at the top of

HALFWAY HOUSE QUARRIES SECTION
(after Richardson, 1930)

				Zone	

Middle Jurassic — Bajocian — Bathonian,

Great Oolite — *fusca* — (early). Traces of Fuller's Earth Clay (in No. 18).

zigzag — 26 ft (seen in No. 17a). The Limestone Beds: rubbly whitish limestone, impure, mingled with irregular brown shaly deposits; the soft material contained abundant crinoid ossicles, scarce foraminifera, rare ostracods, and two minute fish-teeth.

Upper Inferior Oolite

parkinsoni — ? (=*schloenbachi*).
1 ft (=*truellei*). 'The Fossil Bed': shelly limestone with large *Parkinsonia parkinsoni, Strigoceras truellei, Nautilus, Belemnopsis, Natica, Collyrites, Pygorhytis, Pleurotomaria bessina*, etc.

garantiana — 4–8 in '*Astarte obliqua* Bed': rubbly soft, red-brown sandy limestone with limonite veins, crowded with *A. modiolaris* [*obliqua*], mostly encrusted with Serpulae; also *Patoceras, Acanthothyris spinosa*, gastropods, bivalves and a few foraminifera.

subfurcatum — (=*niortensis*) missing.

---------------Vesulian Transgression---------------

Middle Inferior Oolite

humphriesianum — 0–2 in (=*blagdeni*). 'The Irony Bed': local conglomeratic, irony rock.

sowerbyi — 0–3 in (=*sauzei*). Bluish limestone, with *Pleurotomaria amyntos*.
(=*Witchelliae*) missing.
(=*Shirbuirnia*) missing.

-------------------non-sequence-------------------

concavum — 3 ft 8 in (=*discites*). Greenish-yellow ironshot limestone, with *Nautilus, Belemnopsis, Terebratula* 3 spp. and bivalves.

Lower Inferior Oolite

murchisonae — 4 in (=*bradfordensis*). *Rhynchonella ringens* Bed: greenish-brown ironshot limestone; *B. ringens* common, and a good marker.
10 in Limestone.

----------non-sequence; early *bradfordensis* missing----------

2 in Yellowish ironshot limestone, rather soft, with *Cerithium, Cucullaea, Velata*, etc.

scissum — missing.
opalinum — missing.

-----------non-sequence-----------

Lower Jurassic — Upper Lias — Toarcian

(aalensis subzone missing).

levesquei — 2 ft (seen) (*moorei* subzone). 'Dew Bed': hard grey shelly, sandy limestone passing down into yellow sands (Yeovil Sands); with *Entolium, Trigonia navis,* ? *Ostrea*. Sometimes has a very level surface with adherent oysters; with large masses of pyrite, fossil wood and bone, and occasionally large *Ceratomya*, etc.

each was exposed. The bedding is nearly flat. In 1963 road widening led to the filling in of certain of the pits, but cut an improved roadside section through the beds.

This classic and abundantly fossiliferous locality, first described in 1856, was already referred to as 'celebrated' in 1865. The complete composite section is described on the previous page. The new road-section is described in Torrens (1969, p. A.28) and, with emphasis on the fauna, by Whicher (1969: *Proc. Geol. Ass.* **80**, 324).

All three sections are similar, but the Top Limestones are thickest (26 feet) in No. 18; only half as thick in No. 17a; and absent from No. 17.

1856, Wright; 1893, Buckman; 1930, Richardson; 1958, Wilson *et al.*

Louse Hill Quarry, Nether Compton, 1¾ miles W of Sherborne

ST(31)610161	178	312	5SE	P, S
			(G)	PSD, 1963

A shallow disused quarry, No. 21 of Richardson (1930), in nearly flat-bedded strata, and about 100 yards long. It is partly overgrown with brambles. The described section on the following page is mainly after Richardson.

The name Louse Hill appears to be tautological. Buckman (1910, footnote, p. 99) states that 'Louse' is due to folk etymology and the west country habit of adding the sign of the genitive to place names. It is not connected with a louse but is from 'low', Anglo-Saxon (*hlaw*, a hill), as in the place name Ludlow.

Of the section it is important to note that there are no indications of a 'Fossil Bed' of *parkinsoni* (=*truellei*) date immediately above the *Astarte obliqua* Bed.

Of the 'Irony Bed' in this quarry Buckman wrote that it was one of the most remarkable repositories of brachiopod species in this country, having many distinctive and peculiar forms. Some of them are found in the Lower Dogger of Rothenstein in the Vilser Alps of the Northern Tyrol, as *Glossothyris bifida, Waldheimia angustiplecta, W. waltoni;* other species from this Louse Hill bed are *W. brodiei* and *Terebratula lowensis.*

Hudleston had many species of *Pleurotomaria* from this quarry, but all were poorly preserved.

1893, 1910, Buckman; 1930, Richardson.

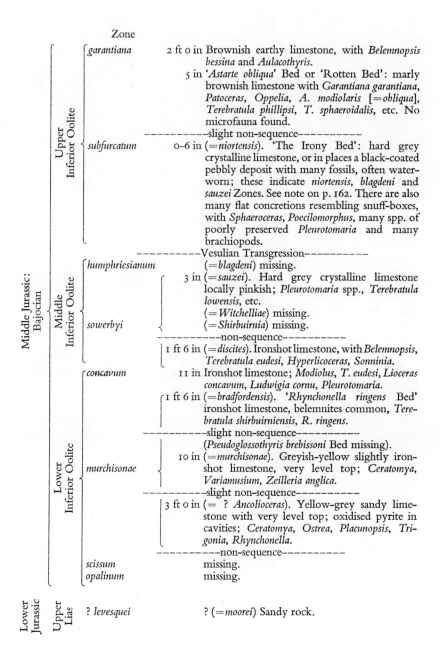

			Zone	

Middle Jurassic: Bajocian

Upper Inferior Oolite

garantiana — 2 ft 0 in Brownish earthy limestone, with *Belemnopsis bessina* and *Aulacothyris*.

5 in '*Astarte obliqua*' Bed or 'Rotten Bed': marly brownish limestone with *Garantiana garantiana, Patoceras, Oppelia, A. modiolaris* [=*obliqua*], *Terebratula phillipsi, T. sphaeroidalis*, etc. No microfauna found.

----------slight non-sequence----------

subfurcatum — 0–6 in (=*niortensis*). 'The Irony Bed': hard grey crystalline limestone, or in places a black-coated pebbly deposit with many fossils, often water-worn; these indicate *niortensis, blagdeni* and *sauzei* Zones. See note on p. 162. There are also many flat concretions resembling snuff-boxes, with *Sphaeroceras, Poecilomorphus*, many spp. of poorly preserved *Pleurotomaria* and many brachiopods.

----------Vesulian Transgression----------

Middle Inferior Oolite

humphriesianum — (=*blagdeni*) missing.

sowerbyi —
3 in (=*sauzei*). Hard grey crystalline limestone locally pinkish; *Pleurotomaria* spp., *Terebratula lowensis*, etc.
(=*Witchelliae*) missing.
(=*Shirbuirnia*) missing.

----------non-sequence----------

1 ft 6 in (=*discites*). Ironshot limestone, with *Belemnopsis, Terebratula eudesi, Hyperlicoceras, Sonninia*.

Lower Inferior Oolite

concavum — 11 in Ironshot limestone; *Modiolus, T. eudesi, Lioceras concavum, Ludwigia cornu, Pleurotomaria*.

1 ft 6 in (=*bradfordensis*). '*Rhynchonella ringens* Bed' ironshot limestone, belemnites common, *Terebratula shirbuirniensis, R. ringens*.

----------slight non-sequence----------

(*Pseudoglossothyris brebissoni* Bed missing).

murchisonae —
10 in (=*murchisonae*). Greyish-yellow slightly iron-shot limestone, very level top; *Ceratomya, Variamusium, Zeilleria anglica*.

----------slight non-sequence----------

3 ft 0 in (= ? *Ancolioceras*). Yellow-grey sandy limestone with very level top; oxidised pyrite in cavities; *Ceratomya, Ostrea, Placunopsis, Trigonia, Rhynchonella*.

----------non-sequence----------

scissum — missing.

opalinum — missing.

Lower Jurassic — Upper Lias

? *levesquei* — ? (=*moorei*) Sandy rock.

163

Owermoigne Heath Pit, 5 miles ESE of Dorchester

SY(30)779880	178	328	48NE (G)	A, S 1963	

A large gravel pit with vertical face, intermittently worked. The section shows:

> Pleistocene: c. 20 ft (seen). Flint gravels of the 100-foot Terrace, with some strings of silt or clay.

The surface of the gravel here lies at about 187 feet above O.D.

In May 1959 the base was flooded with a couple of feet of rain water, indicating the presence of some clayey streaks, since the gravels are presumed to rest upon Eocene Bagshot Sands.

The main interest in the gravels, part of the Moreton or Lower Terrace, lies in the possibility that they may yield Palaeolithic flint implements. I have found no literature specifically referring to this particular pit.

This site and the following one (Parsons' Pit) were chosen as alternatives to the earlier proposed Warmwell and Moreton Railway Station pits, which are now unsuitable for preservation because of their intensive mechanical exploitation.

The drainage of this area, which lies in the syncline of the Dorset–Hampshire Basin, was effected from an early date in the Pleistocene by a longitudinal consequent river, the ancestor of the present River Frome, known as the River Solent. That river flowed through the present-day Solent and Spithead to enter the open sea somewhere east of the present Isle of Wight. An important river, it must have been far greater and swifter than any in this area today, as is shown by the vast spreads of gravel left as its terraces.

Of greatest importance is the 100-foot Terrace (the name referring to its level above the present-day main river), of which a part, the Moreton Terrace, lies to the south of the River Frome. The gravel consists mainly of coarse sub-angular flints, with an admixture of Greensand cherts, quartz, and 'Cornish' rocks derived from the Tertiary beds.

In Dorchester Museum are more than 50 more or less rolled 'Chellian' and early Acheulian flint hand-axes and a few coarse Lower Palaeolithic flake implements from the 100-foot Terrace near Moreton Station. The implements are very rare, and were collected over many years from workmen when the gravels were excavated by hand. Specimens are figured in Arkell (1947a, Figs. 80, 81).

The Bagshot Sands contain an unlimited supply of good builder's sand, and the Plateau Gravels are a principal source of gravel. Where gravel overlies the sand over large areas, as on the heaths between Moreton Station and Warmwell, and can be worked in the same pits, an important industry has grown up. For about a century the centre of activity was just south of Moreton Station, but since 1936 it has spread one and a quarter miles to the south-west.

1947a, Arkell.

Parsons' Pit, Red Bridge, 5 miles ESE of Dorchester

SY(30)789882	178	328	48NE (G)	A, S 1963

A small disused gravel and sand pit, the deepest of the local pits, chosen as one in which the gravel face is most likely to remain unobscured by falling scree. It is to show a second example of the Plateau Gravel spread of the 100-foot (or Lower Palaeolithic or Moreton) Terrace, whose surface here lies at about 175 feet above O.D., with a good exposure of the underlying Bagshot Sands.

The section shows:

HEAD (ON WEST FACE OF PIT)
Pleistocene up to 5 ft Flint gravels of the 100-foot Terrace.
Eocene   c. 20 ft (seen). Bagshot Sands: current-bedded deltaic sands, devoid of fossils.

I have found no literature specifically referring to this particular pit.

For fuller description of the scientific interest see under Owermoigne Heath Pit above.

1947a, Arkell.

Peashill Quarry, Shipton Gorge, 2 miles SE of Bridport

SY(30)495916	177	327	38SE (G)	P, S 1963

A small disused quarry in a field that, according to a local informant, is called 'Great Pistol', of which the name 'Peashill' is said to be a corruption. It lies across the road from the New Inn.

The chief interest is the rich microfauna of its Microzoa Beds, discovered here by Walford in 1885. The following succession has been described in this quarry (Richardson, 1928), almost wholly at the top of the Upper Inferior Oolite, and known as the Top Limestones. In 1963 the lower part was not visible, but the important Sponge Beds were well exposed.

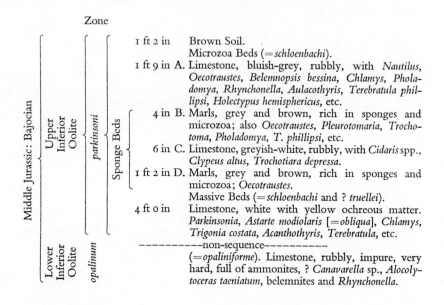

From the two marl beds together, B and D above, the following smaller fossils have been recorded from this quarry.

Fish. Two minute teeth.

Brachiopods. Five of Moore's species of small and uncommon genera: Crustacea. Macrura: two minute crab-claws. Ostracods: species of *laboucheri*.

Crustacea. Macrura: two minute crab-claws. Ostracods: species of *Bairdia* (very abundant), *Cytheridea*, *Cytheropteron* and *Polycope*; probably at least nine species in all.

Polyzoa. Cyclostomata: *Stomatopora* (6 forms), *Proboscina* (4), *Tubulipora* (1), *Idmonea* (4), *Bisidmonea* (1), *Entalophora* (5), and *Pergensia* (9).

Of the above 30 forms, 24 were described as new, including Walford's new genus *Pergensia*, which he erroneously placed in the Cheilostomata.

Sponges. Hexactinellida: *Laocaetis* [*Craticularia*] *foliata* and *Verrucocoelia elegans*. Desmospongia: *Platychonia vagans* and *Reniera oolitica*. Calcarea: *Peronidella* (3 spp.), *Enaulofungia* [*Holcospongia*] (3), *Limnorea* (*Lymnorella*) *pygmaea*, *Oculospongia minuta*, *Eudea pisum*, *Elasmostoma palmatum*, *Leucospongia tinea*. Also recorded is *Leucospongia shiptonensis*, a *nomen nudum*.

Foraminifera. 38 species are recorded by Charles Upton, in Richardson (1928), of which 28 species are of the family Lagenidae. The form identified as *Cristellaria rotulata* was very large and abundant.

The fauna listed from the Microzoa Beds of this quarry, including 19 larger species of ammonites, brachiopods, molluscs and echinoids, thus totals at least 120 species and varieties.

To the student of the Inferior Oolite a most important feature of this section is that the 'Top Limestones' rest directly upon the *opalinum* Zone represented by the calcareous top of the Bridport Sands, which pass down into the Upper Lias. The intervening zones *garantiana* to *scissum*, inclusive, that is, the lower part of the Upper, the whole of the Middle, and all but the lowest zone of the Lower Inferior Oolite, are missing.

Hinde states that the Shipton Gorge sponges are mostly small, and nearly all of the class Calcarea. Of the siliceous sponges there are only a few broken-up fragments, the silica being now entirely replaced by calcite. As the calcareous and siliceous sponges generally have different habitats, it is of interest to find them occurring here in the same beds. Some of the small calcareous forms were found still attached to the surface of the Hexactinellids upon which they grew. Of the 18 species accepted by Richardson and Thacker, nine were described from this area as new, and for at least five of these this quarry is the type site. The evidence of the sponges suggests that these Microzoa Beds were laid down in a shallow sea, probably near the coast.

Walford (1889) states: 'From this one locality I am able to recognise about fifty different forms [of Polyzoa] represented by twelve or more genera.' In his two papers, however, he records only 30 forms divided amongst seven genera, so perhaps more remain to be described. (I have not seen a recent paper: Walter, B., 1967, Revision de la faune des Bryozoaires du Bajocien inferieur de Shipton Gorge. *Trav. Lab. Geol. Fac. Sci. Lyon*, N.S. **14**, 43–52.)

In 1932 the present writer sought the foraminifera and other microfauna from this and other localities collected by Linsdall Richardson and identified by Charles Upton, who died in 1927. No trace of them was to be found in the British Museum (Natural History) or in the Gloucester or Cheltenham Museums, which all preserve other portions of the Upton Collections. Correspondence with his family and others who might have had information yielded no useful results. Mr Richardson finally concluded that Upton may never have mounted any slides of these microfaunas, and the washed material may have been discarded as valueless after his death.

I am indebted to the late Dr H. Dighton Thomas and to Dr H. W. Ball, both of the British Museum (Natural History), for help with the Polyzoa, and for a revised list of the Sponges, respectively.

1889, 1894, Walford; 1893, Hinde (1887–1912); 1920, Richardson and Thacker; 1928, Richardson; 1958, Wilson *et al.*; 1969, Torrens (A).

Portland Bill, 6 miles S of Weymouth

SY(30)678684	178	342	60NW, NE (G)	R, S, P PSD, 1963

A 65-acre coastal strip contains the whole of the Portland Bill Raised Beach, of special interest because of its abundant marine fauna as well as shingle; the fossiliferous loam; and finally a classic example of 'Head' overlying it. It is mapped as stretching from the southern end of the Bill, where it is 350 yards wide, and narrowing northwards along the east coast to Sand Holes, one mile distant. The general section shows (after Prestwich, 1875, and others):

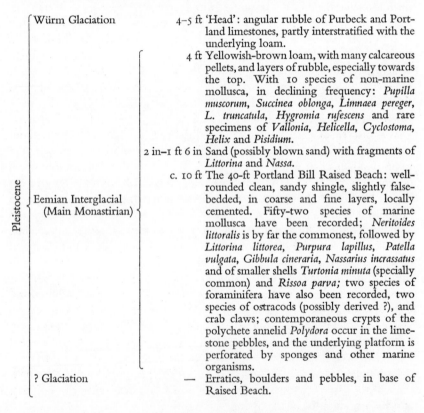

Würm Glaciation — 4–5 ft 'Head': angular rubble of Purbeck and Portland limestones, partly interstratified with the underlying loam.

4 ft Yellowish-brown loam, with many calcareous pellets, and layers of rubble, especially towards the top. With 10 species of non-marine mollusca, in declining frequency: *Pupilla muscorum, Succinea oblonga, Limnaea pereger, L. truncatula, Hygromia rufescens* and rare specimens of *Vallonia, Helicella, Cyclostoma, Helix* and *Pisidium*.

2 in–1 ft 6 in Sand (possibly blown sand) with fragments of *Littorina* and *Nassa*.

c. 10 ft The 40-ft Portland Bill Raised Beach: well-rounded clean, sandy shingle, slightly false-bedded, in coarse and fine layers, locally cemented. Fifty-two species of marine mollusca have been recorded; *Neritoides littoralis* is by far the commonest, followed by *Littorina littorea, Purpura lapillus, Patella vulgata, Gibbula cineraria, Nassarius incrassatus* and of smaller shells *Turtonia minuta* (specially common) and *Rissoa parva;* two species of foraminifera have also been recorded, two species of ostracods (possibly derived ?), and crab claws; contemporaneous crypts of the polychete annelid *Polydora* occur in the limestone pebbles, and the underlying platform is perforated by sponges and other marine organisms.

Eemian Interglacial (Main Monastirian)

Pleistocene

? Glaciation — Erratics, boulders and pebbles, in base of Raised Beach.

The Raised Beach rests upon a wave-cut platform of Jurassic Portland Limestone, dipping gently to the south-east, and is banked against an old cliff of Lower Purbeck tufaceous limestones. The beach must have been formed immediately before a fall in sea level, because it is covered by stratified loam containing fragile land and fresh-water shells, which passes by gradual interstratification into the Head. It has obviously not been subsequently submerged.

The gravel was first recognised by Bristow, who recorded it on his primary Geological Survey one-inch sheet of 1850. After brief notes by Weston (1852) and Damon (1860), the section was described in detail by Whitaker in 1869, Pengelly (1871) and Prestwich (1875).

HEAD

This is one of the classic examples of Head, exhaustively studied by Prestwich. The bulk of the angular rock fragments consists of local Portland and Lower Purbeck Limestones and cherts, but there are also fragments of Cretaceous, Middle Purbeck (such as the Cinder Bed) and other rocks no longer occurring *in situ* on the Island of Portland, and also a few rounded Bunter Pebbles.

Remains of Mammoth, ?Reindeer, Woolly Rhinoceros and other mammals reported in the Head of other parts of the Island in general suggest a sub-arctic land fauna in keeping with our ideas of how Head or Coombe Rock was formed.

LOAM

The loam was probably deposited in a mere enclosed behind the storm beach after a slight retreat of the sea. The abundance of *Succinea oblonga* points to a Pleistocene date, but the assemblage of shells is that of a temperate climate. It is ecologically impossible for all the recorded land and fresh-water shells to have lived together, and the fauna is probably a mixture of marsh and dry terrestrial forms. Eventually marsh and snails were overwhelmed by cold and buried under the Head. At the extreme southern end of the Bill neither Head nor Loam is present.

THE 40-FOOT RAISED BEACH

The platform on which the Raised Beach rests slopes from about 50 feet to 20 feet above present mean tide level. These differences in level are so great as to suggest the presence of two separate beaches, but this is not so. The gradient is such as exists today on any Recent shore, and it is important to note that a '50-foot Raised Beach' can be traced down to 20 feet above O.D. The high-water mark in Raised Beach times was not lower than the present

169

level of 40 feet above O.D. At the northern extremity, at Sand Holes, the upper limit of the deposit reaches 65 feet above sea level.

The beach gravel is well rounded, and 99 per cent is of flint, local limestones and chert. However, at the south-west end the proportion is some 90 per cent of flint, whereas to the north it is only 40 per cent, the rest being mainly local Portland and Purbeck Limestones. The remaining one per cent consists of foreign rocks: porphyries of Dawlish type, quartzites and sandstones of Budleigh Salterton Pebble Bed type; tourmalinised rocks of Cornish type, very rare pink granite, and some rocks perhaps from the Tertiary.

Most of the fauna consists of species still living in the neighbouring sea, but a few show that the sea was slightly colder, cold temperate, equivalent to that on the present shores of the north of England and Scotland, and about three or four degrees Fahrenheit colder than the present sea at Portland. There are no arctic or sub-arctic forms, and, except for one doubtful form, no extinct species. The facies is that of a rocky coast, the commonest shells being periwinkles.

There are marked differences between the exposures. On the south-west the deposits are coarser and thicker but scarcely contain a shell, whereas on the east they are finer, with an abundant marine molluscan fauna, especially towards the northern end. Some of the shells have the original colour bands preserved.

The Raised Beach has been dated to the Monastirian (Baden-Powell) included within the Eemian Interglacial, and the same may apply to the Loam. The Head must therefore be given a date within the Würm Glaciation (Carreck). Correlation of this Raised Beach has been claimed with that at Hope's Nose, near Torquay, 44 miles to the west; with that at Brighton, 100 miles to the east; and with the Langton Herring Terrace, five miles to the north.

? GLACIATION

Two large sub-angular boulders, one of ? Tertiary sandstone and the other a calcareous grit, and some few pebbles of granite, porphyry, etc., are recorded by Prestwich (1875) from the base of the Raised Beach. Although not fully convincing by themselves, their significance is confirmed by the evidence of erratics in comparable beaches in South Wales, Thorney Island and Selsey, and at Sangatte, near Calais. Thus, they may be accepted at Portland Bill as evidence of sea-ice along the Channel coasts, indicative of glacial conditions before the formation of the cold-temperate Raised Beach.

Cryoturbation structures have recently been reported in the deposits and bedrock of the Raised Beach platform along the east side of Portland Bill by Pugh and Shearman (1967).

1875b, Prestwich; 1930a, 1930b, 1953, Baden-Powell; 1947a, Arkell; 1960, Carreck; 1967, Pugh and Shearman.

Portland Coast with Chesil Beach and the Fleet, West of Weymouth

---

SY(30)6478	178	341,	52NE, 53SW,	S, P, Ph
		342	58SE, 60NE	PSD, IGC
			(G)	1963

An 11-mile stretch of varied coast with three distinctive geological features: first, on the Isle of Portland, exposures of the Lower Purbeck and the Port-landian stage of the Upper Jurassic, the most important being the type section of the Portland Stone in the quarries, and a less convenient section of the Portland Sand* in the cliffs; second, the Chesil Beach south of Abbotsbury; and third, the Upper and Middle Jurassic strata of the Weymouth Anticline exposed along the eastern margin of the Fleet (Fig. 19).

ISLE OF PORTLAND COAST

A little more than one mile of coast and cliffs rising to some 400 feet above sea level on the north-western side of the Isle of Portland, comprising West Weare, and West Cliff to Black Nore, with examples of the stone quarries on top.

The Portland Beds mark the last episode of the marine Jurassic, with shallow-water sands and limestones following the thick Kimeridge Clay. Purbeck Beds follow conformably, but with a sharp change to fresh-water conditions.

The 'Island' is a fragment of the southern limb of the Weymouth Anticline, almost detached from the mainland by erosion of the Kimeridge Clay floor; and the Portland Stone and Lower Purbeck form a tableland sloping gently southward with a dip of about one and a half degrees.

The sequence of Lower Purbeck and the Freestone Series of the Port-landian are well known from the quarry exposures, but the thickness of the different beds varies rapidly and very considerably from place to place. The Cherty Series and Portland Sand, however, are not so well known because of less accessible exposures on the coast, vertical cliffs masked by landslips and by waste tips of the stone quarries.

---

* The type section of the Portland Sand is on the mainland at Emmit Hill, near St Alban's Head; see p. 171.

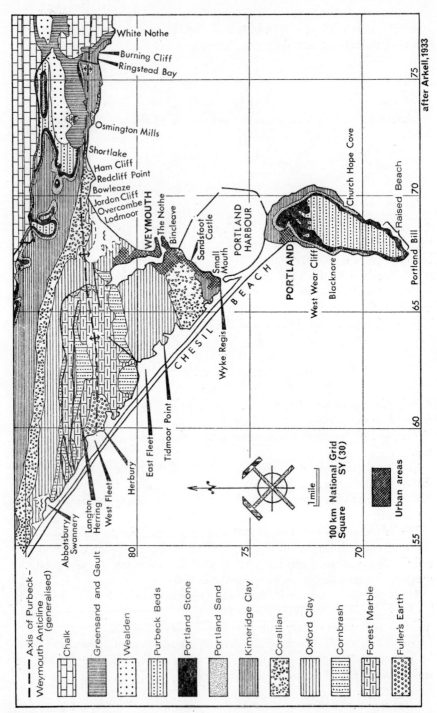

Fig. 19. Portland and Weymouth area, geological sketch-map.

White Nothe
Burning Cliff
Ringstead Bay
Osmington Mills
Shortlake
Ham Cliff
Redcliff Point
Bowleaze
Jordon Cliff
Overcombe
Lodmoor
WEYMOUTH
The Nothe
Bincleave
Sandsfoot Castle
Small Mouth
PORTLAND HARBOUR
Church Hope Cove
PORTLAND
West Wear Cliff
Blacknore
Raised Beach
Portland Bill
Wyke Regis
CHESIL BEACH
Tidmoor Point
East Fleet
Herbury
West Fleet
Langton Herring
Abbotsbury Swannery

after Arkell, 1933

100 km National Grid Square SY (30)

1 mile

Urban areas

— — Axis of Purbeck–Weymouth Anticline (generalised)

Chalk
Greensand and Gault
Wealden
Purbeck Beds
Portland Stone
Portland Sand
Kimeridge Clay
Corallian
Oxford Clay
Cornbrash
Forest Marble
Fuller's Earth

SUCCESSION AT KINGBARROW QUARRIES, SY692729
(Woodward, 1895)

Lower Purbeck

12 ft Three beds of white fissile limestone, called 'slatt', the lowest containing ostracods, interbedded with three beds of marl.

2 ft Bacon Tier: hard streaky limestone with sandy layers.

2 in Greenish clay.

2 ft Aish: fine-grained argillaceous limestone, drying white.

1 ft 4 in Burr: tufaceous limestone with silicified tree-stumps and trunks enveloped in tufa.

up to 1 ft Great Dirt Bed: dark carbonaceous clay with rolled limestone pebbles; well-preserved silicified Cycads and Conifers with trunks up to 23 ft long and 2–4 ft in diameter; some stumps are rooted in position of growth.

c. 9 ft Top Cap: tufaceous limestone, with nearly horizontal holes 2–8 in in diameter, sometimes branching, in places containing fossil wood.

3–6 in Lower Dirt Bed: loamy carbonaceous clay with stones and Cycads.

1 ft 6 in Skull (or School) Cap: hard brown tufaceous limestone overlying a carbonaceous parting, or sometimes a chert band with ostracods.

Upper Jurassic: Portlandian

Portland Stone

Freestone Series

3 ft Roach: creamy oolitic limestone with empty moulds of marine fossils, *Laevitrigonia* [*Trigonia*] *gibbosa* (=Hippocephaloides, or 'Horses Heads'), *Chlamys lamellosa*, *Aptyxiella portlandica* (='the Portland Screw'), *Ostrea expansa*, *Natica*, *Neritoma*, *Lucina*, *Pleurotomaria*.

8 ft Whit Bed: buff oolite, the best Freestone, in places with a layer of chert 3 ft above the base.

2 ft Flinty Bed: limestone full of chert, with *Titanites*.

0–4 ft Curf: soft chalky limestone with much chert.

0–2 ft Base Bed Roach: shelly oolitic limestone with many casts of *Trigonia*, *Pecten*, etc.

6–8 ft Base (or Best) Bed: buff or white oolite; good Freestone with few shells.

(succession continued in the coastal cliffs)

Cherty Series

60–70 ft Cherty Series: hard limestone with bands and nodules of chert throughout; with hollow moulds of *Myophorella* [*Trigonia*] *incurva*, *Protocardia dissimilis*; in places abundant *Serpula gordialis*, with the group of large Behemothian ammonites '*A. boloniensis*', some of which reach 38 in in diameter.

7–8 ft Basal Shell Bed: compacted shells in a matrix of very hard limestone recrystallised as calcite, crowded with *Glomerula* [*Serpula*] *gordialis*, with a total of some 90 species; see note below.

Portland Sand

30–40 ft West Weare Sandstones: brown and grey marly sandstones and sandy cementstones, with *Glaucolithites gorei*.

6–7 ft *Exogyra* Bed: stiff marl packed with *Exogyra nana* (=*E. bruntrutana*) and with *Rhynchonella portlandica*.

35 ft Upper Black Nore Beds: black sands with lines of light grey nodules.

6 ft Black Nore Sandstone: hard black argillaceous sandstone with intensely hard concretions.

40 ft (seen). Lower Black Nore Beds: blue-black sandy clays, with the Russian ammonite *Zaraiskites*.

Upper Kimeridgian   Kimeridge Clay: with *Pavlovia*, *Thracia* and *Protocardia*.

From the Basal Shell Bed, Cox (1925) described a rich and well-preserved fauna of 90 species, comprising Vermes, 2 species; Asteroid ossicles; Echinoids, 6 (3 new); Polyzoa, 2 (*Elaphopora cervina* Lang, new genus and species, and *Berenicea damnatorum* Lang, new species); Bivalves, 54 (18 new); Gastropods, 18 (9 new); Cephalopods, 5 (one new, *Kerberites portlandensis*); Pisces, 1

173

(tooth); Reptilia, 1 (*Cimoliosaurus portlandicus* Owen sp.). The commonest Bivalves are *Myophorella* [*Trigonia*] *incurva*, *T. radiata*, *T. tenuitexta*, *Chlamys lamellosa*, *Pleuromya uniformis*, *Isognomon bouchardi*, *Corbula*, *Barbatia*, *Venericardia*, *Isocyprina* and *Exogyra nana;* and the commonest gastropod *Euspira ceres.*

From the Portland Stone of the Isle of Portland, Delair (1958–60) records seven forms of reptiles, as follows.

Chelonia (Turtles and Tortoises): *Pleurosternum portlandicum* (a water tortoise), *Stegochelys planiceps.*

Plesiosaurs: *Colymbosaurust portlandicus,* ? *C. megapleuron, Cryptocleidus* aff. *richardsoni.*

Pliosaur: *Pliosaurus brachydeirus.*

Dinosaur (Sauropod: giant, herbivorous, amphibious): *Ornithopsis* sp. (represented by a tooth).

Ichthyosaur: ? *Macropterygius thyreospondylus* (represented by the only ichthyosaur vertebra known from the English Portlandian).

Pliosaurus portlandicus is also reported (Savage, 1957: *Proc. Bristol Nat. Soc.,* **29,** 379).

Roughly the top 10 feet of the Portland Sand may represent Damon's Portland Clay; the evidence is conflicting.

Beyond the end of the Chesil Beach at Chesilton, pyritised radial plates of the free-swimming crinoid *Saccocoma* have been collected (Carreck, 1960); they indicate the *Subplanites* Zone of the Kimeridge Clay, but zonal work has yet to be done on the Kimeridge Clay here.

The two beds worked commercially for the famous building stone are the Whit Bed and the Base Bed of the Freestone Series. Both are oolitic limestones of similar quality, with a calcium carbonate content of about 95 per cent, and a crushing strength of over 200 tons per square foot. When a block of stone is cut out in the quarry it is given several quarry marks, one of which shows its cubic content in feet. Some of the marks are very ancient, and the symbols for numerals suggest Roman influence. The total output of commercial stone from Portland Island this century has varied considerably from some 50 000 to more than 115 000 tons yearly.

The earliest known building constructed of Portland Stone is Rufus Castle, Portland, built about A.D. 1080. Inigo Jones selected the stone in 1610 for the Banqueting Hall at Whitehall, London, but its fame and great reputation date from its use on a grand scale by Wren, for the rebuilding of St Paul's Cathedral and other London churches after the Great Fire of 1666. Since that time it has been chosen for large public building all over the country, but especially in London, where it has been used for the British Museum (1753), Somerset House (1776–92), the General Post Office (1829), the Horse Guards, and most of the Government offices in Whitehall. In more recent times it has been used for Imperial Chemical House, Britannic House,

Bush House, Selfridges, London County Hall, the Bank of England and many others. In the provinces conspicuous examples of its use are Cambridge University Senate House (1722–30), Leeds Civic Hall, Southampton Municipal Buildings, and the Manchester Reference Library. The waste stone, roach, and cherty beds, were used for the great Portland breakwaters, the first being completed by convict labour in 1864. The roach has lately been used for the new Cripps building of St John's College, Cambridge (1966).

In early records of the Portland quarries Hooke in 1668 referred to 'Snake or Snail stones of a prodigious bigness', i.e. large ammonites. John Woodward (1728–29), had, among other Portland shells in his collection, 'three vastly large ammonitae dug up in the great quarry, and stone with long slender turbinated shells, and that bivalve that Plot called Hippocephaloides; the shells are perished and the spaces left empty.'

On the Easton Road (B3154), at about SY691725, is the Museum of the Bath and Portland Stone Firms Ltd. Here, set out as a garden rockery, is an excellent collection of the larger fossils from the quarries, especially the giant ammonite *Titanites* from the Whit Bed; and the silicified trunks of coniferous trees and of Cycads (locally known as 'Crows' Nests') from the Lower Purbeck Beds.

1884, Damon; 1895, Woodward; 1897, Seward; 1925, Cox; 1933, 1947a, Arkell; 1958, House; 1960, Carreck; 1969, Torrens (A).

THE FLEET BACKWATER AND ITS EASTERN MARGIN

The Fleet backwater, brackish at the northern end and eight miles long, is thought to represent what remains of the left bank of the hypothetical 'River Fleet', which functioned sometime during the Quaternary period. In its prime it laid down one or more gravel terraces which represent stages in its development, and their fragments remain today. The lowest and latest, the Langton Herring Terrace at 50 feet above O.D., is correlated by Baden-Powell with the Portland Bill Raised Beach, of the Eemian (Last) Interglacial, dated perhaps between 100 000 and 300 000 B.C. Two earlier terraces are more uncertain, the Wyke Wood Terrace at 125 feet, and the oldest, the Fleet Common Terrace, at 220 feet above O.D.

Marine erosion of the coast proceeded until the right bank of the 'River Fleet' was destroyed, including part of the terraces, which are thought to have furnished much of the Chesil Beach gravel. Both the Chesil Beach and the Fleet backwater are believed to have been established during the Neolithic subsidence, about 3000–2000 B.C. The name comes from the Saxon *fleot*, a creek or estuary.

The scenically unattractive low clay coast of the Fleet is irregularly indented where small streams debouch into it. Although some of the sections are poor, they are highly fossiliferous, with beautifully preserved specimens. The series of exposures provides a section across the nearly east–west-trending Weymouth Anticline, and is of great importance for the faunas and correlation of the Upper and Middle Jurassic sequence. The axis of the anticline, which was formed during the Tertiary folding, cuts the Fleet coast near the Coast Guard Station at Langton Herring, at about SY605819, and, in general, the beds dip gently north and south away from it. A number of strike faults are mapped near the axis, and a few dip faults on the southern limb.

Perhaps the first geologist to describe a traverse here was de Luc in 1804. He noticed the *Ludus Helmontii* (septaria) in the Oxford Clay, and, apparently in the Forest Marble, he found quantities of 'gryphites' (probably brachiopods), which, where the valves were separate, 'looked like the shells of pistachio nuts'.

The general sequence of the strata exposed in the eroded Weymouth Anticline along the margin of the Fleet, shows:

```
                 ┌ Kimeridgian          Kimeridge Clay
                 │
                 │                196 ft  Corallian Beds
 Upper      ┌    │ Oxfordian              ┌ Oxford Clay (Upper Clay)
 Jurassic   ┤    │             c. 600 ft  ┤ ─────────────────────────
            │    │                        │ Oxford Clay (Lower Shales)
            └    │ Callovian              └ (Kellaways Beds ? faulted out)
                 │                          ┌ Upper Cornbrash
                 └             c. 30 ft    ┤ ──────────────
 Middle     ┌                              └ Lower Cornbrash
 Jurassic   ┤      Bathonian    c. 100 ft   Forest Marble
            └                               Upper Fuller's Earth
```

The best exposures, described in sequence from north to south, are the following (all from Arkell, 1947a).

SHIPMOOR POINT   SY577836 ⎫
AND CHESTER'S HILL   SY580883 ⎭ CORNBRASH

This splendid natural section gives the complete Cornbrash sequence, the thickest development of the beds in England.

Upper Cornbrash	14 ft 6 in	Alternating hard doggery limestones and cream-coloured marls; at Shipmoor Point the top 6 ft is unfossiliferous, but at Chester's Hill it yielded *Microthyridina* [*Ornithella*] *lagenalis;* the beds below yield *Digonella* [*O.*] *siddingtonensis, Rhynchonella cerealis, Liostrea hebridica,* etc.
Lower Cornbrash	7 ft:	4 ft. Rubbly limestone with many fossils, *Obovothyris* [*Ornithella*] *obovata, Kallirhynchia* [*Rhynchonella*] *yaxleyensis, Meleagrinella echinata, Chlamys vagans, L. hebridica,* etc.
		3 ft. Hard pinkish limestone with *Cererithyris* [*Terebratula*] *intermedia, M. echinata, Nautilus truncatus, Pygurus michelini,* etc.
Forest Marble		Clay.

At Chester's Hill the Lower Cornbrash, repeated by a fault, has thickened by at least 11 feet, and has yielded some additional species.

COASTGUARD STATION, LANGTON HERRING SY607816   FULLER'S EARTH

On the coast west of the Station is exposed:

Upper Fuller's Earth
{
12 ft Oyster Bed, composed mainly of *Liostrea hebridica* var. *elongata*.

3 ft Clay with limestone nodules; many fossils including *Rhynchonella smithi*, *Rugitela powerstockensis*, crushed large Peresphinctid ammonites, *Procerites* and *Hecticoceras*; *Belemnopsis bessina*, *Ostrea undosa*, *Trigonia scarburgensis*, *T. elongata* var. *lata*.
}

HERBURY PENINSULA SY613808   FOREST MARBLE OVER FULLER'S EARTH

Forest Marble
{
Greenish and brown clay with hard shale laminae, and hard flaggy oolitic and shelly limestone; with *Liostrea hebridica*, *Oxytoma costata*, nests of a small Rhynchonellid, sponges, *Apiocrinus* ossicles, etc.

1 ft *Boueti* Bed, a mass of brachiopods, the commonest *Goniorhynchia boueti*, species of *Avonothyris*, especially *A. langtonensis*, and of *Ornithella*, especially *O. digona*; also *Chlamys vagans*, *Placunopsis socialis*, *Apiocrinus* ossicles, and abundant ostracods; many of the fossils are encrusted with Polyzoa, especially *Berenicea*, serpulae, and minute oysters. The microfauna also contains baby brachiopods, foraminifera, ophiuroid ossicles, echinoid spines and minute gastropods (Sylvester-Bradley, 1948).
}

Fuller's Earth
{
c. 8 ft Laminated limestone and sandy shale.
15 ft (seen). Grey shale.
}

At the south-west end of the Herbury Peninsula is found a highly fossiliferous bed of the Forest Marble, known from elsewhere to lie 60 feet above the *Boueti* Bed; in it *Ornithella digona* is common, and there are many other brachiopods.

ABOUT PROMONTORY AT SY638791, SOUTH OF EAST FLEET
OXFORD CLAY, CALLOVIAN: *jason* ZONE

Lower Oxford Clay   *Kosmoceras* Shales   Highly bituminous shales, with many layers of 'turtle stones', large septarian concretions; dip, 5° SE.

TIDMOOR POINT SY642786   OXFORD CLAY, CALLOVIAN: *lamberti* ZONE

Indistinct but highly fossiliferous section. It is one of the most celebrated fossil localities in England, and of the greatest importance because of its remarkable assemblage of pyritised ammonites which correlate it with the *lamberti* Limestone of Buckinghamshire, the Hackness Rock of Yorkshire, and the *lamberti* clay of Normandy and the Jura Mountains. For the classification of the Oxford Clay in north-west Europe 'the importance of this locality could hardly be over-estimated' (Arkell).

This is the type locality of *Quenstedtoceras lamberti* (J. Sowerby), and var. *intermissa* (Buckman), *Q. leachi* (J. Sowerby), *Kosmoceras spinosum* (J. de C. Sowerby), and *K. tidmoorense* Arkell. Arkell records from here more than 40 other species or forms of ammonites, and *Procerithium muricatum, P. damonis, Pentacrinus* ossicles and stems, and casts of *Nucula* and *Grammatodon*.

PROMONTORY OPPOSITE FURZEDOWN FARM SY647783

OXFORD CLAY, LOWER OXFORDIAN: *mariae* ZONE

Indistinct but highly fossiliferous section. Very abundant pyritised ammonites, but of only some 10 species, including abundantly the zone fossil *Quenstedtoceras mariae;* also *Cardioceras, Taramelliceras, Perisphinctes, Belemnopsis hastata, Nucula, Pentacrinus* ossicles, and *Gryphaea*, mainly *G. lituola*.

This is the fauna of the typical Lower Oxfordian of the Jura Mountains, the Boulonnais, and Buckinghamshire, called in France the Marnes à *Creniceras renggeri*.

COAST SOUTH-WEST OF LYNCH FARM, SY648778, WEST OF WYKE REGIS

OXFORD CLAY, LOWER OXFORDIAN: *cordatum* ZONE

The rusty coloured kidney stones of the Red Nodule Beds weather out of the clays and form a bright coloured platform on the beach, mixed with many large oyster shells, *Gryphaea dilatata* and occasional *Lopha* [*Ostrea*] *gregarea*; multitudes of red casts of *Modiolus bipartitus, Thracia depressa* and *Pleuromya alduini*, with ammonites of the genera *Cardioceras, Goliathiceras* and *Aspidoceras* typical of the zone. The assemblage reproduces that of the same zone in Oxfordshire, Cambridgeshire and Normandy.

10–15-FT CLIFF, ON THE FLEET COAST, ABOUT SY650775, SOUTH-WEST OF WYKE REGIS   UPPER AND LOWER OXFORDIAN, CORALLIAN

	Glos. Oolite Series.
	Myophorella [*Trigonia*] *clavellata* Beds. 10 ft (seen).
	2½ ft Chief Shell Beds: rubbly and sandy limestones, with *M. clavellata, Liostrea delta*, etc.
Corallian	7½ ft Sandy Block: sandstones and sandy marl, with casts of *Pleuromya*, and fucoid markings.

Osmington Oolite Series. 52½ ft (seen).

4½ ft Nodular rubble, shelly and pisolitic, with *Nerinea cyane, Bourgetia subelegans, Littorina muricata, Procerithium, Isodonta, Nucleolites, Trigonia blakei,* etc.

48 ft White Oolite, marl, blue clays and sandy limestone; abundant *Isodonta triangularis* in the uppermost oolite, and *Ostrea dubiensis* in the next 4½ ft.

Corallian

(gap)

Berks. Oolite Series. 20 ft (seen).

Bencliff Grit.

10 ft Sands with enormous gritstone lenticles.

Nothe Clay.

10 ft (seen).

Lower Calcareous Grit.

Nothe Grits.

5½ ft (seen). Hard gritstone with *Gryphaea dilatata* and *Chlamys fibrosa* with marl above and below.

FLEET COAST, FROM ABOUT SY653772 TO SMALL MOUTH AT SY665763

KIMERIDGIAN AND CORALLIAN (UPPER OXFORDIAN)

A discontinuous and unspectacular section but one with interesting features. In 1926 it showed the following succession, but by 1939 it was much degraded, and little could be seen *in situ* above the Sandsfoot Grits.

Kimeridge Clay (seen at Small Mouth).

6 ft (seen). Blue-grey shaly clay with layers of *Liostrea delta.*

Kimeridgian

8 in Sandy band of *Exogyra nana* and *E. praevirgula.*

1 ft 8 in *Rhactorhynchia inconstans* band; grey clay with this fossil, *E. praevirgula* and *Modiolus durnovarius.*

Upper Calcareous Grit.

Westbury Iron Ore Beds.

3 in Red claystone nodules.

1 ft Ironshot oolite, with *Ringsteadia anglica, Serpula, Exogyra nana, Goniomya literata,* etc. (=Ringstead Coral Bed).

Ringstead Waxy Clays.

15 ft Brown and red waxy clays, and blue-grey clay with a few red nodules.

Corallian

Sandsfoot Grits.

7½ ft Sands with *Pinna sandsfootensis,* and broken *Liostrea delta;* full of fucoid markings.

15 ft Clay and sand, with *Pleuromya, Liostrea delta,* etc.

2 ft Fossil Bed, gritstone with a mass of *Chlamys midas, Ctenostreon proboscidium, L. delta, Ringsteadia,* etc.

Glos. Oolite Series.

Sandsfoot Clay (total c. 40 ft).

c. 12 ft (seen), much concealed by slips and vegetation.

Myophorella [*Trigonia*] *clavellata* Beds below.

1933, 1947a, Arkell; 1948, Sylvester-Bradley; 1958, 1961, House; 1969, Torrens (A).

THE ABBOTSBURY SWANNERY

It is impossible to leave this account of the Fleet without brief mention of its outstanding biological interest, the Abbotsbury Swannery, which is included in this Site of Special Scientific Interest. Unique in Europe it was referred to in a document of 1393, and later in an Exchequer Court document of 1591: '. . . from all time whereof the memory of man is not to the contrary there was and as yet is a certain flight of wild swans and cygnets called "a game of Wilde Swannes" in that estuary or water and on the banks and soil of the same building, breeding and frequenting . . .' Their numbers have varied from time to time between about 500 and 1500; 24 years ago there were about 800 (Good, 1946).

The Abbotsbury Estate, which includes the Swannery, has had but three owners over nearly a thousand years. Ork the Dane is said to have built a monastery here in 1014. He was granted a charter to the estate by King Cnut in 1023. Later it passed to the Abbey of Abbotsbury, to which about 1165 Henry II confirmed 'all its lands and tenures and all its liberties and free customs and acquittances with Soc and Sac and Toll and Theam and Infang-thef . . .'

After the Dissolution of the Monasteries, Henry VIII in 1541 leased the estate to Sir Giles Strangways for 40 years at a rent of £47 10s. 7½d. and since then it has remained in the same family, whose head is Lord Ilchester (Ilchester v. Raishley, 1888).

THE CHESIL BEACH OR BANK

This grand coast feature is claimed as one of the wonders of the world; it is unique in Europe, and was early described by Leland (1535–43). But for its protection the site of Weymouth would long since have been destroyed by the sea. The name comes from the Saxon word *cisel*, meaning shingle.

The first detailed investigation and description that I have found is that of de Luc (1811), made in 1804, and he arrived at several of the opinions that are held today. He described its extent, then, it seems, mainly a slightly off-shore bar; found that it rested upon clay; located an origin for the pebbles; and believed that their grading was largely caused by two sets of waves of different sizes from different directions. The next considerable investigation, published by Coode in 1853, gave much further detail.

The Chesil Beach is usually considered to begin at Bridport Harbour, where its continuity is broken by an artificial channel for boats to pass into an inner basin at high tide. However, similar graded gravel, although with many more cobbles, extends some three miles farther west towards Golden Cap. Thus, there is no natural end-point on the west.

From Bridport Harbour the beach stretches unbroken for 17·4 miles east and south to the Isle of Portland, where it ends abruptly against the cliffs of Chesilton. Approximately for the first eight miles it now hugs the coast; for the next eight miles it lies from 200 to 1000 yards off-shore, enclosing a shallow lagoon, the Fleet; and for the last two miles it strikes out to sea to join itself to the Isle of Portland. It maintains a beautifully even curve practically unbroken for the whole of its course, by compensatory narrowing or widening as it comes close to the old shore-line or slightly diverges from it.

The outlines of the Chesil Beach and of the Fleet as shown on the six-inch Ordnance sheets surveyed in 1863-4 were compared with those of the 1957-9 survey and found to be sensibly identical. This shows that no appreciable eastward movement of the shingle has taken place within the last century, whatever may have happened earlier.

If de Luc is understood correctly, in 1804 the shingle formed a slightly off-shore bar for nearly the whole distance between Portland and Bridport Harbour; local beaches touched the mainland only for a little way west of Abbotsbury, at Burton Cliff, and at Bridport Harbour, where it finally became a beach. Sixty years later the Ordnance Survey showed that it occupied the position it does to-day; the bar remains only between Portland and Abbotsbury; farther west it is a beach over practically the whole distance, the only vestige of the bar stage being found at Burton Mere, one and a half miles east of Burton Cliff. This might be taken as evidence of movement during the first half of the nineteenth century, or, alternatively, merely of a false premise about the 1804 position.

For the height of the beach there are only Coode's figures of 1853; 22 feet above high-water mark at Abbotsbury, and 42 feet at Chesilton; and a single Ordnance Survey level of 35 feet above Liverpool datum opposite the Fleet village, marked on the 1:2500 plan of 1902. A 25-foot contour line is shown on the beach from Portland to Wyke Regis on the current six-inch Ordnance sheets, but it is not continued on sheets farther to the north-west.

The width of the shingle as measured from the current six-inch Ordnance sheets, between low-tide marks of medium tides, or from low-tide mark to the edge of the mainland, as appropriate, was found as follows.

Approximate width, in feet	Extent, in miles
Beach	
300-400, East of Bridport Harbour for about	$\frac{1}{2}$
120-200, East Cliff to Burton Cliff End	$2\frac{1}{2}$
400-450, Burton Mere to West Bexington	$2\frac{1}{2}$
250-180, West Bexington to Abbotsbury	2
Bank	
500-600, Abbotsbury to Wyke Regis	$6\frac{1}{2}$
660-850, Wyke Regis	$2\frac{1}{2}$
620-320, Chesilton	$1\frac{1}{4}$
Beach	
180-140, Chesilton Esplanade	$\frac{1}{4}$

There has been an exceptional piling up of shingle of the largest size, both in height and width, at Wyke Regis.

From his profile measurements Coode calculated that between Abbotsbury and Portland alone a brief half-gale which blew for four hours on 28 November, 1852, but produced a heavy ground swell, scoured away four and a half million tons of shingle from the bank; however, five days later three and a half million tons had been thrown back.

Measurement from current Ordnance Survey sheets gives the area of the Fleet as about 1275 acres, compared with the estimated 3000 acres in 1630. The only way this reduction could have come about is by the eastern margin of the Chesil Bank advancing eastwards into the Fleet by, I calculate, 1776 feet, assuming the area given in 1630 to have been roughly correct, which is by no means certain.

From Donne's estate map of 1758 (now in the County Archives at Dorchester) two measurements of the width of the Fleet near Abbotsbury suggested that a slight easterly movement of the eastern margin of the Chesil Bank had since taken place. Two other measurements, of the width of the Bank in the same area, gave about 290 and 264 feet, compared with the present-day figures from the six-inch Ordnance sheet of about 528 feet for both sites, suggesting that the width has roughly doubled since 1758. Again, the accuracy of Donne's map cannot be unreservedly accepted, and conclusions from such measurements must remain in doubt.

According to Coode the shingle extends 36 feet below low-water spring tides at Abbotsbury, 42 feet at Fleet, and 48 feet at Portland; at these depths it gives way to sand. On its landward side the shingle has been proved to rest in places on clay at a depth of three or four feet above low-water spring tides. In one excavation between the mainland and Chesilton it rested upon white blown sand (Baden-Powell, 1930).

The lower part of the shingle is generally set in a matrix of grit and sand which makes the beach nearly watertight. The tide therefore seems to gain access to the Fleet only at the southern end, at Small Mouth, and the rise and fall at Abbotsbury does not exceed a foot even at the highest spring tide.

The grading of the pebbles is a particularly remarkable feature. They are described as ranging from the size of peas at Bridport Harbour, through the size of horse-beans about Abbotsbury, to the size of hen's or even swan's eggs near Chesilton. The grading appears to result from the sorting of existing pebbles of different sizes by the sea waves and not to attrition as the pebbles move along the coast. How the extraordinarily regular grading was accomplished is still rather an open question, but de Luc's conclusion that it was caused by two sets of waves was well re-stated by Edmunds (1938):

'The waves reaching Chesil fall into two groups, firstly, large south-westerly waves coming in from the Atlantic, and secondly, easterly and southerly waves generated in the Channel. The former frequently break on Chesil with a slight obliquity from the west, and drive large and small shingle to the east. The latter reach Chesil with a marked obliquity from the south, but owing to their small size, drive only the smaller pebbles west-wards. Thus, although the two sets of waves counterbalance each other with regard to the total quantity of shingle moved, there is a very marked sorting of the material.'

De Luc claimed that the pebbles on the mainland east of the Fleet, remnant terrace gravels of the kind that he believed to have furnished much of the Chesil Beach material, were themselves already roughly graded, decreasing in size to the north like the beach pebbles. I am not aware that this point has since been checked.

Cornish (1898b) gave the average weights of the pebbles he collected from about high-water mark on 12–14 July 1897, at nine sites, with their distances from Bridport Harbour. He took the whole length from there to the Port-land end as 18·2 miles instead of 17·4 miles as measured from the present map. His sites are re-located approximately as follows.

Miles
0·1 Near Bridport Harbour, east side, SY462903
1·7 Cliff End, Burton Bradstock, 486890
5·0 Opposite Coastguard Sta., Puncknowle; (Bexington), 532863
7·8 Opposite west end of Fleet; (Abbotsbury), 569838
10·7 Opposite Coastguard Sta., Langton Herring, 605809
12·6 Opposite Coastguard Sta., Chickerell; (East Fleet), 630790
15·5 Opposite 'Passage'; (Small Mouth), 663757
17·3 200 yards west of the south end of the Beach; (Chesilton), 682805
17·4 East end of Beach; (Chesilton), 683804.

Miles from Bridport Harbour	0·1	1·7	5·0	7·8	10·7	12·6	15·5	17·3	17·4
Weight of average pebble in thousandths of an ounce	16	28	67	111	294	342	783	12 800	9500
	(—)	(38)	(50)	(94)	(—)	(—)	(1100)	(5330)	(—)
Number of pebbles in 1 lb avoirdupois	1000	571	239	144	54	47	20	1·25	1·68
	(—)	(420)	(320)	(170)	(—)	(—)	(14)	(3)	(—)

The figures in parentheses give the results obtained by the present writer from samples taken from near the top of the Beach, on 26–27 October 1963, and suggest the possibility that the size distribution may be changing slightly over the years.

The pebbles are overwhelmingly of Chalk flint of specific gravity about 2·6, with some Greensand chert; but far-travelled rocks are sparingly represented, including black chert, red cherty jasper, vein quartz, purple porphyries and black quartzite. At Chesilton in particular there are many large oval or flat-rounded pebbles of sandstone or quartzite like those of the Triassic Budleigh Salterton Pebble Bed, and in which the same rare fossils

183

have been found; and, here only, white limestone, well-rounded, derived from the Portland quarries refuse tips.

Three features are noteworthy throughout the length of the beach. First, in the lowest part exposed by the sea, the site-average and smaller pebbles are generally set in a matrix of grit and sand. Second, the vast quantity of site-average pebbles alone, brown and often polished, forms the main mass of the beach. Third, there is a slight winnowed scatter of larger pebbles and cobbles up to four inches in diameter, rarely even to 12 inches. These are often of quartzite, are usually light grey in colour when dry, have a distinctively matt surface, and are very well rounded. Experiment showed that under the action of the breaking waves these larger pebbles and cobbles freely travel laterally along the shore in the direction of the wind that happens to be blowing; a test specimen was carried laterally as far as three feet by a single wave in fairly calm sea conditions, sliding easily over the groundmass of smaller pebbles.

The colour of the site-average pebbles, often brown overall, but sometimes only mottled with brown, seems to be caused by ancient iron staining. This feature, and also the high polish of many, is in agreement with the view that they are derived mainly from Quaternary gravels, and swept up from the whole area, particularly the great West Bay, from ancient land that has now vanished, eroded by the sea. They were there probably laid down as terrace gravels of the postulated old 'River Fleet', and in the Raised Beach of Eemian Interglacial age, a relic of which is still preserved at Portland Bill (q.v.).

Finality has not yet been reached in attempts to explain the origin of both the Chesil Bank and the Fleet. According to Baden-Powell (1930) the Bank is just such a deposit as would be expected to result from a blending of the Portland Bill Raised Beach material with pebbles from the Fleet Terraces and the Head, eroded by the sea at a late Quaternary date after the formation of the Head. Arkell (1947a) held that the shingle was a gigantic bay bar pushed back almost on to its hinterland; and that the Fleet is the last relic of a drowned area formerly opening into West Bay (otherwise known as Lyme Bay) and submerged in the Neolithic subsidence. 'From the western cliffs longshore drift brought the pebbles which built up the bar and still form it at the present day, centuries after the supply has been cut off.' That view would suggest the establishment of the Chesil Bank and the Fleet at a date between about 3000 and 2000 B.C.

The depth of the Fleet at low water (Admiralty Chart 3315, of earlier editions than 1954) is shown to range from three-quarters to one-sixth of a fathom. The deepest soundings (four and a half feet) are shown at Abbotsbury, where the Fleet is nearly tideless; with only one foot in the narrows off Wyke Regis.

Perhaps inspired by drainage operations in the Fenland, an attempt was made in the year 1630 to reclaim the Fleet, then estimated to have an area of some 3000 acres. A dam with sluices was built, presumably at the southern end at Small Mouth, but the project failed, apparently because the sea-water continued to penetrate into the Fleet at high tide.

In conclusion, a rough appraisal of what has been achieved by the sea waves, the resultant of many storms, suggests that the whole of the Chesil Beach shingle of something like 50–100 million tons weight of pebbles of different sizes, has been collected together, meticulously graded, and spread over some 18 miles of coast, within a period of 4000–5000 years.

1535–43, Leland; 1811, de Luc; 1853, Coode; 1869, Bristow and Whitaker; 1875a, Prestwich; 1898a, 1898b, Cornish; 1930a, Baden-Powell; 1938, Edmunds; 1946, Good; 1947a, Arkell; 1967, Neate.

Punfield Cove and Ballard Down, Swanage Bay

SZ(40)039810	179	343	57NW (G)	P, S. PSD. 1966

At Punfield Cove in the northern corner of Swanage Bay is found the best section of the Lower Greensand in Dorset, including the type section of the 'Punfield Marine Band', the primary geological interest of this site, although the exposure is usually concealed by vegetation and slipped material (Fig. 20).

The Lower Greensand beds here overlie Wealden Shales etc. and dip very steeply to the north below the Gault, Upper Greensand and Chalk succession of Ballard Down, as far as the Ballard Down Fault, which lies within the *mucronata* Zone of the Upper Chalk and is 'the most puzzling feature of Dorset tectonics' (Arkell).

The succession from Ballard Down to Swanage is as given on page 187 (Lower Greensand from Casey, 1961).

Although the Wealden at Punfield was noted by Fitton in 1824, the interest of the 'Punfield Marine Band' seems to have been first appreciated by Godwin-Austen, who in 1850 laid before the Geological Society a collection of its fossils, and Lyell commented on them in 1851. In 1871 Judd considered that the succession showed a transition from fresh-water Wealden through a fluvio-marine (but still Wealden) 'Punfield Formation' to the marine Lower Greensand. He recognised the strong likeness of the fauna to that described from the province of Teruel in eastern Spain, 60 miles south of Zaragoza, and within 70 miles of the east coast, where equivalent strata more

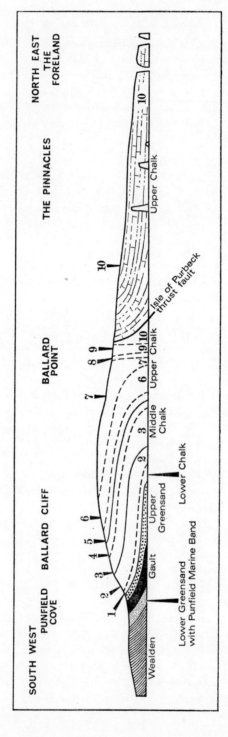

Fig. 20.

BALLARD DOWN–PUNFIELD COVE SECTION

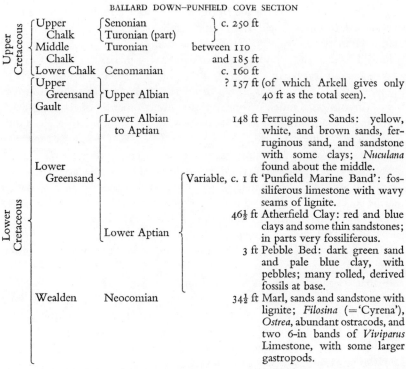

Upper Cretaceous	Upper Chalk	Senonian / Turonian (part)	c. 250 ft
	Middle Chalk	Turonian	between 110 and 185 ft
	Lower Chalk	Cenomanian	c. 160 ft
Lower Cretaceous	Upper Greensand / Gault	Upper Albian	? 157 ft (of which Arkell gives only 40 ft as the total seen).
	Lower Greensand	Lower Albian to Aptian	148 ft Ferruginous Sands: yellow, white, and brown sands, ferruginous sand, and sandstone with some clays; *Nuculana* found about the middle.
		Lower Aptian	Variable, c. 1 ft 'Punfield Marine Band': fossiliferous limestone with wavy seams of lignite.
			46½ ft Atherfield Clay: red and blue clays and some thin sandstones; in parts very fossiliferous.
			3 ft Pebble Bed: dark green sand and pale blue clay, with pebbles; many rolled, derived fossils at base.
	Wealden	Neocomian	34½ ft Marl, sands and sandstone with lignite; *Filosina* (='Cyrena'), *Ostrea*, abundant ostracods, and two 6-in bands of *Viviparus* Limestone, with some larger gastropods.

than 1600 feet thick include the three important productive coal basins of Utrillas, Gargallo and Val d'Ariño. Characteristic are the abundance of gastropods of the genus *Cassiope*, and Judd claimed that there is scarcely a single fossil in the 'Punfield Marine Band' which does not occur in the Spanish beds. Etheridge found 22 species common to both. Meyer (1872, 1873) contended that Judd's 'Punfield Formation' was really part of the Lower Greensand and not of the Wealden, and this view was accepted by Judd. But the strata still provided controversy. Since the large and conspicuous

Fig. 20. (*opposite*). Punfield Cove–Ballard Cliff–The Foreland: coast section showing Chalk overlying Lower Cretaceous. Zones of the Chalk indicated by numbers are:

Senonian	Campanian	10. *Belemnitella mucronata* / 9. *Gonioteuthis quadrata* / — *Offaster pilula* (not shown)	
	Santonian	8. *Marsupites testudinarius* / — *Uintacrinus socialis* (not shown) / 7. *Micraster coranguinum*	Upper Chalk
	Coniacian	6. *Micraster cortestudinarium*	
	Turonian	5. *Holaster planus* / 4. *Terebratulina lata* (=*gracilis*) / 3. *Inoceramus labiatus* (=*R. cuvieri*) / — *Actinocamax plenus* sub-zone (not shown)	Middle Chalk
	Cenomanian	2. *Holaster subglobosus* / 1. *Schloenbachia varians*	Lower Chalk

Cassiope was in 1871 not known from France, and was unlikely to have been overlooked, Judd suggested that Punfield and Teruel must have had a direct marine connection down the English Channel.

Judd described the 'Punfield Marine Band' as 21 inches thick and forming a well-marked feature in the cliff at Punfield Cove, where it dips at 65° to the north. The upper part is a hard laminated micaceous sandstone, more or less calcareous with much carbonaceous matter, sometimes becoming black to resemble an impure lignite. It has few fossils. The middle part is an oyster-bed with a few other dwarfed bivalves, *Corbula*, *Cyrena* (?) and *Cardium*. The lowest six to eight inches is a hard ferruginous shelly limestone, with many pebbles or concretions of clayey limestone, but little carbonaceous matter. The fauna includes ammonites, *Vicarya* [now known as *Cassiope*] and other marine shells, with oysters sometimes large and typical. Judd listed in all 38 species, including teeth of *Lamna* and *Pycnodus*, *Ammonites Deshayesi*, 16 gastropods, 18 bivalves, *Serpula* and many small gastropods and bivalves fragmentary and indeterminate, with much wood and carbonaceous matter. Arkell (1947) gives a revised list that includes five forms of *Cassiope*, and he adds that crustacean remains are common.

The most characteristic forms of the assemblage belong to the striking Turritellid genus *Cassiope*, ponderous thick-shelled gastropods more than two inches in length, ornamented with coarse spiral ribs and spiral rows of tubercles in various combinations. The commonest species at Punfield is *C. pizcuetana*.

The deposit persists as a fossil bed as far west as Worbarrow Bay, and is represented in the Isle of Wight by the Crackers Bed. Casey (1961) has re-investigated this Punfield bed, and concludes that too much has been made of the Spanish affinities of its fauna, which owes much of its distinctiveness to facies. It is a marine-brackish deposit, and the reason its fauna is not found in the Aptian of France is not because sea routes did not exist but because conditions of normal salinity prevented the brackish fauna from using them at that time. He quotes a record of species of *Cassiope* and an assemblage of other shells, comparable with the Punfield fauna, found in the Upper Barremian of the Paris basin.

Casey gives revised determinations of the ammonites as *Deshayesites forbesi* (=*D. deshayesi* in Arkell), *D.* aff. *callidiscus*, *D.* sp. nov., and *Roloboceras hambrovi*. *Deshayesites punfieldensis* has not been found at Punfield. The ammonites are all Crackers species and they confirm Strahan's correlation with the Crackers Bed of the Isle of Wight. On the evidence of the crustacean remains Arkell considered that at least the Upper Lobster Bed was represented in addition to the Crackers.

Comparing the two beds, ammonites are very abundant in the Crackers but exceedingly rare at Punfield. Conversely, *Cassiope* of brackish water affinities is characteristic at Punfield, but is known from the Crackers by only one or two examples. The bivalve *Eomiodon* (=*Astarte obovata* in Arkell), an indicator of marine-brackish conditions, is present at Punfield but unknown in any of the Isle of Wight beds. *Nemocardium (Pratulum) ibbetsoni*, a bivalve of very wide tolerance, is found in both the Wealden and the Atherfield Clay series, and is quite common at Punfield.

Here, then, ironically in the 'Punfield Marine Band' alone is recorded the passage from fresh-water Wealden to marine Lower Greensand, in some measure the vindication of Judd's idea of a transition 'Punfield Formation'.

1824, Fitton; 1871, Judd; 1872, 1873, Meyer; 1947a, Arkell; 1961, Casey.

Putton Lane Brick Pit, Chickerell, 2 miles NW of Weymouth

SY(30)650798	178	342	53SW	S, P
			(G)	PSD, 1963

A large working brick pit, affording one of the best exposures of the Kellaways Beds in the south of England (section from Arkell, 1947a).

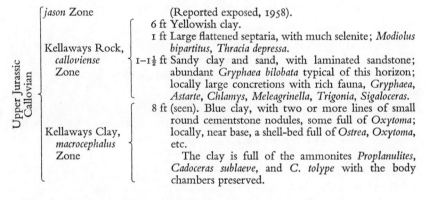

jason Zone	(Reported exposed, 1958).
	6 ft Yellowish clay.
Kellaways Rock, *calloviense* Zone	1 ft Large flattened septaria, with much selenite; *Modiolus bipartitus, Thracia depressa*.
	1–1½ ft Sandy clay and sand, with laminated sandstone; abundant *Gryphaea bilobata* typical of this horizon; locally large concretions with rich fauna, *Gryphaea, Astarte, Chlamys, Meleagrinella, Trigonia, Sigaloceras*.
Kellaways Clay, *macrocephalus* Zone	8 ft (seen). Blue clay, with two or more lines of small round cementstone nodules, some full of *Oxytoma*; locally, near base, a shell-bed full of *Ostrea, Oxytoma*, etc.
	The clay is full of the ammonites *Proplanulites, Cadoceras sublaeve*, and *C. tolype* with the body chambers preserved.

(Upper Jurassic — Callovian)

The *jason* Zone, of whose thickness and strata here I have found no description, is seen in the southern extension of the pit, the older zones in the northern part. The dip is 7° south.

Eight named forms of reptiles are recorded by Delair (1958–60) from the Oxford Clay of the Weymouth area and Chickerell. They are old records for which neither exact locality nor zone may be known, but it is appropriate to give them here as a fragment of the local Oxford Clay fauna.

Crocodiles: *Steneosaurus* sp. (a marine form, with skull about three feet long; fragments found at Putton Lane Brickfield about 1897, together with the remains of a bony fish, *Pholidophorus* sp.).

Plesiosaurs: *Muraenosaurus plicatus* (vertebral centra, Weymouth), *Cryptocleidus eurymerus* (Weymouth), *C. richardsoni* (the type specimen, comprising the major part of a skeleton, Chickerell).

Pliosaur: *Pliosaurus ferox* (typical teeth, Weymouth).

Dinosaurs: Theropoda (carnivorous): *Megalosaurus parkeri* (remains of an animal between 10 and 20 feet in length, running on strong hind-limbs, with three-toed feet; fore-limbs short, with five clawed fingers, only used in resting and feeding; type specimen from near Weymouth). Ornithischia (herbivorous): ? *Cryptodraco* sp. (one incomplete vertebra, Weymouth). Stegosauria: *Dacentrurus* sp. (remains of fibula, Weymouth).

For higher zones of the Callovian see under Crook Hill Brick Pit.
Cope (in Torrens) states that this exposure is now flooded.

1947a, Arkell; 1958, House; 1969, Torrens (A).

Studland Heath, 3 miles N of Swanage

SZ(40)0384	179	343, 329	51SW, 57NW (G)	Ph, S 1963

A coastal area three miles long by one mile wide lying south of the drowned valley of Poole Harbour. It consists of heathland and sand dunes, with strips of marsh and a fresh-water lake called Little Sea. The solid geology is of Eocene Lower Bagshot Beds, with a striking exposure of the upper part of these beds in the Agglestone.

The primary interest (with which this description is not concerned) of this site lies in the ecology of the rich flora and fauna of the recently colonised acid new land bordering Studland Bay, formed since the sixteenth century (Diver).

On this coast sand tends to accumulate on east-facing or lee shore, as at Studland. Here the dunes form three main ridges which trend a little east of north; they are separated from one another and from the Tertiary land on the west by strips of marsh with peat and scattered pools. The evidence of old maps shows that the innermost ridge formed between 1607 and 1721; the next was complete by 1849; and the outer ridge formed between 1849 and 1933. Traces of a fourth ridge had then begun to form.

In 1721 Little Sea was a tidal inlet open to the east; by 1785 it was changing into a brackish lagoon, although still tidal in 1805 (de Luc); it had become a fresh-water lake before 1900.

The dunes are of white silica sand practically devoid of lime and probably derived from nearby Tertiary beds via the sea-bottom. Thus, lime-loving plants and snails are almost absent, and the flora and fauna are distinctively different from those found on more normal calcareous coastal dunes. The original lack of lime may be due to the composition of the parent Bagshot or other Tertiary sand. The sheltered nature of Studland Bay, protected from the prevailing westerly gales, allows Recent molluscan shells generally to remain unbroken, and so may account for the absence of comminuted shell debris to form a calcareous addition to the dunes. In places the sand is of the peculiar quality that gives rise to 'singing sands' when trodden underfoot.

The western shore of Little Sea is the original coastline, then a low sea-cliff. The geological succession shows, at Studland (Arkell, 1947a):

Recent		Blown sand, with strips of peat.
Pleistocene		Plateau gravels, lying around 50 feet O.D.
Eocene, Cuisian	Lower Bagshot Beds	c. 35 ft Agglestone Grit; extremely coarse, pebbly and current-bedded grit.
		135 ft Pipe Clay Series; sands and clays, including seams of pure plastic pipe clay.
		c. 140 ft Redend Sandstone; red and yellow soft sandstone.

The plateau gravels are found in a few small relict patches, e.g. around SZ018845, 025846 and 026839.

The pipe clay has been worked for centuries in Purbeck for pottery-making. Both clays and sands are of deltaic origin, and were probably deposited in a shallow subsiding lake fed by a large river flowing from the granite areas in the west. The pipe clays once yielded from the Corfe Pits a rich tropical flora including large fan palms and ferns; also wing-cases of beetles, and shells (*Unio*); since the clays have been mined, these fossils are no longer generally found.

The Agglestone Grit is locally cemented to a hard irony stone of which relict boulders and slabs litter the heaths, and have been used for building-stone. The largest mass is the Agglestone itself, at SZ024828, likened to an inverted cone or anvil. It is a striking natural rock feature, poised *in situ* upon a large mound 70 feet high of soft Bagshot Sand which has been partially protected from atmospheric weathering by the overlying Agglestone.

In the most detailed measurements published (Hutchins, 1774) the Agglestone is given as about 18 feet high, with a maximum diameter of 36 feet at the top and 18 feet at the base. It was formerly larger, for some considerable portions of it are stated to have fallen and been removed for building prior to 1774. In 1963 the writer paced the top of the mound on which the Agglestone rests as about 60 feet in diameter. The top of the somewhat mushroom-shaped rock was measured as about 28 feet north to south, and 20 feet east to west. Its height was 20 feet at the north end and 15 feet at the south end. It overhangs asymmetrically, most on the west-north-west.

The Puckstone, 380 yards north-west of the Agglestone, is a lichen-covered rock only three feet high, which forms the peak of a mound-like exposure about 50 feet in diameter of many slabs of the rock. A deep pit on the north side suggests that stone has sometime been removed.

191

In the eighteenth century the Agglestone was called the Devil's Nightcap, and tradition told that the Devil threw it from the Isle of Wight with intent to demolish Corfe Castle, but it fell short. It was long popularly connected with the Druids, and sometimes thought to have been erected by them and used as an altar. Even in 1774, however, it was suspected of being really a natural rock feature.

The name 'Agglestone', which has been spelt Adlingstone and Haggerstone, has been variously derived from the Anglo-Saxon *Haelig-stan* = holy stone (Hutchins, 1774), *Hagolstan* = hailstone (Arkell, 1947), the Celtic (Welsh) *Eglwys* = church (Austen, 1856), and Hag's stone = witche's stone (Kingsley, 1858). Puckstone is derived by Arkell from *puca's* (=goblin's) stone. In Anglo-Saxon the name 'Studland' meant 'land where horses were kept' (Arkell). Whatever their derivations, these names were given probably more than a thousand years ago.

1774, Hutchins; 1811, de Luc; 1856, Austen; 1933, Diver; 1942, Baden-Powell; 1947a, Arkell; 1962, Chandler.

Wear Cliffs, with Golden Cap, 3 miles E of Lyme Regis

SY(30)405920	177	327	37SE (G)	S, P PSD, IGC, 1963

One mile of coastal cliffs exposing Upper Greensand and Gault capping standard sections of the Middle Lias and upper part of the Lower Lias, including the type section of the Green Ammonite Beds with the Belemnite Stone at the base, on the shore (Fig. 21). The succession shows (based on Wilson *et al.*, 1958):

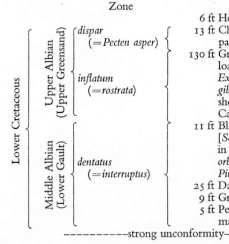

Zone

6 ft Head; flint and chert gravel.

13 ft Chert Beds: slipped and broken chert with pale buff and brown sands.

130 ft Greensand: Foxmould, buff, fine-grained loamy sand, with beekitised shells, mainly *Exogyra obliquata* (=*conica*) and *Neithea gibbosa* (=*quadricostata*); layers of these shells may be seen on the face of Golden Cap, where Cowstones are absent.

11 ft Black and blue loam, with *Rotularia* [*Serpula*] *concava* and other fossils common in lower part, such as *Entolium* [*Pecten*] *orbiculare*, *Lima gaultina*, *Cucullaea glabra*, *Pinna*, *Lingula subovalis*, etc.

25 ft Dark green and yellow loamy sand.

9 ft Green and black loam.

5 ft Pebble Bed: glauconitic sand and pebbles mainly quartz.

----------strong unconformity----------alt. 400 ft O.D.

Lower Cretaceous

Upper Albian (Upper Greensand): *dispar* (=*Pecten asper*); *inflatum* (=*rostrata*)

Middle Albian (Lower Gault): *dentatus* (=*interruptus*)

(*continued on p. 194*)

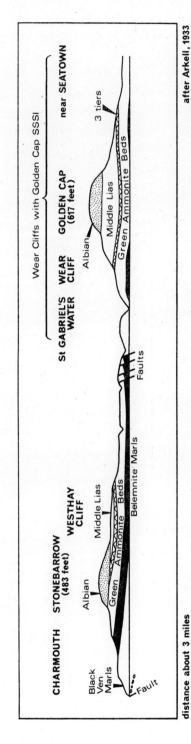

distance about 3 miles

after Arkell, 1933

Fig. 21. Charmouth–Golden Cap: coast section showing Albian (Lower Cretaceous) unconformably overlying Middle and Lower Lias.

13

WEAR CLIFF AND GOLDEN CAP SECTION (*continued*)

Lower Jurassic	Middle Lias—Domerian *margaritatus* Zone	Down Cliff Sands	? 90 ft Sandstone band (14 ft thick) at top, overlying shaly sandy clays and calcareous lenticles; a local conglomerate 62 ft above, and a bed of *Isocrinus* ossicles 8–14 ft above, the Starfish Bed. Some nests of small cuboidal Rhynchonellids, and Amaltheid ammonites, *Gryphaea cymbium*, *Pseudopecten equivalvis*, etc.

Eype Clay

3–4½ ft Starfish Bed: flaggy, calcareous, micaceous sandstone, with a two-inch-thick limestone with *Ophioderma egertoni* and *O. tenuibrachiata* on the under side. Strong spring-line at base.

3 ft Shaly clay or marl.

0–½ ft Day's Shell Bed: sporadic marly limestone locally full of shell fragments, fish remains, crinoid stems and belemnites. (Not well developed on Golden Cap.)

62 ft Micaceous marls, with some bands and nodules of grey earthy limestone.

c. 3 ft Hard calcareous sandstone with *Amaltheus stokesi*, *Montlivaltia*, *Furcirhynchia*, *Lingula*, *Spiriferina*, *Oxytoma*, *Chlamys*, *Lucina*, *Procerithium*, etc.

to 2 ft Eype Nodule Bed: grey clay, sandy at top, with one or two bands of small hard grey limestone nodules, with *A. stokesi*, *A. compressus*, *Tragophylloceras loscombi*, *Leptaleoceras* aff. *pseudoradians*, *Oxytoma inequivalve*, *Isocrinus*, *Belemnites*, *Lucina*, etc.

110 ft Blue micaceous marls, with local calcareous sandstone bands, limestone and ironstone nodules, iron pyrites; two thin, hard bands with *A. stokesi*, *T. loscombi*, etc.

The Three Tiers

c. 2½ ft Upper Tier: hard calcareous sandstone with *A. stokesi*, *T. loscombi*.

7½ ft Sandy shale.

c. 2 ft Middle Tier: flaggy micaceous sandstone, with *A. stokesi*.

8–14 ft Sandy shale.

c. 2½–5 ft Lower Tier: flaggy micaceous sandstone with *A. stokesi* common, *T. loscombi* and *Lytoceras fimbriatum* rare; small gastropods; saurian bones; the most fossiliferous tier.

11 ft Clay and shale, grey to brown, with pyrite includes six-inch band of shelly clay, with *A. stokesi* and *T. loscombi*.

Lower Lias — Lower Pliensbachian or Carixian

davoei (Beds 130c–122a)

100 ft Green Ammonite Beds: bluish-grey micaceous marly clays, some indurated beds, limestone nodules, and ferruginous layers; with the Red Band (Bed 26) c. 52–58 ft above the Belemnite Stone.

ibex
jamesoni

— Belemnite Marls: Belemnite Stone (Bed 121) at top, with many ammonites, and full of belemnites.

Owing to various difficulties in measurement, the individual beds are assigned slightly different thicknesses by almost every author.

Golden Cap, rising 618 feet above O.D., and the highest cliff on the South Coast, owes its name to the Cretaceous yellow sands capping its summit, 'gilded' on a sunny day in strong contrast to the dark Lias clays beneath. The grandest of the Dorset Lias cliffs, it is claimed to afford the finest Lias sections in England. The site also includes the best Gault section in the area, although it is difficult of access. It shows a practically unbroken succession from the Foxmould down to the unconformable junction with the Middle Lias. Here, too, is found the thickest development of the Middle Lias in the British Isles; but it is all in the *margaritatus* Zone, the *spinatum* Zone not being present.

The Eype Clay was so named by the Geological Survey as a locality name more distinctive than the Blue Clay (Woodward, 1893), or the *margaritatus* Clay (Lang, 1913), or the Micaceous Beds (Arkell, 1933). The Eype Nodule Bed of the Survey was called the *pseudoradians* Bed by Spath (1936).

The bands of calcareous sandstone known as the Three Tiers and named by Day (1863) form great buttresses along the lower portions of the cliffs, a tripartite ledge some 30 feet thick here, at their greatest development.

There is difficulty in drawing a boundary between Middle and Lower Lias here; it is defined as the lowest and strongest line of soak or change of slope observed as occurring above strata which are known or presumed to be Green Ammonite Beds not containing *Amaltheus*. This lies at about 11 feet below the Lower Tier.

This site provided Lang's (1936) type section of the Green Ammonite Beds of the Lower Lias. Replete with Liparoceratid ammonites, the beds are here at their thickest, and thin rapidly to the west. The sections in them are most accessible between the Ridge Faults to the west, at SY392924 and the main promontory at Golden Cap.

From about 300 yards east of St Gabriel's Mouth nearly to the west buttress of Golden Cap the Belemnite Bed (129), Crumbly Bed (120c–e), and the top of the Pyrite Marls (120a–b), are well exposed at the cliff foot; but the cliff is much concealed by talus. Beds as low as the *jamesoni* Zone are exposed at extremely low tides on the foreshore of The Corner (SY410919).

The Belemnite Stone at beach level beneath Golden Cap forms an excellent collecting ground as far west as St Gabriel's Water, and yields ammonites of the genera *Beaniceras*, *Liparoceras*, and *Lytoceras*, and the large bivalve *Parainoceramus ventricosus*. The Crumbly Bed and Pyrite Marls below show very many belemnites weathering out, and pyritised remains of *Acanthopleuroceras*.

1863, Day; 1933, Arkell; 1936, Lang; 1957, Howarth; 1958, Wilson *et al.*; 1969, Torrens (A).

GLOUCESTERSHIRE

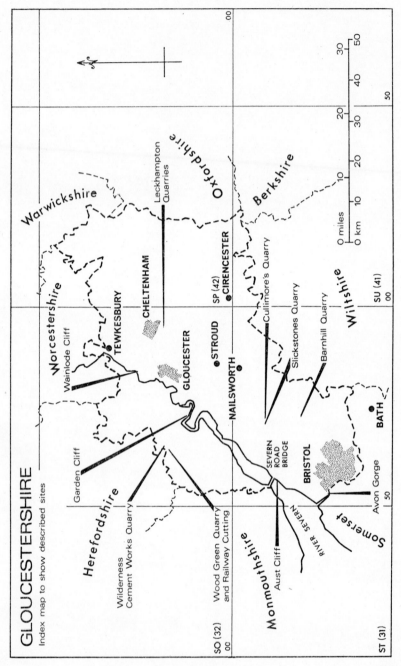

GLOUCESTERSHIRE
Index map to show described sites

Warwickshire

Worcestershire

Herefordshire

Oxfordshire

Berkshire

Wiltshire

Monmouthshire

Somerset

Wainlode Cliff

Garden Cliff

Wilderness Cement Works Quarry

Wood Green Quarry and Railway Cutting

Aust Cliff

Leckhampton Quarries

Cullimore's Quarry

Slickstones Quarry

Barnhill Quarry

Avon Gorge

TEWKESBURY

CHELTENHAM

GLOUCESTER

STROUD

NAILSWORTH

CIRENCESTER

BRISTOL

BATH

SEVERN ROAD BRIDGE

RIVER SEVERN

SO (32) 00

ST (31) 00

SP (42) 00

SU (41) 00

0 miles
0 km

Fig. 22

Aust Cliff, Severn Estuary, 10 miles N of Bristol

National Grid Reference	One-inch Sheet Pop. and 7th ed. Ordnance Survey	One-inch Sheet, Geol. Survey	Six-inch Sheet, Ordnance Survey	Main Site Interests
ST(31)566898	156	Bristol	62NE (–)	S, P, T, U IGC, PSD 1967

A classic river cliff section, one mile long, exposing the base of the Lower Lias, the Rhaetic and the Keuper, first described by Buckland and Conybeare in 1824. Since the cliff is largely inaccessible, the beds have to be studied in fallen blocks (Fig. 23).

Aust Cliff is the truncated face of a ridge of Trias and Lower Jurassic rocks surrounded by alluvium. A very gentle anticlinal structure is shown, cut by five small faults with throws ranging from three to 16 feet to the south, which give rise to small springs on the cliff face. Both flexing and faulting have been explained by compaction of the Keuper Marls (Whittard, 1949). The Trias rests unconformably on the upturned edges of a Carboniferous Limestone ridge exposing the Lower Dolomites, which dip about 15° southwest. This platform is one of erosion, cut in Trias times. Viewed from the cliff top at low water, it affords an excellent example of a rather rare feature, an unconformity seen in plan.

The section shows (after Reynolds, 1946):

Lower Lias, 3 ft 0 in		Shale and thin-bedded limestone divisible into: b. *planorbis* beds, with *Psiloceras planorbis* only; a. *Pleuromya* beds, with *Liostrea hisingeri*, [=*Ostrea liassica*], *Pleuromya, Modiolus, Dimyodon, Macrodon, Protocardium, Lima, Oxytoma.*
Upper Rhaetic, Cotham Beds, 11 ft 6 in	6 in	Cotham Marble equivalent, 'Crazy Cotham'; pale limestone; a thin limestone at the top is full of *Meleagrinella* [*Pseudomonotis*] *fallax;* with the complete fish *Pholidophorus higginsi* and *Legnonotus cothamensis* and fish remains of *Gyrolepis, Birgeria* [='*Saurichthys*'] and insects.
	4 ft 2 in	Yellow clay with thin limestone bands, with the ostracod *Darwinula liassica.*
	2 ft 10 in	'Insect Bed': grey or cream-coloured argillaceous limestone, with a rather uncommon liverwort, *Naiadites lanceolata*, the phyllopod *Euestheria minuta* var. *brodeiana* and insects.
	4 ft 0 in	Yellow thin-bedded argillaceous limestone; no fossils recorded.

(continued on p. 201)

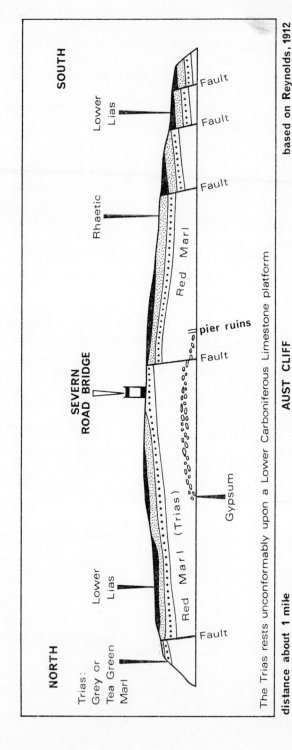

NORTH

Trias:
Grey or
Tea Green
Marl

Lower
Lias

SEVERN
ROAD BRIDGE

Red Marl (Trias)

Gypsum

Fault

Fault

pier ruins

Fault

Red Marl

Rhaetic

Fault

Lower
Lias

Fault

Fault

SOUTH

distance about 1 mile

The Trias rests unconformably upon a Lower Carboniferous Limestone platform

AUST CLIFF

based on Reynolds, 1912

Fig. 23. Aust Cliff: coast section showing basal Lower Lias, Rhaetic and Trias resting unconformably upon Carboniferous Limestone.

	1 ft 0 in	Greenish-black shales, *Myophoria*, *Palaeocardite* [*Cardium*] *cloacinum*.
	5 in	'Upper *Pecten* Bed': hard grey limestone; *Chlamys valoniensis*, *Pleurophorus elongatus*, *Acrodus*, *Gyrolepis*.
Lower Rhaetic Westbury Beds, 14 ft 4 in	8 ft 0 in	Black shales, *Rhaetavicula* [*Pteria*] *contorta*, *Protocardia rhaetica*, with selenitic '*Pullastra*' (= *Schizodus*) Bed.
	8 in	'Lower *Pecten* Bed': hard dark, pyritous limestone, *Myophoria*, *Ch. valoniensis*, *P. rhaetica*, *Schizodus* sp.
	4 ft 0 in	Black shale, the upper part fissile paper-shale, very barren, with selenitic '*Pullastra*' Bed.
	1–6 in	*Ceratodus* Bone Bed: impersistent, includes fragments of hard Tea-Green marlstone.
	22 ft 9 in	Grey or Tea-Green sandy marl with 1 ft of hard sandstone 3–4 ft from the top.
	52 ft	Red Marl.
Trias, Keuper, 119 ft 9 in	25 ft	Gypsiferous marl, including a primary deposit of masses of white and pink gypsum, with secondary strings and veins of the fibrous form 'satin spar' with traces of celestine and barytes.
	20 ft	Red sandy marl; a thin-bedded sandstone near the base contains abundant clay pseudomorphs after rock-salt.

----------unconformity----------

Carboniferous Limestone	Limestone platform, exposing the Lower Dolomite of zones Z and C_1.

The Cotham (or Landscape) Marble equivalent at Aust is a pale limestone in a highly brecciated condition, known as 'False (or Crazy) Cotham'. The insects figured by Brodie (1845) from Aust have been largely renamed by Handlirsch (1906–8).

In the Westbury Beds thin-shelled bivalves are abundant in some bands of the black shales, and '*Pullastra*' beds occur in them both above and below the 'Lower *Pecten* Bed'. It is in the '*Pullastra*' beds that the remains of 'Labyrinthodonts' and small Dinosaurs are found most frequently.

The Bone Bed is the most interesting deposit at Aust, and probably its best development in England. It is a conglomerate or breccia of water-worn fragments of Palaeozoic rocks, chiefly Carboniferous Limestone mingled with bones, teeth, spines, scales and coprolites of reptiles and fish, with some well-rounded quartz pebbles. The most characteristic remains are the palatal teeth of the fish *Ceratodus*, from about two to four inches in length, and very variable. There are also found *Mytilus cloacinus* and *Schizodus* sp.

Coprolites are extremely abundant; they contain from 25 per cent to 50 per cent of calcium phosphate, and some remains of crustacea.

Teeth of the Carboniferous fish *Psephodus magnus*, *Psammodus porosus* and *Helodus* are found in the Bone Bed, probably derived from the Carboniferous Limestone, and *Ctenoptychius* from the Carboniferous Limestone or the Coal Measures.

The quartz pebbles are generally regarded as stomach stones swallowed by fishes and saurians. Hitherto considered as an aid to the trituration of their food, the stones may, plausibly, have been taken as ballast to improve a desired equilibrium in the water, as recently found in crocodiles (Cott, 1961, p. 236).

VERTEBRATE REMAINS

The contemporary vertebrate remains recorded from Aust, mainly from the Bone Bed, include the following (some others are recorded by Davis, 1881).

REPTILES

Ichthyosauria: Mainly as vertebral centra up to seven and a half inches in diameter.

Sauropterygia: Including *Plesiosaurus rugosus*, *P. costatus* and the commonest form here, *P. rostratus*.

Dinosaurs: Extremely rare very large bones, some assigned to the carnivorous *Avalonia*, and some smaller bones to *Zanclodon;* and others to *Rysosteus oweni*.

'AMPHIBIANS'

Mandibles, usually very fragmentary, have been referred in the literature to the Labyrinthodonts, under the queried name *Metopias diagnosticus*. However, Savage and Large (1966) have re-examined these and consider that they belong to large specimens of the Lepidosteid fish *Birgeria acuminata* (Agassiz). They conclude that no evidence remains for the presence of amphibians in the Rhaetic of the Bristol Channel area.

FISHES

Remains are plentiful, commonest being the primitive sharks.

Selachians: *Hybodus cloacinus*, *H. minor*, *H. laeviusculus* and *H. punctatus*; *Acrodus minimus* (probably the commonest fossil in the Bone Bed); *Nemacanthus monilifer* (as fin spines) and pieces of skin (shagreen).

Chondrosteous 'Ganoids', primitive bony fishes: *Gyrolepis alberti*; *Sargodon tomicus*; the Lepidosteid (bony pike) *Birgeria acuminata* (some material previously recorded under the name *Saurichthys acuminatus* now re-determined by Savage and Large).

Dipnoi (lung fishes): *Ceratodus* very variable and abundant, with the species *C. latissimus*, *C. parvus* and *C. disauris*; the last two species are probably invalid (Donovan, *in lit.*).

A cartilaginous fish now lives in Queensland, Australia, whose teeth correspond in every respect with those of the Rhaetic *Ceratodus*. It is a vegetable-feeder and lives in muddy creeks, burying itself in the mud when hot weather sets in, until the return of the wet season (Richardson, 1906b).

KEUPER

These beds are formed of fine quartz dust, much of desert origin, distributed by the wind, some deposited on land, some in salt lakes; sun cracks are seen with clay pseudomorphs after rock-salt crystals near the north end of the section. No fossils are recorded from the Keuper beds.

The colour difference between the Tea Green Marls and the Red Marls is considered to be due to the reduction of part of the iron oxide from the ferric to the ferrous state by decaying organic matter. Both marls contain about 13 per cent of iron oxide; in the green marl it is wholly in the ferrous state, but in the red marl about nine per cent is ferric and four per cent ferrous (de la Beche.)

From early times a ferry has served as conveyance across the River Severn at this place, which was known as Aust Passage. In 1966 a suspension bridge was constructed to carry a new motorway. The masonry abutments resting on Aust Cliff inevitably conceal part of the geological exposure, but much will remain available to geologists.

1824, Buckland and Conybeare; 1845, Brodie; 1846, de la Beche; 1881, Davis; 1903, Short; 1906b, Richardson; 1906–8, Handlirsch; 1921a, 1946, Reynolds; 1949, Whittard; 1961, Cott; 1966, Dreghorn; 1966, Savage and Large.

Barnhill Quarry, Chipping Sodbury

ST(31)725830	156	Bristol	ST78SW (G)	S, T, P PSD, IGC, 1965

The notified site stretches over some 1200 yards as a narrow strip along the eastern, disused, side of the vast quarry, and is bounded by the main Wickwar road, B4060, as far as the first milestone from Chipping Sodbury.

The geological interest lies largely in the enormous slabs of Carboniferous Limestone, which dip about 40° west, and their finely displayed bedding planes, which are often beautifully ripple-marked. Three small thrusts of little moment coincide with the bedding planes; they are of post-Rhaetic date since they cut the Rhaetic strata. Considerable slickensiding is present, both on the thrust planes and also on the dip-joint faces.

The Carboniferous Limestone comprises mainly the S Zone, with a little of the underlying C Zone, as follows (after Tuck, 1925):

Carboniferous Limestone, Viséan

S₂
- Thick band of oolite, and thin-bedded grits and shales, often ripple-marked.
- Massive limestones.
- Gritty limestone bed.

S₁ Concretionary algal limestone and calcite mudstones, with some oolite, and shale partings.

C Mainly *laminosa:* dolomite, dark crinoidal dolomitic limestones and fine grey dolomite, including an algal band; the sub-oolite occurs as a fossiliferous band at the top of the quarry.

Fossils in this section include the typical corals and brachiopods of the Avon section, with abundant *Cyrtina carbonaria.*

In this south-eastern part of the quarry Rhaetic strata lie unconformably upon a horizontal platform of the Carboniferous Limestone, which has apparently been planed by the Rhaetic sea. Where this platform has been protected by Rhaetic clay its surface remains smooth, but where it has been covered only by a porous soil layer the surface is deeply griked by solution channels.

When this part of the quarry was in active work, about 1936, the Mesozoic succession was very well exposed. It was recorded by Reynolds (1938).

Lower Lias
- Blocks in the surface soil of pale grey limestone, with *Pleuromya tatei, Modiolus minimus* and *Liostrea hisingeri* [=*Ostrea liassica*].

Upper Rhaetic, 5 ft 6 in
- (Sun Bed and Cotham Marble wanting).
- 2 ft Pale shale.
- to 1 ft *Estheria* Bed, irregular pale argillaceous limestone, with no recognisable fossils.
- 2 ft 6 in Greyish shale.

Lower Rhaetic, Westbury Beds, 14 ft 9 in
- 6 in Irregular bed of grey argillaceous limestone, in places sandy, with many *Chlamys valoniensis, Schizodus* cf. *ewaldi, Atreta* [*Dimyodon*] *intusstriata.*
- 3 ft 6 in Dark selenitic shale with impersistent ferruginous layer full of vertebrate remains; *Cardium, Myophoria emmrichi, Protocardium rhaeticum*; many *Acrodus minimus, Birgeria acuminata* [=*'Saurichthys acuminatus'*], *Sargodon tomicus, Gyrolepis alberti, Plesiosaurus* sp.
- to 1 ft *Pecten* Bed, argillaceous limestone with prominent layers of beef (fibrous calcium carbonate).
- 7 ft 6 in Dark-grey and black shales with abundantly fossiliferous bands, with *Rhaetavicula* [*Pteria*] *contorta, Schizodus ewaldi, Ch. valoniensis, P. rhaeticum*, pyrite.
- 1 ft 3 in Black paper-shale without fossils.
- 1 ft (or less). Bone Bed, irregularly developed; vertebrate remains as in the higher bed, together with teeth of *Hybodus*, vertebrae termed *Rysosteus*, and quartz pebbles up to 1 in long.

------------unconformity----------------------------

Carboniferous Limestone
Steeply dipping, with a smooth and level planed-off surface.

1925, Tuck; 1938, Reynolds.

Cullimore's Quarry, Charfield, 14 miles NNE of Bristol

ST(31)720926	156	Bristol	56SW	P, S, Ig
			(–)	IGC, PSD
				1964

Within the Tortworth Inlier, this is a small but classic section on the west side of an old quarry where working ceased many years before 1908. It was first described and fossils were first collected by Weaver in 1819.

The section now shows:

Lower Silurian ⎰ 2¼ ft Dark grey, very fossiliferous ashy limestone, with a grit
⎱ parting in the middle.
Upper Llandovery ⎱ 1 ft (seen). Black highly amygdaloidal basic lava ('trap').

The ashy limestone lies in a hollow in the lava. It contains minute angular particles of glassy basalt thickly scattered, with some lapilli up to half an inch or more in diameter.

This limestone outcrop has yielded more than 60 species of invertebrate fossils, mainly corals and brachiopods, but also trilobites, polyzoa, gastropods, etc. The commonest are the coral *Favosites gothlandicus*; the brachiopods *Orthis calligramma*, *Leptaena rhomboidalis*, *Atrypa reticularis* and *Coelospira hemispherica*; and the trilobites *Phacops weaveri* and *Encrinurus punctatus*. The *Palaeocyclus* (coral) Band is also present. Two of the brachiopods, *Stropheodonta compressa* and *Stricklandia lirata*, are of special interest since they link the Tortworth fauna with that of Pembrokeshire and distinguish it from the Shropshire facies.

The lava, known to be about 60 feet thick, is a pyroxene andesite, highly amygdaloidal at the top, with abundant calcite, which fills many of the cavities; it has a few large but altered and eroded feldspar phenocrysts. The lava is compact below. Originally termed the 'Upper Trap Band', being one of two that are known in the Tortworth area, it is of particular interest since here and in the Eastern Mendips are found the only contemporary Silurian volcanic rocks in Great Britain.

1824, Weaver; 1901, Morgan and Reynolds; 1908, Reed and Reynolds; 1924, Reynolds; 1934, Smith.

Garden Cliff, Westbury-on-Severn, 8 miles SW of Gloucester

SO(32)718128   143   OS43SE   32SW   S, P
                                    (–)   PSD, 1963

This is the middle one of three classic river-cliff sections of the Rhaetic and underlying Keuper Marls in Gloucestershire, located over 28 miles along the River Severn in a north-east to south-west direction, with Aust Cliff (q.v.) 17 miles to the south-west and Wainlode Cliff (q.v.) 11 miles to the north-east. Each section exposes the complete thickness of Rhaetic development at the site, but each has its own distinctive peculiarities, ascribed to land barriers of Palaeozoic rocks separating each from the others at the time of deposition.

The half-mile-long Garden Cliff is also known as Golden Cliff. It was termed 'Gould Clyffe' by the English people according to Giraldus Cambrensis writing in the mid-twelfth century; the name referred to the golden colour of the rocks there, due to the abundant pyrite. With its highly fossiliferous Bone Bed it was mentioned by Townsend in 1813 (p. 277), so the section was probably known to William Smith. It was first described in detail by Conybeare and Phillips in 1822, and recognised by later authors as one of the finest exposures of the Rhaetic. Offering the most accessible exposure of the three sites mentioned above, it includes the type section of the Westbury Beds, named from the locality; and was the first place in England to yield brittle-stars (*Ophiolepis damesii*) from the black shales. The beds dip about 9° south-south-east and the section is as follows (condensed from Richardson, 1903).

Lower Lias,* 2 ft 5 in	2 in 'Bottom Bed': hard grey and blackish-blue limestone, *Liostrea hisingeri* [=*Ostrea liassica*], *Modiolus minimus*, *Pleuromya* and *Gervillia*.
	1 ft 10 in Brown and grey calcareous laminated shales with some insect remains.
	2½ in *Pseudomonotis* Bed ('Insect Limestone'): bluish grey-centred limestone; the top inch is fissile, with many elytra of beetles, and traces of wings and abdomens, and small plants resembling reeds.
	2½ in Light-brown limestone with many bivalves.

* On the difficult question of where to draw the boundary between the Lower Lias and the Upper Rhaetic the late Mr Linsdall Richardson informed me (*in lit.*, 24 March 1962) that he now supported Brodie's correlation of the Insect Bed of Wainlode Cliff with the *Pseudomonotis* Bed of Garden Cliff, and he concluded that the Insect Limestone and overlying shales at Wainlode, and the *Pseudomonotis* Bed and overlying shales and underlying limestone with many bivalves at Garden Cliff, represent Moore's Lower Liassic Insect and Crustacean Beds of Somerset. Accordingly, the beds referred to above at Wainlode Cliff and Garden Cliff require classifying with the Lower Lias and not with the Rhaetic.

Upper Rhaetic, Cotham Beds, 14 ft 1 in	7 ft 7 in	Laminated shales with sandstone and shale bands.
	c. 1 ft	'Estheria Bed' ('Cypris' Bed, or Naiadites Bed): feature-forming limestone with many fossils including bivalves Palaeocardita cloacinum, Chlamys valoniensis, etc.; a phyllopod, Euestheria minuta var. brodieana; an ostracod, Darwinula liassica; the liverwort, Naiadites lanceolata; and fish-scales.
	5 ft 6 in	Grey, marly, non-laminated shales, with many bivalves etc.

----------non-sequence----------

Lower Rhaetic, Westbury Beds, 19 ft 8½ in	2 ft 11 in	Black shales, with a sandstone band; with selenite and pyrite; many bivalves, Rhaetavicula contorta, Modiolus minimus, Ch. valoniensis, etc. 1-in limestone band at base with fish remains in addition, Birgeria acuminata [= 'Saurichthys acuminatus'], Gyrolepis alberti and coprolites.
	4 ft 1½ in	Alternating black shales with thin sandstone bands; many bivalves and fish remains.
	1 in	Hard blackish-blue limestone with bivalves and fish remains.
	4 ft 4 in	Black shales, with some thin sandstone layers and one fossiliferous horizon.
	2 in	Sandstone, with 'Pullastra' (= Schizodus) and fish-scales.
	1 ft 7 in	Black and grey shales, with a ½-in sandstone.
	1 in	Bone Bed, sandstone in 1–4 layers, very pyritic with small quartz pebbles. Many vertebrate remains. Fish: Acrodus minimus, Sargodon tomicus, Hybodus cloacinus, H. minor. Saurians: Ichthyosaurus, Plesiosaurus; and 'Pullastra'.
	1 ft 6 in	Black laminated shales with several grey calcareous sandstone layers; no fossils.
	1 ft 0 in	'Upper Pullastra Sandstone': feature-forming with ripple-marked upper surface, lower part a bone bed; small quartz pebbles; Schizodus, R. contorta, etc. and fish (B. acuminata, S. tomicus, G. alberti, H. minor, Nemacanthus), coprolites.
	2 ft 0 in	Black shales, coarsely laminated, with thin, irregular sandstone layers.
	5 in	'Lower Pullastra Sandstone': feature-forming lower part a bone bed; with Schizodus (several species), R. contorta, and fish (B. acuminata, G. alberti, S. tomicus).
	1 ft 4 in	Black laminated shales, no fossils recorded.
	2 in	Arenaceous deposit with fish (G. alberti, A. minimus, B. acuminata), coprolites, small quartz pebbles and occasional pebbles of red marl.

----------non-sequence----------------------------

Trias, Upper Keuper	18 ft 0 in	Tea-Green Marls: greenish-grey marls, no fossils.
	73 ft 6 in	(seen). Red Marls with zones of grey; with very thin veins of gypsum; no fossils.

Micro-organisms preserved in minute grains, or 'micro-berries', of pyrite have been found in the Westbury Beds of this section, as well as in other strata elsewhere. These little-known fossils, perhaps comparable with bacteria, are still being investigated (Love, 1962.)

The fauna of the Cotham Beds is impoverished and stunted, and the 'Estheria Bed' is a fresh- or brackish-water deposit, as is shown by its fossils.

Prominent cornice-like ledges are formed by the 'Estheria Bed' and by both the Upper and Lower *Pullastra* Beds, these *Pullastra* Beds being peculiar to the Garden Cliff section.

The Bone Bed here differs from that at Aust, in that *Ceratodus* teeth, abundant at Aust, are extremely rare at Garden Cliff, only two having been reported, of which one is certain (Richardson, 1906).

Although the main part of the Rhaetic is typically developed at Garden Cliff, the sequence at both top and bottom, above the Cotham Beds and below the Black Shales, is less complete than in Somerset and Glamorgan.

1822, Conybeare and Phillips; 1903, 1906b, 1947, Richardson; 1934, Gardiner *et al.*; 1966, Savage and Large.

Leckhampton Quarries, 2½ miles S of Cheltenham

SO(32)950185	143	OS44	26SE, 34NE	S, P
			(–)	PSD, IGC
				1964

Disused quarries are found on the hill-top, and hill scarps with old quarries extend for more than a mile round the west and north sides of Leckhampton Hill. They afford magnificent sections of the highly fossiliferous Cotswold Inferior Oolite, from which in 1853 Lycett recorded 184 species. The dip is about 7° south-east. The quarries were worked for 'many centuries' until, in 1793, the Lord of the Manor developed them on a large scale, and their famous freestone was extensively used for buildings in Cheltenham.

The first geological description is by Murchison (1834), and is followed by a large literature. The thicknesses of the beds differ considerably according to different authors; the figures given below are substantially those of Richardson, 1933.

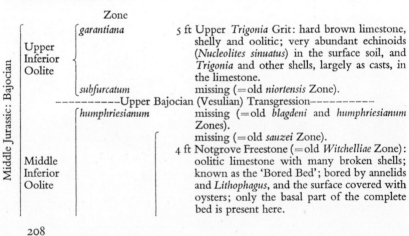

Zone

Middle Jurassic: Bajocian

Upper Inferior Oolite

garantiana — 5 ft Upper *Trigonia* Grit: hard brown limestone, shelly and oolitic; very abundant echinoids (*Nucleolites sinuatus*) in the surface soil, and *Trigonia* and other shells, largely as casts, in the limestone.

subfurcatum — missing (=old *niortensis* Zone).

----------Upper Bajocian (Vesulian) Transgression----------

humphriesianum — missing (=old *blagdeni* and *humphriesianum* Zones).

missing (=old *sauzei* Zone).

Middle Inferior Oolite

4 ft Notgrove Freestone (=old *Witchelliae* Zone): oolitic limestone with many broken shells; known as the 'Bored Bed'; bored by annelids and *Lithophagus*, and the surface covered with oysters; only the basal part of the complete bed is present here.

Middle Inferior Oolite	*sowerbyi*	5 ft Gryphite Grit (=old *Shirbuirnia* Zone): the 'Bottom Bed', hard brown, coarse gritty iron-shot limestone, with a vast accumulation of *Gryphaea;* fauna of 42 species listed; formerly quarried for road metal and walls.
		12 ft *Buckmani* Grit (=old *discites* Zone): shelly ragstone, upper surface bored; overlying yellow sandy limestone and marls; many fossils, including large ammonites, *Gryphaea* and brachiopods, many *Lobothyris buckmani.*
		6 ft Lower *Trigonia* Grit (=old *discites* Zone): light-brown and grey thin-bedded ironshot and rubbly limestone, conglomerate near base; yields good bivalves, a few ammonites, and corals.

----------Middle Bajocian Transgression--------------------

Middle Jurassic: Bajocian

	concavum	0–1½ ft Snowshill Clay: yellow clayey marl.
Lower Inferior Oolite		33 ft Upper Freestone (=old *bradfordensis* Zone): false-bedded oolitic and marly and rubbly limestone, with corals, *Pentacrinus,* brachiopods, and bivalves; formerly quarried for lime-burning.
		10 ft Oolitic Marl (=old *bradfordensis* Zone): cream-coloured marl famous for its beautifully preserved fossils, large brachiopods, especially *Plectothyris* [*Terebratula*] *fimbria* in profusion, and large species of bivalves and gastropods; locally rich in corals; the crustacean *Eryma Guisei* was described from here; 120 species listed; formerly quarried for lime-burning.
	murchisonae	78 ft (see *Note*). Lower Freestone: massive yellowish oolitic limestone with comminuted fossils; the lower part, the 'Roestone', is false-bedded, with many fossils. The upper part forms the famous freestone, formerly extensively quarried for building; when freshly mined it is easily cut by the saw, but it hardens on exposure.
		21 ft Pea Grit: rubbly ochreous shelly and coarsely pisolitic beds, with masses of flattened concretions up to half an inch in diameter, described from here as the calcareous alga *Girvanella pisolitica;* 75 species listed, with 19 species of echinoids, and crinoids, many corals, polyzoa, brachiopods, gastropods, bivalves, ammonites, belemnites and serpulae. Formerly quarried for gate-posts and other rough work.
		9 ft Lower Limestone (=old *sinon* Zone; see *Note*): brown and grey oolitic and marly limestones.
	scissum	10 ft *Scissum* Beds: ferruginous oolite, dark shelly crystalline limestone, and at base 3 ft of sandy marls with broken shells, chiefly *Pecten* and *Cidaris* spines, some ammonites, many belemnites, brachiopods, bivalves and fish remains; some *Lithophagus* boring.
	opalinum	missing.

(*continued overleaf*)

LECKHAMPTON QUARRIES SECTION (*continued*)

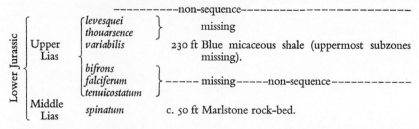

			----------non-sequence-------------------
Lower Jurassic	Upper Lias	{ *levesquei* *thouarsense* }	missing
		variabilis	230 ft Blue micaceous shale (uppermost subzones missing).
		{ *bifrons* *falciferum* *tenuicostatum* }	----- missing-----non-sequence-----------
	Middle Lias	*spinatum*	c. 50 ft Marlstone rock-bed.

Leckhampton Hill, rising to 978 feet above O.D., is part of the bold escarpment of the north-west flank of the Cotswolds, formed by excavation of the Severn Valley. Exposures of the upper strata are found in the extensive shallow quarries on top of the hill, the Gryphite Grit forming the floors. The rest of the succession is to be seen in the old quarries on the hillside overlooking Cheltenham, although the Upper Lias and the Marlstone (Middle Lias) are not normally exposed.

The isolated pinnacle known as the Devil's Chimney may have begun to be separated from the main outcrop only some 400–500 years ago. In the early nineteenth century it was just possible for the adventurous to jump on to it from the main hill.

The earliest traces of man on Leckhampton Hill comprise many flint-flakes, arrow-heads and scrapers found in the fields round about, and dated to the Neolithic and Bronze Age. On the hill-top are ancient earthworks. A tumulus and camp are dated to the early Iron Age, about 500 B.C., and may have been constructed by the Dobuni, a powerful tribe here shortly before the Roman invasion. The hill is in the direct line of most of the ancient trackways from the south, and a path leads from the camp to a British track, the Greenway. Silver pennies of A.D. 835–901 suggest that the camp may also have been occupied during the turbulent times of Dane and Saxon conflict.

Note. The Lower Freestone: other authors give different thicknesses, up to 130 feet (Arkell, 1933).

The Lower Limestone: Wilson *et al.* (1958) p. 69, state that '. . . there is no evidence in Britain of a subzone of *Cylicoceras sinon* above the *costosum* Sub-zone.'

This is the type locality of the Buckmani Grit.

1834, 1845, Murchison; 1850, Brodie; 1857, Hull; 1893, 1897, Buckman; 1904, 1906a, 1933, Richardson; 1920, Gray; 1933, Arkell; 1965, Dreghorn; 1969, Torrens (B).

Slickstones Quarry, 1 mile NE of Cromhall, 13 miles NE of Bristol

ST(31)705915	156	Bristol	64NW	P, S
			(–)	PSD, IGC
				1964

A large working quarry in Carboniferous Limestone, also known as Crom-hall Quarry. The special scientific interest lies in a vertebrate-bearing fissure-deposit of Trias, the visible part of which is given as about 250 feet long by up to 30 feet wide. The mass was exposed during quarrying operations, and is now left as unwanted refuse in the quarry.

The material is a cave deposit, Trias landwash laid down in an old solution cavern in the Carboniferous Limestone, traces of whose walls and roof are to be seen, in places coated with stalagmite. It varies from a soft pale-green silt to a hard calcite-cemented sand, with Carboniferous crinoid ossicles and Triassic bones. It is nearly horizontally bedded, with some small-scale current-bedding, and contains fragments of stalagmite and rare cave pearls. The only invertebrate fossil found is an occasional Phyllopod, *Euestheria minuta*.

The deposit contains an important vertebrate fauna of land animals, considered to be of Keuper, Upper Trias, age, originally washed by land drainage through some swallow-hole into the cavern and deposited therein, at a time when it was functioning as an underground drainage channel through the Limestone. Although many of the bones are fragmentary, they are beautifully preserved if fragile, and a few specimens are articulated; part of the skin was preserved on one specimen. A new genus and species, *Glevosaurus hudsoni*, was described from here by Swinton in 1939, a Rhyn-chocephalian about one foot long; it is considered to be ancestral to the modern Tuatara, *Sphenodon*, a lizard-like creature only known from islands off New Zealand. Further work has yielded some nine other species of reptiles, ranging in stature from about six feet to six inches. One is a Theco-dont, about 20 inches long, the first known from England. Four others may be Rhynchocephalians.

This Upper Trias fauna forms an important sample of the vertebrates then living on dry land at a time when reptilian evolution was at a particularly interesting stage. As found at Slickstones and in other somewhat similar fissure fillings in the Bristol Channel area, this is stated to be the only known example of a Mesozoic reptile fauna which lived on ground higher than swamp. The fauna, which is still in process of description, may double the number of known forms and contribute valuable information about the early history of the group. It has been described as 'the most exciting work in vertebrate palaeontology being carried out in this country today'.

The Carboniferous Limestone quarried here is of Tournaisian, Z_2, age, well exposed, with a good fauna of Zaphrentid corals and brachiopods, and worthy of study and collection in its own right.

1924, Wallis; 1939, Swinton; 1957, Robinson.

Wainlode Cliff, 5 miles N of Gloucester

SO(32)845257	143	OS44	18SE	S, P
			(–)	1963

A fine natural cliff about 400 yards long cut by the River Severn, exposing a complete section of the Rhaetic, dipping from 2 to 9° south-south-west. The section, whose upper part is inaccessible except by ropes, was first described by Strickland in 1842, and is as follows (condensed from Richardson, 1903).

Lower Lias, 2 ft 0 in	4 in	Hard blue limestone crowded with *Liostrea hisingeri* [=*Ostrea liassica*] and *Modiolus minimus*.
	1 ft 3 in	Shales with intercalated limestone with *Pseudomonotis* and *Darwinula*.
	5 in	'Insect Limestone' (*Pseudomonotis* Bed): hard blue-grey limestone with *Modiolus*, *Ostrea*, insects, fish-scale.
Upper Rhaetic, Cotham Beds, 11 ft 8 in	5 ft 2 in	Brown and blue laminated shales.
	6 in	*Estheria* ('Cypris') Bed: hard yellow nodular limestone, with *Estheria*, *Naiadites*, ostracods, fish remains, shell fragments. (A widespread and constant horizon marker.)
	6 ft 0 in	Pale greenish-yellow laminated shales, with *Modiolus*, *Schizodus*, *Cardium*.
Lower Rhaetic, Westbury Beds, 14 ft 7 in	14 ft 7 in	Black shales, with selenite crystals, and thin limestone and sandstone bands; in the lower part the shales contain vertical veins of baryto-celestine; many fossils, including *Rhaetavicula contorta*, *Chlamys valoniensis*, *Modiolus*, *Gervillia*, *Protocardium*, '*Pullastra*' (=*Schizodus*), *Ostrea*, fish remains. A 3-in-thick Bone Bed 2 ft above the base, is composed of up to four thin sandstone bands separated by clay partings, and in places very pyritic; it contains '*Pullastra*', *R. contorta*, fish remains and coprolites, including *Birgeria* [='*Saurichthys*'], *Gyrolepis*, *Acrodus*, *Nemacanthus*, *Hybodus*.
		----------non-sequence----------
Trias Upper Keuper	23 ft	Tea-Green Marls: light green-grey marls with conchoidal fracture, weathering blue-grey and white.
	75 ft	(seen). Red Marls: variegated, rarely gypsiferous with zones of grey-green and bluish marl, with angular and conchoidal fracture.

Two new species of foraminifera, *Spirillina* [*Orbis*] *infima* and *Polymorphina liassica*, were described by Strickland (1846) apparently from the basal Lower Lias here.

The passage of a true bone-bed into an unfossiliferous sandstone may be studied in this section.

Note. On the boundary between Lower Lias and Upper Rhaetic, see foot-note under Garden Cliff (p. 206).

1842, 1946, Strickland; 1903, 1947, Richardson.

Wilderness Cement Works Quarry, 600 yards SW of Mitcheldean Church, 10 miles W of Gloucester

---

SO(32)659183	143	OS43SE	23SE (G)	S, P PSD, IGD 1963

A disused quarry, 300 yards long, which seems to have been in work in 1886, but working ceased before 1945. It lies in the north-east corner of the Forest of Dean Coalfield, on the eastern flank of the sharply folded and pitching Wigpool syncline. Here, dipping about 58° west-south-west, is finely exposed the complete succession of the Carboniferous Lower Limestone Shales, by far the best section of these beds in the Forest of Dean. Duplicate exposures of the beds up to and including No. 7 are afforded by the upper and lower working faces of the quarry.

The attenuated development here (190 feet) differs considerably from that of the group in the Bristol area—in the Avon section it is more than twice as thick—and in the South Wales area. Here is a lower moiety of limestones with subordinate shales, and an upper moiety of shales with subordinate limestones, which become dolomitised and conspicuous towards the top. Sibly and Reynolds did not consider it possible to subdivide them into K_1 and K_2 Zones. The junction with the Old Red Sandstone is sharp, and there is an exceptional current-bedded oolite at the base; a well-marked Bryozoa bed is found 90 feet above the base. However, the basal *Modiolus* phase (the Shirehampton Beds) and the 'true Bryozoa Bed' near the base of K_1, both features found in the Avon section, are lacking.

Sibly and Reynolds described this quarry section in great detail as follows, but their ninefold subdivision has proved of only limited local validity.

		Bed	Thickness	Description
Lower Carboniferous: Tournaisian			c. 10 ft (seen).	Lower Dolomite (Zones Z–C_1); base of lowest member of the Main Limestone.
	Lower Limestone Shales—Zone K	9.	40 ft	Micaceous dolomite (some beds simulating a sandstone) interbedded with shales. The upper part has yielded no recognisable fossils, but a band 12 ft above the base contains an abundant large *Syringothyris*, other brachiopods, and *Euomphalus*, and below this are richly fossiliferous dolomites.
		8.	30 ft	Micaceous shales and thin dolomitised limestone bands. Fossils include abundant *Camarotoechia mitcheldeanensis*, and some eight other genera of brachiopods.
		7.	25 ft	Dark calcareous shales, weathering yellow, with a few thin, plentifully fossiliferous limestone bands, with *Fenestella* and *Rhabdomeson*, *Productus* and other brachiopods and Spirorbids.
		6.	6 ft	Bryozoa Bed: a coarse crinoidal limestone and ferruginous dolomite, the fossils replaced by iron oxides; abundant *Rhabdomeson* and Fenestellids and plentiful brachiopods.
		5.	10 ft	Shales overlie 6 ft of highly fossiliferous grey crinoidal limestone, with abundant brachiopods of some six genera.
		4.	30 ft	*Modiolus* phase of calcareous shales full of ostracods, with calcareous mudstone and nodular limestone; with calcareous algae (*Mitcheldeania*, *Solenopora*, and *Spongiostroma*), ostracods, annelids, and abundance of broken bivalves. On the bedding planes of the higher beds are *Modiolus* and large ostracods.
		3.	4 ft	Compact, grey, crinoidal limestone, with ostracods and abundant calcareous algae; in three bands, with shale partings, forming a conspicuous feature in the quarry face. A chemical analysis showed about 92% $CaCO_3$ and 0·57% MgO.
		2.	20 ft	Richly fossiliferous thin-bedded sandy limestone and shale, with crinoids; *Fenestella* and *Rhabdomeson*; and *Camarotoechia mitcheldeanensis* in great abundance; this overlies 4 ft of grey crinoidal limestone with brachiopods, bivalves, ostracods, foraminifera and calcareous algae; and at base 6 ft of flaggy limestones and shales.
		1.	25 ft	Massive limestones; the upper beds are grey crinoidal limestones packed with brachiopods, and some current-bedded oolite; the lowest beds are gritty oolites current-bedded and ripple-marked.
Old Red Sandstone; Tintern Sandstone Group			c. 20 ft (seen).	Red and yellow-brown thin-bedded sandstones, with red, purple and green clays.

From the Lower Limestone Shales of this quarry or nearby Wethered described the calcareous alga *Mitcheldeania nicholsoni* as a new genus and species, and recorded, among other fossils, six species of ostracods: *Kirbya variabilis*, *K. plicata*, *Cytherella extuberata*, *Bythocypris sublunata*, *Darwinula berniciana*, *Leperditia okeni*.

In the Tintern Sandstone in this quarry, a few feet below the base of the Lower Limestone Shales, has been found an interesting fauna and flora including Athyrid, Camarotoechid, Orthotetid, and Spiriferid brachiopods, bivalves, gastropods, and—especially remarkable in this position—a goniatite, probably referable to the genus *Imitoceras* Schindewolf; also fish remains, moulds of *Ctenacanthus* spines and *Orodus* teeth; and Lycopodiaceous plant-stem moulds nearly two inches in diameter, provisionally identified as

Cyclostegia kiltorkense Haughton. In general, these remains have a Carboni-ferous aspect (Stubblefield, in Pringle, 1937).

1886, Wethered; 1937, Sibly and Reynolds; 1937, Pringle; 1942, Trotter.

Wood Green Quarry and Railway Cutting, Blaisdon, 8 miles W of Gloucester

---

SO(32)694168	143	OS43SE	32NW	S, P.
			(–)	1964

Old red sandstone marls (seen at the north end of the cutting).

Downton Series — 10 ft Clifford's Mesne Beds: false-bedded yellowish micaceous shales, in places yielding *Lingula minima;* at base a phosphatised pebble bed up to 1 in thick, looking, when weathered, like dirty gingerbread. This is the Ludlow Bone Bed; it is not now seen in the cutting, but may be found nearby (see *Note* below).

50 ft Upper Longhope Beds: flaggy and shaly slightly calcareous siltstones. Typical Upper Ludlow fossils include *Protochonetes ludloviensis* [='*Chonetes striatellus*'], *Salopina* [*Dalmanella*] *lunata, Camarotoechia nucula,* fragments of *Ceratiocaris,* and *Beyrichia torosa;* but there is no Lower Ludlow element. (Discontinuous exposures are seen in the cutting, but the highest beds are not seen here.)

10 ft Lower Longhope Beds: flaggy calcareous siltstones with some shelly limestone bands; fauna shows an important and dis-tinctive horizon of transitional character, a typical Upper Ludlow assemblage flourishing alongside a waning Lower Ludlow fauna; *Shaleria ornatella* and *Whitfieldella canalis* are common only in these beds. (Seen in cutting.)

Silurian;
Ludlow Series

25 ft Upper Blaisdon Beds: flaggy and shaly siltstones with many bands of shelly limestones; polyzoa are common and the bedding planes are crowded with *Dalmanella orbicularis, Dayia navicula, Sphaerirhynchia wilsoni, Camarotoechia nucula, Fuch-sella amygdalina, Howellella elegans.*

10 ft Lower Blaisdon Beds: feature-forming conglomeratic lime-stone bands (which partly represent the Aymestry Limestone) more abundant in basal part, separated by poorly fossiliferous shaly siltstones; polyzoa, with *Ptylodictya lanceolata* common; brachiopods include *Atrypa reticularis,* mollusca, *Poleumita globosa,* and uncommon simple corals. (Typical section seen in quarry, but more accessible section in cutting.)

c. 50 ft Upper Flaxley Beds: calcareous silty mudstones and siltstones with subordinate limestone bands and nodules; Stropho-menids characteristically abundant; many brachiopods, simple corals common, mollusca, *Beyrichia,* and crinoid ossicles. (Best seen in southern part of cutting.)

A small disused roadside quarry and adjacent railway cutting extending over 75 yards. The cutting was made for the Hereford, Ross and Gloucester

Railway in 1852, and was described by Strickland (1853) with the locality name of Velt-house.

Here are displayed important sections in the remarkably thin development of the Silurian Ludlow Beds of the May Hill Inlier, which lies at the junction of the Malvern and Woolhope axes. The Ludlow Beds, dipping up to 65° west, exhibit a shelly calcareous shelf facies, and reach no more than a quarter of their thickness in the Malvern and Woolhope areas, and the massive Aymestry Limestone is missing, represented only by thin bands of conglomeratic limestone in the Lower Blaisdon Beds. The succession is as shown on p. 215 (Lawson, 1955).

Note. The Ludlow Bone Bed recorded in the cutting by Strickland is not now exposed within the Site of Special Scientific Interest, but may be examined in the stream section 70 yards south-south-east of the railway–road crossing, together with the Clifford's Mesne Beds.

I am much indebted to Dr J. D. Lawson for help with this section.

1853, Strickland; 1955, Lawson.

SOMERSET

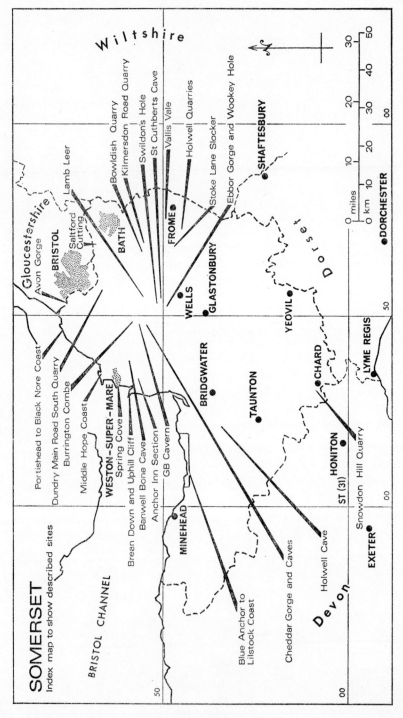

SOMERSET
Index map to show described sites

Fig. 24

Anchor Inn Section, Purn Hill, Bleadon, 2 miles S of Weston-super-Mare

National Grid Reference	One-inch Sheet, Pop. and 7th ed. Ordnance Survey	One-inch Sheet, Geol. Survey	Six-inch Sheet, Ordnance Survey	Main Site Interests
ST(31)333570	165	OS20	16SE (–)	P, A 1965

A small gravel quarry under a cliff east of the main A370 road, at the base of Purn Hill and opposite the Anchor Inn; largely destroyed during road-widening in the early 1930s but still worthy of study.

Purn Hill lies on the southern flank of the Black Down uplift. Its western margin exposes Zone Z_2 and Horizon Gamma of the Carboniferous Limestone, dipping about 20° south-south-west. The rock is mainly highly crinoidal limestone, the hill showing prominent scars formed by cherty bands. In the quarry some small-scale faulting was recorded.

The west side of Purn Hill formed a sea-cliff during late Palaeolithic times and after, and the quarry probably included the remains of a sea-cave. The section exposed was as follows (after Palmer, 1934).

Recent {
At surface, reaching 80 feet above O.D. was a kitchen midden site, with British or Romano-British pottery, domestic animal bones, and edible mollusc shells, resting upon

Pleistocene {
12–15 ft Upper sandy breccia of small sub-angular limestone fragments, with much sandy matrix, grading into stony loam.
More than 15 ft Lower breccia of larger limestone fragments with less matrix, and with basal boulders; includes a fossiliferous bed 1 ft thick, full of shed Reindeer antlers, a few fragmental metapodal bones, and remains of Brown Bear; at the base of this bed were found bones and teeth of Voles (*Microtus ratticeps*), fish vertebrae, and the marine shells *Macoma balthica* and *Littorina littorea*.

The section apparently rests upon the platform of a 50-foot Raised Beach. The terrestrial breccias are considered to be of Late Pleistocene, Upper Palaeolithic, date, and are claimed to have been formed under arctic blizzard conditions.

1924, Bamber; 1931, 1934, Palmer.

219

Avon Gorge, Bristol

ST(31)560745	155	Bristol	ST57SE	S, P, Ph
			(G)	IGC, PSD,
				1967

This spectacular gorge lies on the western outskirts of Bristol, the River Avon down to its mouth on the Severn estuary five miles distant marking the county boundary. It forms an important line of communication between Bristol and Avonmouth, with the river, navigable by ocean-going steamships of limited tonnage, a railway, and the main A4 road.

On both right (Gloucestershire) and left (Somerset) banks practically the same geological succession is seen. Even in 1811 the right bank had the better exposures, owing to quarrying (Bright, 1817). It was later claimed as probably the most complete section of the Carboniferous Limestone in the British Isles. Studied since before 1811, it has been taken as the type section of the Carboniferous Limestone Series of the South-western Province (Fig. 25).

In the early nineteenth century the Black Rock was quarried for paving stones, and the limestone of St Vincent Rocks (Observatory Hill) was burnt to give a very pure lime that was exported to the West Indies for sugar refining (Bright, 1817). Much more limestone must have been quarried for other local purposes.

The Avon gorge traverses the southern limb of a greatly denuded anticlinal uplift of Carboniferous Limestone and Old Red Sandstone. With the strata dipping about 30° south-east, the section stretches for one and a quarter miles downstream from the Clifton Suspension Bridge. Overlain by the Millstone Grit, it passes down into the Upper Old Red Sandstone. The Clifton (or Great or Observatory) reversed faulting, of Armorican date, crosses the gorge some 150–300 yards downstream of the bridge, downthrowing about 1100 feet to the north and duplicating the upper part of the succession, the Upper Cromhall Sandstone, the Hotwells Group and the Clifton Down Limestone. A small reversed fault in the Great Quarry duplicates about 50 feet of Clifton Down Limestone.

Bright (1817) and Cumberland (1821), whose accounts are somewhat intermingled, give a detailed description of the lithology and measurement of the succession, divided into 294 beds, but identify few fossils. Stoddart (1867) gives a similarly detailed description and also names the fossils he found in each bed, a very valuable paper. In 1905 Vaughan published his zonal system based on the corals and brachiopods. In this area, it was claimed,

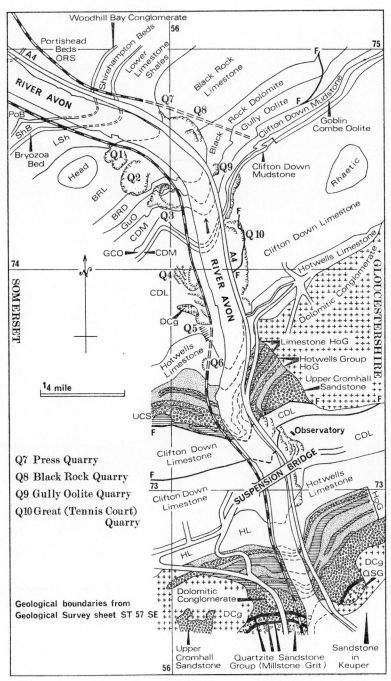

Fig. 25. Avon Gorge: geological map. (Based on Crown Copyright Geological Survey map. Reproduced by permission of The Controller, H.M. Stationery Office.)

these are the only two classes sufficiently abundant throughout the sequence to render a system of zones founded upon faunal variation both reliable and useful; either corals or brachiopods alone are not satisfactory (Vaughan and Reynolds, 1935). However, bryozoa (polyzoa) are suggested as having possibilities; and foraminifera are plentiful and extend certainly from the *Zaphrentis* Zone to the top of the succession. Both these groups might prove of zonal value if they were studied here.

Since Vaughan's classic paper attention has been focused upon the corals and brachiopods to the apparent neglect of the other elements of the very rich fossil fauna, which includes algae, plants, foraminifera, sponge spicules, corals, worms, crinoids, echinoids, bryozoa (=polyzoa), brachiopods, bivalves, gastropods, pteropods, cephalopods (*Orthoceras*), trilobites, crustacea, ostracods, conodonts and fishes; these last furnished Agassiz with some of his types of Lower Carboniferous fishes. At least 10 of them from Zone Z_2 and two more from Zone K are still preserved in the Bristol City Museum, after its partial destruction by enemy action in 1940.

Vaughan's zonal classification, however, proved to be not readily usable as a basis for geological mapping in the region, and officers of the Geological Survey (Kellaway and Welch, 1955) propounded an alternative scheme based on the lithology. The general correlation of these two schemes is given below.

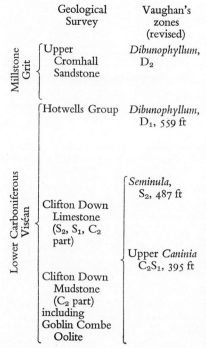

	Geological Survey	Vaughan's zones (revised)	
Millstone Grit	Upper Cromhall Sandstone	*Dibunophyllum,* D_2	Red grits, one horizon with quartz pebble, and shales, coarse oolites, foraminiferal limestones; corals (*Lonsdaleia floriformis, Lithostrotion irregulare,* etc.), large crinoids, bryozoa.
Lower Carboniferous — Viséan	Hotwells Group	*Dibunophyllum,* D_1, 559 ft	Grey foraminiferal limestones and pseudobreccias, shales, grits and massive white foram and coral limestones; coarse red oolite; large *Productus hemisphaericus;* first appearance of *Dibunophyllum* and *Cyathophyllum,* which are abundant.
	Clifton Down Limestone (S_2, S_1, C_2 part)	*Seminula,* S_2, 487 ft	Massive foram limestones and concretionary beds, with algal bands (*Mitcheldeania* and *Spongiostroma*), pisolites and oolite; *Chonetes, Productus cora, Lithostrotion martini, Diphyphyllum, Carcinophyllum,* bryozoa, ostracods.
	Clifton Down Mudstone (C_2 part) including Goblin Combe Oolite	Upper *Caninia* C_2S_1, 395 ft	*Seminula* oolite with abundant *S. ficoides* and massive foram limestones; first appearance of *Lithostrotion* (*L. martini*); *Mitcheldeania, Spongiostroma;* ostracods, *Spirorbis,* bryozoa (*Heterotrypa*), ? *Palaeechinus* spines, trilobites (*Phillipsia*), *Dictyoclostus* [*Productus*] *semireticulatus, Caninia bristolensis.*

Lower Carboniferous Tournaisian	Gully Oolite (C₁ part)	Lower *Caninia*, C_1, 245 ft	*Caninia* Dolomite, Gully Oolite (90 ft) and *Laminosa* Dolomite (93 ft) with Horizon Gamma at base; fish beds with some large specimens and types of Agassiz; first appearance and abundance of *Caninia*; *Syringothyris*, *Michelinia*, *Orthotetes*, *Chonetes*.
	Black Rock Limestone and Dolomite (C₁ part, Hor. Gamma, Z_2, Z_1, Hor. Beta)	*Zaphrentis*, Z, 536 ft	Dark limestone, sometimes dolomitised, with fluorite and strontium sulphate crystals, and bitumen; first appearance and great abundance of *Zaphrentis*; forams first become plentiful; crinoids, bryozoa, *Spirifer*, *Chonetes*, *Orthotetes*, *Schizophora resupinata*; fishes (some Agassiz types).
	Lower Limestone Shales / Bryozoa Bed (12 ft)	*Cleistopora*, K, 380 ft	Shales and limestones, with rich fauna, trilobites (*Phillipsia*), *Spiriferina octoplicata*, *Productus bassus*, *Camarotoechia mitcheldeanensis*, *Vaughania vetus* [= *Cleistopora* aff. *geometricus*]; calcareous conglomerates; fish (some Agassiz types); at base Horizon Alpha = Bryozoa Bed, red crinoidal limestone with bryozoa (*Rhabdomeson*), ostracods, very abundant microfossils.
	Shirehampton Beds	*Modiolus* phase, Km, 83 ft	Shales with limestones, essentially passage beds; *Spirorbis*, ostracods, crinoid fragments, bivalves (*Modiolus latus, Sanguinolites*), gastropods (*Bellerophon*), bryozoa (*Rhabdomeson, Rhombopora, Fenestella*); plant remains; shallow-water brachiopods (*Lingula, Discina*), and deeper-water forms (*Athyris, Chonetes*), Spiriferids and algae (*Mitcheldeania, Ortonella*).
Upper Old Red Sandstone	Portishead Beds	—	Sandstones, with Fish Beds at top.

At the Clifton Suspension Bridge the Avon Gorge is more than 700 feet wide and some 300 feet deep. A little way upstream the rock-bottom falls to a minimum of 39·3 feet below O.D. (Donovan, 1960) and in the Avonmouth area the buried channel is at least 65 feet below O.D. (Hawkins, 1962). Hawkins suggests that the channel was eroded during the first or second phase of the Last (Würm) Glaciation. Although it is agreed to be an erosion channel, not along a fault line, and cut mainly since Pliocene times, its formation is difficult to understand. It may have been due to ice damming the probably earlier Flax Bourton valley during the Pleistocene, the resulting lake over the Bristol area finding a spillway to the north-west. However it was formed, the Avon Gorge was cut across the grain of the hard Carboniferous Limestone to a depth that ensured the continuance of this drainage line after the ice melted.

The Clifton Suspension Bridge was built between 1836 and 1864 with long delays owing to lack of funds, to plans by Isambard K. Brunel. There were used in it the chains of the dismantled Hungerford (London) suspension bridge completed in 1845, also by Brunel.

Geology: 1817, Bright; 1821, Cumberland; 1867, Stoddart; 1905, 1906, Vaughan; 1921a, 1921b, 1921c, Reynolds; 1935, Vaughan and Reynolds; 1955, Kellaway and Welch; 1956, Dineley and Rhodes; 1958, George. Physiography: 1907, Harmer; 1938, Trueman; 1960, Donovan; 1962, Hawkins; 1966, Bradshaw.

Banwell Bone Cave, Banwell, 4 miles ESE of Weston-super-Mare

| ST(31)382588 | 165 | Bristol and 280 | 17NW (G) | C, P (1959) |

A Carboniferous Limestone solution cave on the northern flank of the Blackdown uplift. The succession in the hill shows a selvage of Clifton Down Limestone bordering the main northern slopes, which are of Burrington Oolite (Viséan); the southern slopes are in the underlying Black Rock Limestone (Tournaisian), which dips 40–60° to the north. The cave was discovered about 1800 by miners who broke into it while seeking mineral ores, galena (lead sulphide) and 'calamine' (zinc carbonate, now known as smithsonite). These ores are sometimes found in veins and fissure fillings in the limestone. Lead has been worked in the Mendips since Roman times; there was some production in A.D. 49. Zinc was extracted from Mendip calamine early in the eighteenth century, probably the first operation of this kind in Europe.

The cave chamber is about 30 feet across, and when found it was full nearly to the roof with sand, loose stones and a great quantity, estimated at 'several wagon loads', of bones of animals. The roof showed a choked fissure, presumably once a natural pitfall, through which the detritus and animal remains reached the cave.

Seventeen species of mammals are recorded by Wilfred Jackson in Cullingford (1953) as: Mammoth, Woolly Rhinoceros, Reindeer (abundant), Bison (abundant), Horse (rare), Cave Lion, Leopard, Lynx, Wild Cat, Hyaena, Cave Bear (but see footnote on p. 58 above), Brown Bear, Badger, Otter, Wolf, Fox and Glutton. These constitute a Late Pleistocene assemblage. All the bones are well preserved and never gnawed; the rarity of Horse is remarkable, since it is usually abundant in Mendip caves. Balch (1948) reported that quantities of the limb-bones still remained stacked round the walls.

The cave was explored by Beard early in the nineteenth century, and by later investigators, and it is now being re-worked. Specimens of the bones are reported in nine museums throughout the country and in private collections.

1829, Rutter; 1948, Balch; 1953, 1962, Cullingford; 1957, Barrington; 1965, Green and Welch.

Blue Anchor to Lilstock Coast, Watchet district

ST(31)1143	164	OS20	35SE, 36SW, SE	S, P, T
			(–)	IGC, PSD,
				1965

After an early description by Horner (1816) and commendation by De la Beche (1839) and Richardson (1911), this nine-mile stretch of coast was largely neglected in the pre-automobile age because of remoteness. It includes some of the best and most complete Rhaetic sections in the West Country, especially of the Sully Beds, and the thickest development of the Westbury Beds. The relation of the White Lias to the Lower Lias may be studied in many sections.

The general succession here shows (Lias from Woodward, 1893; Rhaetic and Trias from Richardson, 1911):

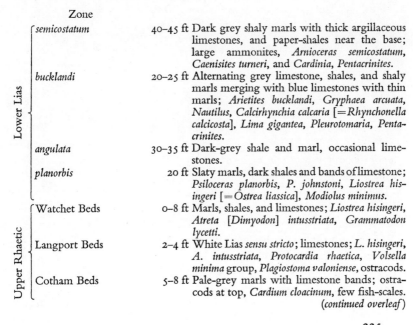

	Zone		
	semicostatum	40–45 ft	Dark grey shaly marls with thick argillaceous limestones, and paper-shales near the base; large ammonites, *Arnioceras semicostatum*, *Caenisites turneri*, and *Cardinia*, *Pentacrinites*.
Lower Lias	*bucklandi*	20–25 ft	Alternating grey limestone, shales, and shaly marls merging with blue limestones with thin marls; *Arietites bucklandi*, *Gryphaea arcuata*, *Nautilus*, *Calcirhynchia calcaria* [=*Rhynchonella calcicosta*], *Lima gigantea*, *Pleurotomaria*, *Pentacrinites*.
	angulata	30–35 ft	Dark-grey shale and marl, occasional limestones.
	planorbis	20 ft	Slaty marls, dark shales and bands of limestone; *Psiloceras planorbis*, *P. johnstoni*, *Liostrea hisingeri* [=*Ostrea liassica*], *Modiolus minimus*.
Upper Rhaetic	Watchet Beds	0–8 ft	Marls, shales, and limestones; *Liostrea hisingeri*, *Atreta* [*Dimyodon*] *intusstriata*, *Grammatodon lycetti*.
	Langport Beds	2–4 ft	White Lias *sensu stricto*; limestones; *L. hisingeri*, *A. intusstriata*, *Protocardia rhaetica*, *Volsella minima* group, *Plagiostoma valoniense*, ostracods.
	Cotham Beds	5–8 ft	Pale-grey marls with limestone bands; ostracods at top, *Cardium cloacinum*, few fish-scales.

(continued overleaf)

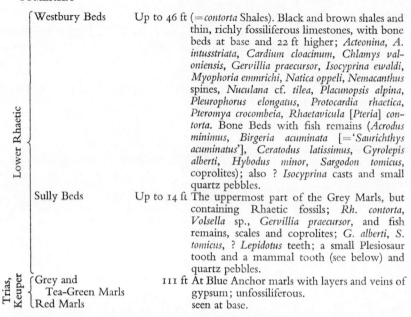

| | Westbury Beds | Up to 46 ft (=*contorta* Shales). Black and brown shales and thin, richly fossiliferous limestones, with bone beds at base and 22 ft higher; *Acteonina, A. intusstriata, Cardium cloacinum, Chlamys valoniensis, Gervillia praecursor, Isocyprina ewaldi, Myophoria emmrichi, Natica oppeli, Nemacanthus* spines, *Nuculana* cf. *tilea, Placunopsis alpina, Pleurophorus elongatus, Protocardia rhaetica, Pteromya crocombeia, Rhaetavicula* [*Pteria*] *contorta*. Bone Beds with fish remains (*Acrodus minimus, Birgeria acuminata* [='*Saurichthys acuminatus*'], *Ceratodus latissimus, Gyrolepis alberti, Hybodus minor, Sargodon tomicus,* coprolites); also ? *Isocyprina* casts and small quartz pebbles. |
| | Sully Beds | Up to 14 ft The uppermost part of the Grey Marls, but containing Rhaetic fossils; *Rh. contorta, Volsella* sp., *Gervillia praecursor,* and fish remains, scales and coprolites; *G. alberti, S. tomicus,* ? *Lepidotus* teeth; a small Plesiosaur tooth and a mammal tooth (see below) and quartz pebbles. |

Where the left bracket labels read: **Lower Rhaetic** (bracketing Westbury Beds and Sully Beds), and **Trias, Keuper** bracketing:

Grey and Tea-Green Marls   111 ft At Blue Anchor marls with layers and veins of gypsum; unfossiliferous.

Red Marls   seen at base.

The best sections, from west to east, are found at:

Blue Anchor Point	ST039437 Lower Lias basement overlying the Rhaetic and Keuper succession down to the Red Marls.
St Audries	ST105431 Lower Lias and Rhaetic, down to the Keuper Red Marls.
Kilve	ST144444 Lower Lias only.
Lilstock (Little Stoke)	ST176451 Lower Lias and Rhaetic.

For stratigraphy the Lilstock section is by far the most satisfactory, because the beds are undisturbed by faulting, landslips or talus. The Westbury Beds are particularly well displayed and easily accessible (Woodward, 1893; Richardson, 1911).

The St Audries section has been recorded by Bristow and Etheridge in Vertical Sections, Geological Survey, Sheet 47, No. 6, showing about 40 feet of strata.

The cliff and foreshore sections are generally clearly exposed, although in places so faulted as to be of more interest for the tectonics than the stratigraphy, and dips are often high.

The reason for the clear separation of both lithology and fauna of the subdivisions of the Rhaetic is that they are generally non-sequentially related (Richardson).

Below East Quantockshead are the finest Lias cliffs on this coast, although they are not very accessible.

On the east side of Blue Anchor Point, in the Sully Beds exposed on the intertidal reefs, and 10½ feet stratigraphically below the basal Bone Bed of the Westbury Beds, Boyd Dawkins found in 1861 a single mammalian tooth. It was associated with fish remains of *Acrodus*, *Gyrolepis* and *Sargodon*; *Chlamys valoniensis*, and fragments of wood. Dawkins described it as a new genus and species, *Hypsoprymnopsis rhaeticus*, the earliest known mammal, similar to the Kangaroo Rat. The crown of the tooth was apparently about four millimetres across. By 1894, however, this unique specimen could no longer be found. It was considered by Lydekker (1906, Victoria County History of Somerset) to be referable to the same genus as the Frome specimens (*Microlestes*) if, indeed, it was specifically distinct although it seems to have been larger (see below under Holwell Quarries).

The types of the ammonites *Psiloceras planorbis* and *P. johnstoni* from the Lower Lias were described by J. de C. Sowerby from specimens obtained near Watchet.

Old limekilns, mostly disused, are recorded hereabouts by Woodward (1893), the stone being sometimes burnt for building-lime; and certain of the beds yielded a hydraulic cement capable of setting under water. At Blue Anchor the gypsum used to be worked for the Watchet paper mills and for agriculture.

1816, Horner; 1818, Sowerby; 1839, De la Beche; 1864, Dawkins; 1893, Woodward; 1911, 1948, Richardson; 1969, Torrens (B).

Bowldish Quarry, 1½ miles NE of Radstock

ST(31)669558	166	Bristol	20SW (G)	P, S PSD, 1966

This disused quarry, No. 6 of Tutcher and Trueman (1925), and Kilmersdon Road Quarry (q.v.), No. 10, are taken as examples of the condensed and broken sequence of the Lower Lias found in the Radstock district.

At the time of deposition, on the northern flank of the east–west-trending Mendip anticline, conditions were unstable, for there was repeated movement about this axis intermittently from post-Carboniferous times until well into the Mesozoic. As a result, there is only a little Upper Lias known in the district, and no Middle Lias. The Lower Lias is attenuated, partly from paucity of sediment, and interrupted by non-sequences, brought about by deposition, uplift, erosion, re-sorting, and deposition again; a cycle several times repeated.

Deposition of the Lower Lias took place from a shallow—but not littoral— sea, teeming with marine life, so that from the abundantly fossiliferous beds of the Radstock district as a whole Tutcher and Trueman record some 445 species. The most numerous are ammonites (169 species), followed by bivalves (120), brachiopods (55), gastropods (50), belemnites (22), and a few forms of saurians, fishes, nautiloids, polyzoa, annelids, crinoids, echinoids and corals. Microfossils are not included in their list, but foraminifera and a few ostracods are found in the clays of the *raricostatum*, *turneri*, *bucklandi* and probably *angulata* Zones, although not as yet, so far as they have been sought, in the *striatum* and *planorbis* clays. Foraminifera were first recognised here a century ago by Charles Moore, and a few have been recorded by Macfadyen (1941) from the first three zones noted above, in this quarry and elsewhere hereabouts. They consist overwhelmingly of forms of the family Lagenidae.

The eight-foot thickness of Lower Lias found in this quarry may be compared with the succession deposited under more tranquil conditions seven miles to the north at Saltford Railway Cutting (q.v.), where 57 feet of strata are found up to the top of the *semicostatum* Zone only; and also with the more-than-300-foot thickness of the same zones in Dorset (Black Ven, q.v.) and across the border into Devon (Axmouth–Lyme Regis, q.v.) 50 miles to the south-west.

The following section is exposed at Bowldish Quarry (Tutcher and Trueman, 1925).

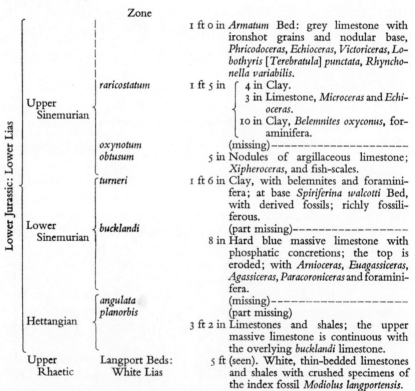

		Zone		
Lower Jurassic: Lower Lias	Upper Sinemurian	*raricostatum*	1 ft 0 in	*Armatum* Bed: grey limestone with ironshot grains and nodular base, *Phricodoceras, Echioceras, Victoriceras, Lobothyris* [*Terebratula*] *punctata, Rhynchonella variabilis.*
			1 ft 5 in	4 in Clay.
				3 in Limestone, *Microceras* and *Echioceras.*
				10 in Clay, *Belemnites oxyconus*, foraminifera.
		oxynotum		(missing)--------------------
		obtusum	5 in	Nodules of argillaceous limestone; *Xipheroceras*, and fish-scales.
	Lower Sinemurian	*turneri*	1 ft 6 in	Clay, with belemnites and foraminifera; at base *Spiriferina walcotti* Bed, with derived fossils; richly fossiliferous.
		bucklandi		(part missing)----------------
			8 in	Hard blue massive limestone with phosphatic concretions; the top is eroded; with *Arnioceras, Euagassiceras, Agassiceras, Paracoroniceras* and foraminifera.
	Hettangian	*angulata*		(missing)------------------
		planorbis		(part missing)
			3 ft 2 in	Limestones and shales; the upper massive limestone is continuous with the overlying *bucklandi* limestone.
	Upper Rhaetic	Langport Beds: White Lias	5 ft	(seen). White, thin-bedded limestones and shales with crushed specimens of the index fossil *Modiolus langportensis.*

1875, Tawney; 1925, Tutcher and Trueman; 1941, Macfadyen.

Brean Down and Uphill Cliff, 2 miles SW of Weston-super-Mare

ST(31)295588   165   OS20   16NW   A, P
                               (–)   PSD, 1965

Brean Down is a striking coastal rock trending east–west, about 2600 yards long by up to 470 yards wide, rising to a maximum height of 321 feet above O.D. Before the Somerset levels were embanked about the tenth century, it must have been an island rising from shallow estuaries, reed swamps or saltings. It is formed by a northerly-dipping (and sloping) mass of Carboniferous Limestone faulted along its southern margin; the C Zone constitutes the main, northern, part, with a smaller outcrop of the Z Zone on the south. The beds contain large corals (*Caninia*) and brachiopods. The southern face is a cliff of marine erosion, the final stage formed when the sea level was lower than it is today, and represented by the Howe Rock Platform Stage.

Banked against the eastern one-third of the southern face up to about 120 feet above O.D. is found an unusually complete archaeological succession, traces of which were first noted by Knight in 1902. A recent paper (ApSimon, Donovan and Taylor, 1961) shows the main section (perhaps the finest of its kind) to provide a most important reference standard, with information about the climatic and environmental conditions of Late Glacial and Post-Glacial times. On top of the down are the remains of a fourth century Roman temple and, close by, of an early fifth century small building and shell midden. On the west and east summits are Round Barrows, together with old fields possibly of mediaeval date or even later; and there is an ancient camp on the eastern ridge.

The general succession at Brean Down has been pieced together and dated as follows (after ApSimon, Donovan and Taylor, 1961).

Layer	Period and date B.C.		
1	Present day		Growing turf and humus.
2	Recent		Modern pebble beach and blown sand: storm beach with limestone shingle; in places 1 ft of marine grey clay is intercalated.
3	Recent	c. 2 ft	Grey stony sand: sometimes reddish-brown, much burrowed by rabbits; with cemetery burials, perhaps early Christian of 5th–10th century; remains of building dated to late 17th century, with fragments of window glass, pottery, nails, etc.

(continued overleaf)

Holocene Post-Glacial	4	Iron Age 'A': Sub-Atlantic, c. 300	$\frac{3}{4}$–1 ft	Iron Age Sand: reddish, loamy, stony; Iron Age 'A' potsherds; bones of Ox, Sheep, Pig, Fish; traces of ruined structures.
	5	Bronze Age: Sub-Boreal, 450–950	To 3 ft	Blown Sand: yellow, with two loamy bands, and thin iron pan at base.
	6	Middle Bronze Age: Sub-Boreal, 450–950	c. 2 ft	Bronze Age Sand: grey-green stony clay and reddish loamy sand, with hearths; Middle Bronze Age and 'A' Beaker potsherds, charcoal, bones of Horse, Sheep, Wild Boar, Ox, Goat, and limpet shells; large blocks of limestone may be ruined structures.
	7	Bronze Age: Sub-Boreal, c. 1800	To 6 ft	Beaker Sand: yellow blown sand with 'A' Beaker potsherds, bones of Sheep, jaw of a Pike, molluscan shells.
	8A	Atlantic to Boreal, c. 5000	to $1\frac{1}{2}$ ft	Red Loam: a mature soil, in which is a Bronze Age burial; rare 'B' Beaker potsherds, also a flint scraper, bone fragments of Ox, etc., charcoal, limpet shells. Climatic optimum.
	8B, C	Pre-Boreal to Upper Dryas, c. 8000	To 20 ft	Upper Breccia: angular limestone fragments in reddish matrix, earthy above, sandy below; unfossiliferous. Climate ameliorating.
Pleistocene: Würm Glaciation	9	Upper Dryas, c. 8300	50–60 ft	Main Sand: orange to brown blown sand, with abundant comminuted marine shell fragments. Frost-weathering.
	10A	Upper Dryas, c. 8600	3 in	Breccia Band: discontinuous, of limestone fragments. Frost-weathering.
	10B, C	Alleröd Interstadial, c. 8900–10 000	2–3 ft	Silty Sand: reddish brown and laminated. Climatic optimum.
	11A, B	Alleröd–Lower Dryas, c. 10 000	To $3\frac{1}{2}$ ft	Middle Breccia and Bone Bed: reddish limestone breccia and earthy sand; Bone Bed at top has yielded worked bone, the earliest evidence of human activity at Brean; also Reindeer and Horse (predominant) and Ox or Bison, Mammoth, Giant Deer, Wolf, Arctic Fox, Lemmings (2 species), Arctic Hare, birds (several species, including Owl);

Pleistocene: Würm Glaciation	12	Lower Dryas c. 10 000–15 000	c. 12 ft	Stony Silt: clayey, red-stained in the middle, and laminated at the top; has angular limestone fragments, and an intercalation of barren stiff clay; with Reindeer antlers, Lemming, Vole (*Microtus agrestis*), fragments of large bones, Ox or Bison, etc., birds (including Owl and Duck). Climate ameliorating.

also 5 species of non-marine mollusca, of which 92% are *Pupilla muscorum* and 6 species of marine mollusca and *Balanus*, mostly as fragments. Climate ameliorating.

12 Lower Dryas c. 10 000–15 000 — c. 12 ft — Stony Silt: clayey, red-stained in the middle, and laminated at the top; has angular limestone fragments, and an intercalation of barren stiff clay; with Reindeer antlers, Lemming, Vole (*Microtus agrestis*), fragments of large bones, Ox or Bison, etc., birds (including Owl and Duck). Climate ameliorating.

13A, B, C Upper Würm c. 15 000–20 000 — c. 5 ft — Lower Breccia: top $\frac{1}{2}$–1 ft and base stained red; a coarse limestone breccia not now exposed in this section; farther west it rises above beach level, with Ox or Bison, Reindeer, Arctic Fox, Vole (*Microtus anglicus*) and small fish-bones. Frost-weathering.

13D Upper Würm, c. 22 000 — Boulder pile: limestone blocks up to 3 ft long, without cement, attributed to intense frost action in an extreme periglacial climate.

Paudorf Interstadial, c. 25 000 — Howe Rock Platform: wave-cut; ?–20 ft below O.D. on seaward side, rising to Ordnance Datum at its inner limit.

Other strand-lines are visible at Brean at 40–47 ft, 70 ft and 120–140 ft above O.D.; with traces of higher and still older strands.

Five other small sites in caves or along the south of the cliff have been excavated and have yielded information; some of them had fossil material.

Nearly a mile to the south-east, Uphill Cliff is formed of Carboniferous Limestone in two bands similar to Brean Down, Zone C over Z, also dipping north but here reduplicated by a strike fault.

In a large quarry facing west, and almost below Uphill Church, are three small caves, 30 feet above the sea, discovered during quarrying, the first in 1826. That cave yielded a Pleistocene and later fauna, including palaeolithic (Proto-Solutrean) flint artifacts and some 23 mammal species, including Mammoth, Woolly Rhinoceros, Cave Lion, Cave Bear,* Hyaena, Urus,

* But see footnote on p. 58.

Bison, Horse, Reindeer etc. and bird remains.

Another cavity found about 1863 yielded human skulls.

1791, Collinson; 1829, Rutter; 1924, Bamber; 1934, Palmer; 1948, Balch; 1961, ApSimon, Donovan and Taylor; 1969, Savage.

Burrington Combe and Caves, 11 miles SW of Bristol

ST(31)478585	165	Bristol and 280	18NW (G)	S, P, C, Ph, B IGC, PSD, 1966

Burrington Combe, one of the major dry valleys of the Mendips, is a deep ravine cut for a mile and a quarter in the northern flank of the Blackdown uplift. Between ST477591 and 476580 it exposes a magnificent section, the most complete and readily accessible of the Carboniferous Limestone Series in the Mendips. The steep dip (48–65° north-north-east) exposes the sequence within some 1300 yards. It was first described in detail by Lloyd Morgan in 1890.

At the northern end the Dolomitic Conglomerate, exposed astride and then west of the road, conceals the uppermost part and rests on the lower beds of the Hotwells Limestone, Zone D_1. For half a mile along road B3134 in the Combe the western side and crags mostly consist of the Dolomitic Conglomerate (Fig. 26). On the eastern side, however, the Carboniferous succession is exposed down to the Black Rock Limestone, Zone Z. Where the road turns sharply east, the section is continued straight up the Western Twin Stream, where the lowest part of the Black Rock Limestone is seen. South of Goatchurch Cave the Lower Limestone Shale, Zone K, is poorly exposed for 250 yards, up to its junction with the Old Red Sandstone. These last two formations have been studied in an adit here by Hepworth and Stride (1950). After turning to the east the main road continues uphill along a fine strike section in the Black Rock Limestone, in this part of the gorge ascending a pre-Triassic valley.

The section is given as follows (after Green and Welch, 1965).

Triassic Keuper	Dolomitic Conglomerate	Red conglomerate, mainly of dolomitised Carboniferous Limestone fragments ranging from large boulders downwards, cemented in a matrix of sandy marl and fine limestone debris.

232

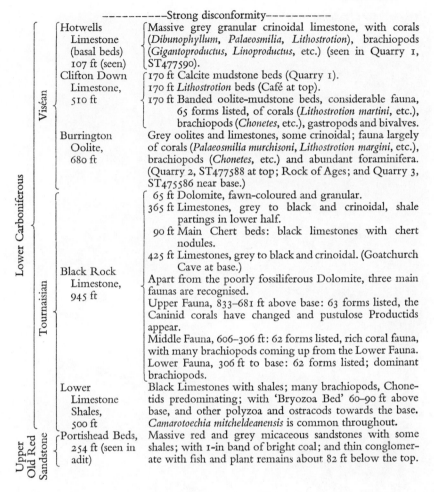

----------Strong disconformity----------

Lower Carboniferous	Viséan	Hotwells Limestone (basal beds) 107 ft (seen)	Massive grey granular crinoidal limestone, with corals (*Dibunophyllum, Palaeosmilia, Lithostrotion*), brachiopods (*Gigantoproductus, Linoproductus*, etc.) (seen in Quarry 1, ST477590).
		Clifton Down Limestone, 510 ft	170 ft Calcite mudstone beds (Quarry 1). 170 ft *Lithostrotion* beds (Café at top). 170 ft Banded oolite-mudstone beds, considerable fauna, 65 forms listed, of corals (*Lithostrotion martini*, etc.), brachiopods (*Chonetes*, etc.), gastropods and bivalves.
		Burrington Oolite, 680 ft	Grey oolites and limestones, some crinoidal; fauna largely of corals (*Palaeosmilia murchisoni, Lithostrotion margini*, etc.), brachiopods (*Chonetes*, etc.) and abundant foraminifera. (Quarry 2, ST477588 at top; Rock of Ages; and Quarry 3, ST475586 near base.)
	Tournaisian	Black Rock Limestone, 945 ft	65 ft Dolomite, fawn-coloured and granular. 365 ft Limestones, grey to black and crinoidal, shale partings in lower half. 90 ft Main Chert beds: black limestones with chert nodules. 425 ft Limestones, grey to black and crinoidal. (Goatchurch Cave at base.) Apart from the poorly fossiliferous Dolomite, three main faunas are recognised. Upper Fauna, 833–681 ft above base: 63 forms listed, the Caninid corals have changed and pustulose Productids appear. Middle Fauna, 606–306 ft: 62 forms listed, rich coral fauna, with many brachiopods coming up from the Lower Fauna. Lower Fauna, 306 ft to base: 62 forms listed; dominant brachiopods.
		Lower Limestone Shales, 500 ft	Black Limestones with shales; many brachiopods, Chonetids predominating; with 'Bryozoa Bed' 60–90 ft above base, and other polyzoa and ostracods towards the base. *Camarotoechia mitcheldeanensis* is common throughout.
Upper Old Red Sandstone		Portishead Beds, 254 ft (seen in adit)	Massive red and grey micaceous sandstones with some shales; with 1-in band of bright coal; and thin conglomerate with fish and plant remains about 82 ft below the top.

Cummings (1958) has published some of his work on the foraminifera and records seven genera of the family Palaeotextulariidae, finding the greatest variety of these in the D_1 zone, Hotwells Limestone.

In the now dry Burrington Combe Dolomitic Conglomerate is found along the whole of the western side from the mouth of the Western Twin Stream to the right-angled elbow at ST477591 and beyond to the east, showing that the Combe was a narrow valley in early Trias times, and nearly as deep as at present, 300–400 feet. Strong river erosion under periglacial conditions during the Pleistocene has removed much of the Triassic deposits from the old drainage line. Collapse of the roofs of caverns (which probably also date back to the Trias) in the Carboniferous Limestone may have played a small part.

Fig. 26. Burrington Combe, Somerset; on road B3134, looking south (1957). Valley in Carboniferous Limestone previously infilled with Triassic Dolomitic Conglomerate; crags of the latter form the skyline and underlie the bracken-covered hillside. The figure on the road is looking towards the cleft of the 'Rock of Ages' in a cliff of Burrington Oolite. (Crown Copyright Geological Survey photograph A.9154. Reproduced by permission of The Controller, H.M. Stationery Office.)

CAVES AND POTHOLES

Some 18 caves or pot-holes are known in Burrington Combe, and the first four noted below were excavated by Boyd Dawkins in 1864.

Aveline's Hole, ST476587, discovered in 1794, has two chambers, one of them 78 feet long. In 1820 about 50 human skeletons were found in it, laid side by side with their weapons, a stalagmite crust sealing bones and weapons to the floor. These fossils disappeared before they were adequately studied. Three more human crania were found in 1914, and later other human remains and implements referred to the local latest Palaeolithic (Cheddarian). A fine harpoon or fish spear was identified as Magdalenian VI in type. A total of 116 species of animal remains have been found, and two if not three faunas are represented. They include three cold forms, Reindeer, Pika and Lemmings; and *Bos longifrons*, Boar, Bear, Lynx, Wolf, Badger, Fox, Field Mouse, Bats, and domestic animals, Dog and Sheep; 49 species of birds and 31 of non-marine mollusca (Savage, 1969).

Plumley's Den (Foxes Hole), ST483583, the alleged hiding place of John Plumley in 1685. The Lord of Locking Manor he had joined the Monmouth rebellion. Escaping after Sedgemoor he went into hiding, but was later caught and hanged (Glennie, 1957). This cave contained bones of mammals and birds, including the extinct Pleistocene mammals. During World War II it was occupied by the Home Guard.

Whitcomb's Hole, ST477583, probably discovered before 1848, is a small shelter; it contained some pottery and iron objects attributed to the Early Iron Age (Balch, 1948).

Goatchurch Cave or 'The Goatchurch', ST476582, has long been known. Its entrance is about 40 feet above the floor of the Western Twin Stream valley, and the system consists of a series of galleries developed mainly along the strike. The Late Pleistocene fauna comprises Cave Bear (but see footnote on p. 58), Brown Bear, Wild Cat, Pig, Horse, Badger, and some unidentified rodents. In another part of the cave were found a Mammoth tusk, and remains of Hyaena and Cave Lion (Tratman, 1963b). The living fauna has been studied, and two species of bats and at least 50 forms of invertebrates have been collected and identified. The commonest are three species of Diplopoda (Millipedes) *Brachydesmus superus*, *Polymiorodon polydesmoides*, and *Blaniulus guttulatus*, the same as the common species in Great Oone's Cave, Cheddar; and also the beetle *Quedius mesomelinus*.

Plumley's Hole (not to be confused with Plumley's Den) lies east of the main road at ST476588. Joseph Plumley died here in an accident in 1874. This hole is now blocked and inaccessible under a concrete slab in the north-east corner of a quarry used as a car park (Glennie, 1957).

Sidcot Swallet, at ST476582, near Goatchurch (of which it may be considered a deeper part), has its entrance a few feet above the valley floor and was discovered in 1925. It has yielded at least 29 named species of living

235

invertebrates, of which 15 have also been recorded from Goatchurch (Balch, 1948).

Rock of Ages. On the west side of the road, at ST476588, is a plaque by a fissure in a great rib of limestone (Burrington Oolite) in which the Anglican divine Augustus Montague Toplady is traditionally believed to have been inspired to write his hymn 'Rock of Ages' while sheltering from a storm about 1762.

1865, Dawkins; 1890, Morgan; 1907, Sibly; 1911, Reynolds and Vaughan; 1921a, Reynolds; 1931, Palmer; 1933, Welch; 1948, Balch; 1957, Glennie; 1957, Barrington; 1956–9, Cave Research Group; 1958, Cummings; 1963b, Tratman; 1965, Green and Welch; 1969, ApSimon; 1969, Savage; 1969, Donovan.

Cheddar Gorge and Caves, 14 miles SW of Bristol

ST(31)470542	165	Bristol and 280	18SW, 27NW (G)	A, B, C, P, Ph, S, IGC, 1967

Cheddar Gorge is cut for one and a half miles in the southern limb of the Blackdown uplift in the the Mendip Hills and provides a general strike section in the Viséan portion of the Carboniferous Limestone Series. It lies almost wholly within the Clifton Down Limestone, although a little of the Hotwells Limestone, D_1 Zone, is exposed in the cliffs south of the main road B3135, which runs through the gorge, in the area of Gough's Cave and farther east. Burrington Oolite is exposed only in the district north of Black Rock Gate (ST482545).

The succession (after Green and Welch, 1965) shows:

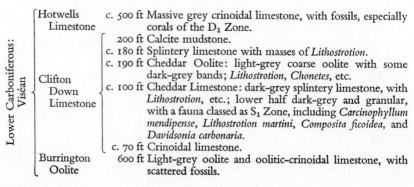

Lower Carboniferous: Viséan	Hotwells Limestone	c. 500 ft	Massive grey crinoidal limestone, with fossils, especially corals of the D_1 Zone.
	Clifton Down Limestone	200 ft	Calcite mudstone.
		c. 180 ft	Splintery limestone with masses of *Lithostrotion*.
		c. 190 ft	Cheddar Oolite: light-grey coarse oolite with some dark-grey bands; *Lithostrotion, Chonetes*, etc.
		c. 100 ft	Cheddar Limestone: dark-grey splintery limestone, with *Lithostrotion*, etc.; lower half dark-grey and granular, with a fauna classed as S_1 Zone, including *Carcinophyllum mendipense, Lithostrotion martini, Composita ficoidea*, and *Davidsonia carbonaria*.
		c. 70 ft	Crinoidal limestone.
	Burrington Oolite	600 ft	Light-grey oolite and oolitic-crinoidal limestone, with scattered fossils.

The S_1 Zone is claimed to have a localised fauna of corals and brachiopods present also at Weston-super-Mare but lacking at Burrington.

Cheddar has perhaps the most magnificent limestone gorge in the British Isles. Its southern side has a vertical scarp face up to nearly 500 feet high, and the northern side is conditioned by the dip slopes of around 20° south. The sheer cliffs are due to strong vertical joints in the limestone. Although there have been differences of opinion as to its origin, it has probably been formed mainly by normal stream erosion during the Pleistocene. Youthful characters are shown in its narrowness, steep sides with cliffs, scree slopes and steep uneven valley floor gradients. That the gorge was complete by late Pleistocene times is proved by the caves, such as Gough's Cave with a Late Pleistocene fauna, in the valley bottom.

In the periglacial conditions that then obtained here the old underground drainage system of solution caverns and passages must have been out of action owing to freezing of the groundwater or other blocking of its flow, so that the limestone behaved as an impervious rock. In any case, the underground drainage system would have been incapable of taking the enormous flow from the relatively large catchment of about 12 square miles. There may have been minor effects caused by roof collapse of existing caverns as the river erosion proceeded, but from the quantitative angle this cannot have been the significant factor it has sometimes been claimed to be. After the general melting of the ice, the old pot-holes and solution channels again became open and competent to take the lessened flow, leaving the gorge and upper reaches dry as they are today. However, quite exceptional rainstorms in August 1930 and July 1968 gave rise to surface floods which rushed down the gorge.

It is impossible to define an exact surface catchment in these Mendip karst conditions, and the Cheddar water now travels from many swallets in underground channels probably largely below the four dry valleys which formed the headwaters of the Cheddar River in its Pleistocene prime. Patches of Dolomitic Conglomerate on the sides of Velvet Bottom, Charterhouse, show that the Cheddar drainage line was established in pre-Triassic times.

The water now resurges in Cheddar village at ST413538 as the head of the River Yeo, with a discharge estimated to range from three million as a minimum to more than 78 million gallons per day once measured, with a normal flow of about 15 million gallons per day. Collinson (1791) derived the name 'Cheddar' from this impressive spring issuing below towering cliffs, *ced* signifying a brow or conspicuous height, and *dwr* water.

Cheddar Gorge has been known for its score of caves, which are relict solution channels, since the late Palaeolithic, when they were occupied at times by both men and animals; later some of them became refuges in the Romano-British period. An early account of 'Cheder Hole' was given by Henry of Huntingdon in his *Historia Anglorum*, A.D. 1125-30.

237

The Cheddar caves have yielded human, animal and bird remains, and artifacts dated as Upper Palaeolithic. The earliest culture has been termed 'Proto-Solutrean' (at Soldier's Hole) and there is an extensive later industry identified as Cheddarian. This includes remains of 'Cheddar Man', found at Gough's Cave, Soldier's Hole and Sun Hole. From the lower layers, but not in Gough's Cave, remains of the Late Pleistocene mammals have been obtained, including Mammoth, Rhinoceros, Bison, Cave Lion and Hyaena, which all disappeared from the area about the incoming of the Cheddarian. The remainder of the mammal fauna persisted, and included Reindeer, Irish Great Deer, Red Deer, Roe Deer, Horse, Brown Bear, Arctic Fox, Common Fox, Variable Hare, Boar, Wildcat, Voles, Lemmings (*Dicrostonyx* and *Lemmus*), other small mammals and birds, including Ptarmigan, and mollusca. The Cave Pika is confined to the Cheddarian levels.

Material has also been collected of the Neolithic, Bronze Age, early Iron Age, and Romano-British periods, including pottery and Roman coins. Some of all this is preserved in a small museum on the site.

The front part of *Gough's Cave* (ST467539) was commercialised by R. C. Gough, who lighted it by gas about 1880. He started exploratory digging in 1893 and broke into the main caves in 1898. The show part is now 800 feet in length, floodlit by electricity first introduced about 1904 to illuminate the beautiful stalactite formations. The skull and skeleton of 'Cheddar Man' were found in Gough's Cave in 1903, and after later excavations, which exposed the succession and yielded quantities of flint artifacts, remains of animals and a large piece of Baltic amber, this cave has been claimed as probably the most important Upper Palaeolithic station in Britain.

There are colonies of Bats in *Gough's Old Cave* (ST467539).

Cox's Cave (ST465539), discovered in 1837, is also floodlit to show this small but exquisite stalactite cavern, described as a gem of fantastic architecture. Conybeare wrote to Dean Buckland in 1843 '. . . It is really the only graceful cave fit for ladies to visit which we have.'

Roman Cave, which is practically a part of Gough's Cave, yielded bones, a quantity of Roman pottery, and Roman coins of gold, silver and bronze dating from the Emperor Nero, A.D. 64, to Gratian, A.D. 367–83; and a living fauna of Great Horseshoe Bats, many moths and large spiders.

The living fauna of *Great Oone's Cave* (ST468540), high up in the cliff and 442 feet long, has been studied and at least 17 forms of invertebrates have been collected and identified. The commonest are three species of Diplopoda (Millipedes) the same as those found in the Goatchurch Cave of Burrington Combe (q.v.).

Sun Hole. Tratman (1963a) records a total of some 41 species from the various levels excavated down to eight feet deep. Most of the deposits, dated by the Cheddarian flint industry, must have been accumulated during the Alleröd Interstadial and the Upper Dryas, the latest Pleistocene, between about 10 000 and 8400 B.C. His list includes the remains of Man, Mole,

Varying Hare, Cave Pika, Lemmings (2 or 3 species), Mouse, Voles (5 species), Beaver, Horse, Reindeer, Boar, Wolf, Weasel, Brown Bear, Wild Cat; Birds (8 species); Snails (12 species) and Slug. It was intermittently occupied by Man from Neolithic to Roman times (T. R. Shaw, 1969: *Proc. Univ. Bristol Spelaeol. Soc.*, **12**, 18).

1681, Beaumont; 1791, Collinson; 1811, de Luc; 1921a, Reynolds; 1931, Palmer; 1947b, Balch; 1953, 1962, Cullingford, Ed.; 1955, Donovan; 1957, Barrington; 1956-9, Cave Research Group; 1963a, Tratman; 1965, Green and Welch; 1968, Ford and Stanton.

Dundry Main Road South Quarry, 5 miles S of Bristol

ST(31)567655	155	Bristol	12NW	P, S
			(–)	PSD, 1965

A small disused quarry on the east side of the Dundry–Chew Stoke road, one of the remaining exposures of the abundantly fossiliferous Inferior Oolite, present as a thick cap on Dundry Hill, overlying Lias. The quarry ceased work about 1928, but a pit in the floor was dug by geologists of the Bristol Naturalists' Society in 1955. Unfortunately, this has now become filled in and only the Upper Inferior Oolite is still well exposed in the walls of the old quarry.

The succession originally showed (after Buckman and Wilson, 1896):

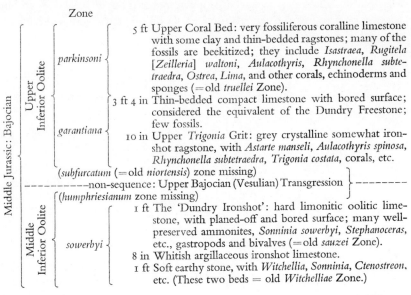

Zone

Middle Jurassic: Bajocian

Upper Inferior Oolite

parkinsoni — 5 ft Upper Coral Bed: very fossiliferous coralline limestone with some clay and thin-bedded ragstones; many of the fossils are beekitized; they include *Isastraea, Rugitela [Zeilleria] waltoni, Aulacothyris, Rhynchonella subtetraedra, Ostrea, Lima,* and other corals, echinoderms and sponges (=old *truellei* Zone).

garantiana — 3 ft 4 in Thin-bedded compact limestone with bored surface; considered the equivalent of the Dundry Freestone; few fossils.

10 in Upper *Trigonia* Grit: grey crystalline somewhat ironshot ragstone, with *Astarte manseli, Aulacothyris spinosa, Rhynchonella subtetraedra, Trigonia costata,* corals, etc.

(*subfurcatum* (=old *niortensis*) zone missing)

——————non-sequence: Upper Bajocian (Vesulian) Transgression

(*humphriesianum* zone missing)

Middle Inferior Oolite

sowerbyi — 1 ft The 'Dundry Ironshot': hard limonitic oolitic limestone, with planed-off and bored surface; many well-preserved ammonites, *Sonninia sowerbyi, Stephanoceras,* etc., gastropods and bivalves (=old *sauzei* Zone).

8 in Whitish argillaceous ironshot limestone.

1 ft Soft earthy stone, with *Witchellia, Sonninia, Ctenostreon,* etc. (These two beds = old *Witchelliae* Zone.)

Dundry Hill is a classic site in the annals of the Inferior Oolite, because it is the type locality of many species of ammonites, gastropods, bivalves and brachiopods described particularly by Sowerby, Tawney, Hudleston, Whidborne and Buckman; many of them came from the present quarry. These, together with belemnites, echinoids, corals, sponges and other fossils abound, and some foraminifera have been recorded. Stoddart (1879) lists 256 species from Dundry.

Dundry has another special interest. Whereas the Upper Inferior Oolite is of the Cotswold type in both lithology and fauna, the Middle Inferior Oolite, judging from the brachiopods and ammonites, is of the Dorset–East Somerset, or Anglo-Norman, type and differs from that of the nearby Cotswolds, although the beds are strictly contemporaneous. At that time Dundry must have had good connection with the Dorset sea, but little or none with that of the Cotswolds.

The famous Dundry Freestone attained a thickness of 27 feet very locally, at the western end of the outlier, near the church. The quarries in it are stated by Reynolds (1921a) to have been worked since Roman times, and they are certainly mentioned in the Pipe Rolls of Edward I and Edward II as yielding revenue in 1293, 1303 and 1310. Large-scale working ceased about 1914. Dundry church, and St Mary Redcliffe and other Bristol churches, are built of it. It was analysed as containing 94·5 per cent calcium carbonate, and had a crushing strength of more than 2000 pounds per square inch (Stoddart).

Dundry village lies 500 feet above the low ground to the north on the edge of a spectacular scarp. This feature formed part of a north-facing fortified boundary, the Wansdyke, which, at least partly of Romano–British date, is said to run from east of Marlborough in Wiltshire for some 60 miles, with Maes Knoll, Dundry, as the terminal strong point at the western end, within 10 miles of the Bristol Channel. At Maes Knoll it partly overlies and incorporates an Iron Age fort.

1822, Conybeare and Phillips; 1879, Stoddart; 1896, 1897, Buckman and Wilson; 1921a, Reynolds; 1948, Kellaway and Welch.

Ebbor Gorge and Wookey Hole, 2 miles NW of Wells

ST(31) 527487	165	Bristol and 280	27SE (G)	A, C, P, Ph, S 1967

As a picturesque gorge Ebbor is second only to Cheddar, from which it differs in being beautifully wooded and wild, with no road through the dry valley. For nearly a mile there is a winding passage of the narrow defile, the rocks on each side rising from 50 to 400 feet, many of them nearly perpendicular. The gorge with 116 acres of surrounding woodland and pasture is being presented to the National Trust in memory of Sir Winston Churchill.

On the southern flank of the North Hill uplift in the Mendips Ebbor Gorge is eroded chiefly in Carboniferous Limestone, mainly the Clifton Down Limestone, Viséan, which includes splintery limestone and oolite, with silicified *Lithostrotion* in places; the dip is around 30° south-south-west.

At Ebbor Wood (ST524485) is found the Ebbor Fault, trending north-west–south-east, with a maximum downthrow of about 1000 feet to the south-west. To the south is exposed the Hotwells Limestone, Viséan, followed by Upper Carboniferous rocks—the Quartzitic Sandstone Group (Millstone Grit), which forms crags in the gorge near the Ebbor Rocks; and the Lower Coal Series, mainly shale, with the Ashton Vale Marine Band containing the goniatite *Gastrioceras*, defining its base. Some 250 yards south-west of the Ebbor Fault lies the parallel Ebbor Thrust, whose greatest development is found at Ebbor Rocks, where Black Rock Limestone, Tournaisian, is thrust against Lower Coal Series giving a stratigraphical downthrow of some 2000 feet. The pressure came from the south, and the adjacent rocks are shattered and recrystallised. Farther south again is badly exposed Burrington Oolite, Viséan, and Triassic Dolomitic Conglomerate.

East of the Ebbor Gorge lies Wookey Hole (ST532480) in the midst of a large mass of the reddish Dolomitic Conglomerate 300–500 feet thick, infilling an early Triassic valley in the Carboniferous Limestone.

The Mendips are exceptional among ancient hill ranges in their anticlinal structure. The folding, and the above-mentioned fault and thrust, are all of Armorican (Permo-Carboniferous) date. The topography is of quite extraordinary antiquity. During Permian and early Triassic times tremendous erosion removed a thickness of some 9000 feet of Carboniferous strata, and valleys were cut in the exposed Carboniferous Limestone. These were later filled with the Dolomitic Conglomerate, and the area was then covered by marine Mesozoic strata. Erosion has removed much of the Mesozoic rocks and has now reached the Carboniferous Limestone again and the remnants of the Dolomitic Conglomerate. This last is being removed by present-day erosion to reveal once more the landscape of 200 million years ago.

The ancient Wookey ravine, still filled with the Dolomitic Conglomerate, can be traced from Wookey Hole northwards to Higher Pitts Farm (ST535491). The winding Ebbor Valley lies some 700 yards to the west and has a short tributary running to Higher Pitts Farm. Both valleys have at different times formed part of the headwaters of the River Axe. The Wookey Valley was cut first, in pre-Trias or early Trias times, and in the Keuper period was filled with the Dolomitic Conglomerate.

When the present topography was re-exposed during the Pleistocene, periglacial conditions obtained here and the groundwater must have been frozen, so that the underground channels already existing in the conglomerate were out of action. Then, apparently, instead of the Wookey Valley being re-excavated, the Ebbor Gorge was cut in the limestone. The Ebbor screes, the largest in Mendip, may reflect the frigid conditions. On the general melting of the ice at the end of the Ice Age, the pot-holes and Wookey solution channels became open and active, and the water again flowed through them, leaving the Ebbor valley dry.

The headwaters of the River Axe now continue to flow below ground from pot-holes to the north-north-east, including St Cuthbert's and East-water Swallets, both two miles distant. The average flow at the cave mouth is given as about 23 million gallons per day, although after prolonged rain this figure may be doubled.

The history of the River Axe headwaters is recorded in the relics of the five levels of the Wookey channels which have been in use at different times, the highest being the oldest. Cross-sections of them are seen in the face of the nearly 200-foot-high cliff. The *Badgers Hole* (see below) occupies a channel at the first (top) level. The second is not many feet lower. The third is the entrance now used by visitors to the main Wookey Cave and is about one-third of the cliff's height above the base. The fourth is represented by a blocked entrance and the fifth is the present outlet of the River Axe at the base of the cliff. The river formed part of the frontier between the British and Saxon kingdoms from A.D. 577 to 658.

Wookey Hole is one of the finest Mendip cave systems, but differs in being mainly dissolved out of the Dolomitic Conglomerate instead of the Carboni-ferous Limestone. Access is gained by a path along a rocky terrace on the left, which leads to the cavern's mouth. The entrance is narrow but opens into a spacious chamber 80 feet high 'the whole roof and sides of which are in-crusted with sparry concretions of whimsical forms' (Collinson, 1791).

The cave has been commercialised and the interior floodlit by the private owner since 1927, and the electric lights allow the growth of green plants close to them within the 2000 feet of the show cave's length. The walls of the lofty chambers are mostly coloured red, and contrast beautifully with the many fine pearly stalactite and stalagmite formations. From the west side of the 75-feet-high second chamber some of the fine roof stalactites were removed before 1744 to decorate Alexander Pope's artificial grotto at Twickenham, near London.

Beaumont (1681) found eels and trout in the river within the cave; frogs and 'other little animals' in the cave and its waters; and a multitude of bats on the roof.

The main show portion comprises three large chambers; the fourth and fifth are not seen by the general public. Behind these lie a further series of chambers that are permanently flooded and only accessible by diving. In 1966 the sixteenth and seventeenth were reached for the first time by a dive of 480 feet from the ninth (*The Times*, 2 May 1966). These flooded chambers were formed at a time when Mendip stood higher in relation to the sea level.

Wookey Hole is claimed to be the only cavern in Britain in which curious loud noises are occasionally heard. They have been likened to a great pot boiling, to galloping horses, and to loud hammering. Once there was heard a confused murmur growing to a thunderous roar which abruptly ceased (Balch, 1947). Some of the sounds are heard only when the river in the cave is running strongly and the flow is rapidly increasing. These are thought to be caused by the archway between the third and fourth chambers becoming submerged, and the air imprisoned within the cave being expelled through the water in great bubbles. The sounds thus produced are magnified by the resonance of the chambers and can be quite terrifying.

Noises made by comparable movements of air and water may be simply studied in one's bath. Take a bath in not much more than four inches of water, rest the shoulders above water level against the sloping end, with the arms lying loosely at your sides to form a shallow 'box', of which your back makes the top. This air-water bellows is then operated by alternately raising and lowering the small of the back. With a little practice surprisingly loud roars can be achieved, reminiscent also of waves entering a sea-cave.

In the Ebbor and Wookey areas are known nine other small caves and rock shelters, mostly disused sections of old underground water channels, the mouths having been inhabited by wild beasts and sometimes by Man. These caves were discovered by Palaeolithic Man and have never since been lost.

In the Wookey area, of *Wookey Hole* itself the earliest proved occupation is dated to the Iron Age. The first written mention, based on still earlier accounts, is thought to be by Titus Flavius Clemens (Clement of Alexandria) in the second century, this particular cave being identified by the comment that strange noises were sometimes heard in it. 'Wokyhole' is mentioned in charters of 1065 and 1294; it was described by William of Worcester in 1478, and thereafter figures in the literature to the present day.

The *Hyaena Den* was discovered accidentally in 1852, 100 yards south of the entrance to Wookey Hole. The entrance is 36 feet wide, and the cave has a maximum height of nine feet. It was filled with cave earth containing a bone bed, and was excavated by Boyd Dawkins from 1859 to 1874. The first of the Mendip caves to yield traces of Palaeolithic Man, now dated as Proto-Solutrean, the Hyaena Den was among the first in this country to prove Man's co-existence with the extinct Pleistocene mammals. Although Hyaenas were the normal inhabitants, some 26 species of mammals have been recorded from it, including Mammoth, Rhinoceros, Horse, Bears, Deer and Cave Lion, some of them the remains of Hyaenas' meals. Apart from abundant Hyaena remains there was a great preponderance of Horse and Rhinoceros. Man, as a Palaeolithic hunter, was an occasional visitor who left implements of flint, chert, bone and antler, and the ashes of his fires near the mouth of the cave.

The *Badger Hole*, in the highest of the five old river levels, was excavated by Balch from 1938 onwards, and yielded nearly the same fauna as the Hyaena Den except for Cave Lion. Birds and many non-marine mollusca were also found here, and the bones and artifacts of Palaeolithic Man.

The Ebbor caves and shelters have also yielded much, particularly the *Bridged Pot Shelter* at ST526488. The finds ranged from artifacts now assigned to the Proto-Solutrean, to Bronze Age pottery, and some 20 species of mammals including Bear, Wolf, Horse, Great Deer and Roe Deer, with abundant remains of Reindeer, Fox, Hare, Cave Pika, Lemming and five species of Voles (Balch, 1947).

From the evidence of the finds, Man lived in these Ebbor and Wookey caves during the last stages of the Würm Glaciation, more than 15 000 years ago. Much later Bronze Age folk are represented by their pottery, and have left their round barrows and weapons on the hills above. They were followed by Celts during the Iron Age. Roman coins found in Wookey Hole include a silver denarius of Marcia (124 B.C.) but others to a total of 105 coins ranged from Vespasian (A.D. 70–79) to Valentinian II (A.D. 375–92) indicating that the cave was inhabited from the Roman Conquest to the end of the Roman occupation of Britain.

Certain of the local place-names are held to derive from the Brythonic or P-Celts who dwelt here during the Iron Age, about the beginning of the Christian era. They include Mendip, from *maen*, stone; Ebbor, connected with the modern Welsh *aber*, mouth or entrance; Wookey, earlier spelt Woky, Woki, Ochie, Owky and Okey, connected with the modern Welsh *ogof*, cave; and *Axe*, the Goidelic or Q-Celt word for water.

Iron ore was mined on a small scale up to 1891 near Higher Pitts Farm (ST535491) in pockets and veins in the Dolomitic Conglomerate, and Kingsbury (1941) records finding in the old spoil heaps many mineral species, some of them rare. They comprise those of iron (hematite, goethite, and various brown and yellow iron oxides); manganese (pyrolusite, manganite, wad, psilomelane, manganocalcite, rhodochroisite and pink manganiferous dolomite); copper (malachite, chessylite, crednerite and tetrahedrite); copper plus lead (chloroxiphite and diaboleite); and lead (cerussite, hydrocerussite, mendipite, mimetite, wulfenite and pyromorphite). In addition, there were also found quartz, calcite, aragonite and baryte.

In 1904 many fragments of a black igneous rock were found in a field at ST528483, near Ebbor Rocks. Identified as Clicker Tor Picrite from Cornwall, nearly 100 miles to the south-west, its adventitious presence here was explained by Evens only in 1958 (see p. 22).

1681, Beaumont; 1791, Collinson; 1862, 1863a, b, Dawkins; 1914, 1947a, Balch; 1929, Welch; 1957, Barrington; 1958, Evens; 1965, Green and Welch.

G. B. Cavern, Charterhouse, 8 miles NW of Wells

ST(31)475561	165	Bristol and 280	18SW (G)	C, B (1959)

A typical Mendip influent cave in the lowest beds of the Carboniferous Limestone; it is well documented. It was discovered in 1939, largely through the efforts of Mr. F. J. Goddard and Dr C. Barker, for whom it is named, after attempts extending over 20 years.

The altitude of the swallet is 830 feet above O.D., the explored length (1957) 4500 feet, and the depth 430 feet. The water is said to resurge at Cheddar (Barrington).

The entrance system is unusually complex and mud-choked, in contrast with the spacious dimensions of the main chambers. Of these the largest is slightly more than 120 feet long, 60 feet wide and 120 feet high, with magnificent roof formations. These include one splendid cluster of five enormous stalactites growing from a common origin and giving rise to a five-pointed stalagmite beneath it.

Of special interest are the helictites, once abundant but now largely looted, formed of aragonite or alternate deposition of calcite and aragonite; and both dried and moist moon-milk. The aragonite may be of significance in assessing palaeoclimatic temperature change (see under Holwell Cave below).

A living invertebrate fauna of at least 12 species has been identified. The skeletons of a number of bats were found at one spot 1500 feet in from the entrance and some 400 feet below the surface; some of the bones were cemented into the stalagmite. Three species were identified, all of living forms found in Britain: Bechstein's Bat (*Myotis bechsteini*), one of the rarest of British bats; the Whiskered Bat (*M. mystacinus*); and the Long Eared Bat (*Plecotus auritus*), very common in Somerset (Gilbert, 1963).

The cave entrance is protected by a small building with a locked steel door, and is under the control of the University of Bristol Spelaeological Society, whose Proceedings of 1943 and 1951 contain descriptions of the results of survey. This cave is not normally accessible; it is not open to the public.

1948, Balch; 1957, 1969, Barrington; 1956–9, Cave Research Group; 1963, Gilbert; also see D. Savage, 1969: *Proc. Univ. Bristol Spelaeol. Soc.*, **12**, 123–6.

Holwell Cave, Quantock Hills, 6 miles SW of Bridgwater

ST(31)211340	165	295	60NE	C, B
			(–)	(1959)

A small isolated cave system in Devonian Limestone, similar in style to the complex South Devon caves. It opens at the base of a disused quarry, the working of which has destroyed its outer part. It lies at an altitude of 620 feet, has a reported length of 250 feet and was discovered about 1800.

As a distinctive feature it contains anthodites—branching, flower-like aragonite clusters, often tinged with pink—found on the roof near the pool. Anthodites are known from several American and other foreign caves, but Holwell is the only British cave where they occur. It is claimed that aragonite is not deposited north of the 60°F mean annual isotherm. As past deposits occur north of the present position of this line, the material has been suggested as an indicator of palaeoclimatic change.

The cave is also distinguished as the only known British station of two blind shrimps, *Niphargus aquilex aquilex* and *N. kochianus kochianus*, which inhabit a stagnant pool 140 feet in from the entrance; and at least five other species of invertebrates have been identified as living within it.

1907, Baker and Balch; 1953, 1962, Cullingford, Ed.; 1956, Moore; 1956-9, Cave Research Group; 1958, Warwick.

Holwell Quarries, Nunney, 3 miles SW of Frome

ST(31)725450	166	OS19	42NE	P, U
			(–)	PSD, 1966

Large quarries straddling the main road A361, in part actively working Carboniferous Limestone which dips about 25° south; north of the road in C (*Syringothyris*) Zone, and south of the road in S (*Seminula*) Zone. The limestone surface is extraordinarily level, with a *Lithophagus*-bored surface, the final planing and boring dated to Upper Inferior Oolite time. Thin shore-deposits of Rhaetic, Lias, and Inferior Oolite overlie the Carboniferous Limestone with striking unconformity.

In both quarries are found fissures, now mainly filled, which once gaped, probably owing to repeated small earth movements along the Mendip axis. The main interest lies in the fossiliferous fissure deposits of Rhaetic, Lias, or Inferior Oolite age, differing from fissure to fissure. In the quarries north of the road the fillings appear to be of Inferior Oolite and Lias age, which yield fish-teeth but no mammals. In Moore's classic *Microlestes* Quarry, south of the road at ST727450, the fissures seem to be Neptunian dykes, the fossils most probably being derived from Rhaetic strata. The fissures may be from one to one and a half feet wide or may widen downwards to 10–15 feet at the quarry floor. The filling may be a peculiar greenish clay, or hard clay with pebbles, etc., or friable sand.

From fissures in the *Microlestes* Quarry Moore, and later Kühne, obtained in all 48 mammalian teeth of the highest scientific interest, some of the earliest known mammalian remains. They are now identified as follows. Microcleptids: *Microcleptes* [*Microlestes*] *Moorei* (Owen), *M. fissurae* Simpson, *Thomasia anglica* Simpson. Triconodonts: *Eozostrodon parvus* Parrington, *E. problematicus* Parrington.

All the teeth, mainly molars, measure about two millimetres across and were collected from this quarry, Moore discovering the first in 1858. If the size of these animals may be judged from their teeth, then all were about the size of a mouse. Two problematical small vertebrae were also found by Moore and described by Owen. Some of his material is housed in the Bath Museum (Richardson, 1911, p. 63).

To collect 19 mammalian teeth Kühne washed 2250 kilograms of the material through sieves, which yielded 121 kilograms of concentrate which he hand-picked, the yield being one tooth to 7·5 kilograms, or approximately, he says, one tooth in three million particles of similar size. The same fissures yielded Moore some 70 000 specimens of fish remains. Kühne identified his own collection of fish as: *Hybodus minor* and *H. cloacinus*, *Acrodus minimus* (of which Moore got 45 000 perfect teeth), *Saurichthys acuminatus*, *Palaeobates* sp., *Sargodon tomicus*, *Birgeria* sp., *Gyrolepis alberti*, and *Colobodus* sp.

There were found also eight or nine genera of Reptiles: a Crocodilian; a Placodont (the first record in Britain of this sub-order); and the Dinosaurs *Thecodontosaurus* and *Palaeosaurus*.

For comparison Kühne treated material from the Bone Bed at Aust (q.v.) with acetic acid and obtained teeth or other remains of about a dozen species of fishes identical with those in his collection from Holwell.

1858, 1860, Moore; 1860, Owen; 1911, Richardson; 1928, Simpson; 1947, Parrington; 1948, Kühne; 1957, Robinson; 1966, Savage and Waldman.

Kilmersdon Road Quarry, ¾ mile S of Radstock

ST(31)689542   166   Bristol   20SE   P, S
                                  (–)   1967

A disused quarry, well known and often visited, Tutcher and Trueman's No. 10; it exposes a second example (the first being Bowldish Quarry, q.v.) of the richly fossiliferous but much condensed and broken succession of the Lower Lias on the northern flank of the Mendip Anticline. These two sections epitomise the chequered history of this small area in Lias times, where every exposure has its own individual features reflecting the rapid and very local changes, and contrasting with the quiet deposition to the south of the Mendips.

There appears to be no detailed description published of this quarry, Tutcher and Trueman having limited themselves to the thicknesses of the beds and the zones recognised. Details of the succession below are therefore provisional. The fossils found in this quarry are not separately given, the very large fauna of the Radstock district, mainly from the Lower Lias, being recorded as a whole by the same authors. Nevertheless, from this Kilmersdon Road Quarry they record some 22 species of ammonites, including 10 Echioceratidae, and also *Peripleuroceras rotundicosta* and ? *Platypleuroceras bituberculatum*, both described from the *jamesoni* Limestone as new species and the former as a new genus.

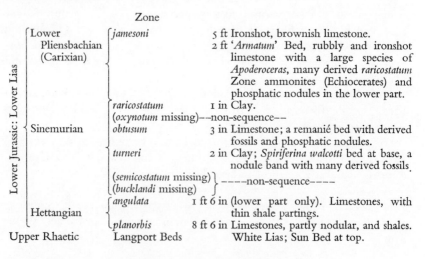

		Zone	
Lower Jurassic: Lower Lias	Lower Pliensbachian (Carixian)	*jamesoni*	5 ft Ironshot, brownish limestone.
			2 ft '*Armatum*' Bed, rubbly and ironshot limestone with a large species of *Apoderoceras*, many derived *raricostatum* Zone ammonites (Echiocerates) and phosphatic nodules in the lower part.
	Sinemurian	*raricostatum*	1 in Clay.
		(*oxynotum* missing)--non-sequence--	
		obtusum	3 in Limestone; a remanié bed with derived fossils and phosphatic nodules.
		turneri	2 in Clay; *Spiriferina walcotti* bed at base, a nodule band with many derived fossils.
		(*semicostatum* missing) (*bucklandi* missing)	----non-sequence----
	Hettangian	*angulata*	1 ft 6 in (lower part only). Limestones, with thin shale partings.
		planorbis	8 ft 6 in Limestones, partly nodular, and shales.
Upper Rhaetic		Langport Beds	White Lias; Sun Bed at top.

1925, Tutcher and Trueman; 1969, Torrens (B).

248

Lamb Leer, West Harptree, 5 miles N of Wells

ST(31)543550	165	Bristol	19SW (G)	C, B (1959)

Lamb Leer rather than Lamb Lair is stated to be the correct form of the name, meaning the leer (an old local name for cavern, cognate with the German *leer*, empty) at Lamb Bottom, Lamb Hill or Lamb Down, old names for this part of Mendip.

Hereabouts is found a curious thin semi-littoral facies of the basal Lower Lias, containing the ammonites *Psiloceras planorbis*, *P. johnstoni*, and other fossils. The strata consist of brown, grey or white chert resting on ochreous sand with seams of clay once worked for ochre. Traces of these beds here overlie the S Zone of the Carboniferous Limestone. The east–west-trending Lamb Leer Fault is exposed in the cavern, downthrowing 200 feet to the south.

Lamb Leer cave was discovered by lead miners, who broke into it some time before 1676, by which date it was well known. It was first described in 1681, but the site was later forgotten and was re-discovered only about 1880. Its altitude is 820 feet, length 800 feet and depth 200 feet (Barrington, 1957). No natural opening to the surface has been discovered.

Lamb Leer is an exceptional cavern in Carboniferous Limestone. Its chief feature is the very large Beehive Chamber, 110 feet high and of similar dimensions in length and breadth, with a rock-strewn floor. It is entered by a series of decorated passages 80 feet above the floor. There are well-marked avens in the roof and a remarkable mass of aragonite flowstone near the entrance. The Beehive Chamber is so called from an enormous mass of stalagmite, probably covering a heap of fallen boulders, 12 feet high and of regular rounded shape recalling that of the old straw beehive. Its sides are streaked with yellow and white bands. The vaulted roof is covered with a profusion of stalactites, and over the walls are masses of stalactite formations whose varying tints are a special feature of this cave. The Beehive Chamber is the largest cave chamber in Somerset and it is claimed that it is unequalled for beauty in England.

There are three main chambers, the Beehive, Great and Final Chambers. Between the first two is an artificial passage cut through aragonite, which mineral also covers boulders on the floor, and a small inlet. For the possible significance of aragonite deposits see under Holwell Cave. Unsuccessful search by geophysical methods has been made for another chamber, apparently described in 1681 and hitherto unidentified.

249

At least nine named forms of invertebrates have been recorded as living in this cave system.

The entrance to Lamb Leer is through a trap-door leading from the surface to a vertical ladder. It is not normally open to the public.

1681, Beaumont; 1933, Arkell; 1948, Balch; 1953, 1962, Cullingford, Ed.; 1956-9, Cave Research Group; 1957, Barrington; 1962, Shaw.

Middle Hope, 3 miles N of Weston-super-Mare

ST(31)335663	165	263	9NE, 10NW	Ig, P, R, S
			(G)	IGC, 1965

Coast sections extending over one mile on the north side of Middle Hope, otherwise known as Woodspring or Swallow Cliff, an east–west ridge rising to 100 feet above O.D. The country rock is Carboniferous Limestone dipping about 30° south, with four exposures of interbedded contemporary volcanic rocks, predominantly tuffs but with a submarine lava flow. The volcanics were first mapped by Sanders between 1840 and 1864. Repetition of the exposures over a distance of two miles is due to dip faults; all four lie at the same horizon. The comparable Spring Cove section (q.v.) consists of more lava than tuff, and lies about 350 feet higher stratigraphically. Both are probably connected with the start of earth movements. Elsewhere in England, Lower Carboniferous volcanicity is known only in Derbyshire, and also in the Isle of Man.

Of the outcrops, No. 1 is found at ST325661. Volcanic material extends over a thickness of about 100 feet. It is mainly dull-green tuff, but includes 12–14 feet of olivine basalt (with traces of copper carbonate). Pillow structure occurs in places and there are abundant calcareous veins, some with many Productids and corals in growth positions. The overlying limestone contains lapilli scattered through the lower seven feet, with a band of vesicular lapilli up to two inches long at the base. No. 1 is the best igneous exposure, and the only one with lava.

No. 2 is at ST337664: about 95 feet thickness of igneous material with fewer limestone bands but thick beds of calcareous tuff. A sandstone lying a few feet below the upper limit of the igneous material contains peculiar vertical pencil-like bodies of a controversial nature. The most massive bed of tuff contains bivalves (*Edmondia*) and a few gastropods.

No. 3 is at ST339665: about 80 feet thickness of beds closely resembling No. 2, including the sandstone with vertical bodies, and a band of tuff with

Michelinia favosa abundant; but the beds are considerably disturbed and even contorted in the upper part. Crinoidal bands with bivalves and a few gastropods occur at intervals.

No. 4 is at ST349669: only about four feet thickness of volcanic material, consisting of ash and ashy limestone.

The best section of the Carboniferous Limestone is seen at the east end of the ridge and south of St Thomas's Head, where a dip section along the western shore of Woodspring Bay exposes a complete succession as follows (Bush, 1930):

	Zone		
Lower Carboniferous: Tournaisian	C_1 296 ft	Caninia (Gully) Oolite: upper beds especially rich in *Chonetes* and *Orthotetes;* no corals.	
		Laminosa Dolomite: fossils rare, no corals.	
	Hor. Gamma Z_2 202 ft	Limestone Volcanics Limestone	The limestones both over- and underlying the volcanics contain abundant *Caninia* and *Zaphrentis* together, and crinoid ossicles. The fossils are generally silicified and weather out beautifully.

The coast between outcrops No. 1 and No. 2 affords perhaps the finest collecting ground for Zone Z_2 and Horizon Gamma in the Bristol district.

Two Raised Beaches are recorded on Middle Hope. The lower is well seen at Swallow Cliff, about ST325661, near No. 1 igneous exposure. The base lies at 35–40 feet above O.D., about 15 feet above the present beach. It was described by Sanders (1840) as consisting of rolled limestone pebbles, fine gravel, sand and shells of species of mollusca found living in the adjacent sea. It is associated with a prominent inland cliff and shore platform. The higher Raised Beach, described in 1956, is seen farther east about No. 3 igneous exposure at ST339665. Its base lies at about 50 feet above O.D., or 30 feet above the present beach.

The higher beach is assigned to the Eemian (Last) Interglacial or Late Monastirian, and the lower beach to the still later Epi-Monastirian, correlated with the Gottweig Interstadial (which has been dated to about 33 000–46 000 B.C.) of the Würm Glaciation. Both are thus held to be of Pleistocene age.

Two other nearby phenomena unconnected with the main geological interests are worthy of note.

At ST320657, on the saltings of the adjoining northern part of Sand Bay, is found a remarkable example of patterned vegetation, a large stand of *Spartina* growing in a series of parallel lines running seaward from the beach. Air photos show that this patterned ground has developed since 1946.

At Sand Bay also was found on 4 November 1959 a distinctively coloured light-grey sand lying in small drifts on the surface of the normal brownish quartz sand of the beach. It proved to be fly ash, minute glassy globules not more than a quarter of a millimetre in diameter of the mineral mullite, blown out of the chimneys of some electricity generating station fired by pulverised fuel. The mullite is believed to result from the fusion of the clay constituent

251

of the fuel. The globules contain gas bubbles; thus they float on the sea and may become concentrated in patches on the beach when they are blown ashore.

1898, Geikie and Strahan; 1903, 1904, Morgan and Reynolds; 1905, Sibly; 1907a, 1907b, Reynolds; 1930, Bush.
Raised Beaches: 1840, Sanders; 1956, ApSimon and Donovan.

Portishead to Black Nore Coast, 7 miles WNW of Bristol

ST(31)4677	155	Bristol	2SW, NW (G)	S, P, T PSD, IGC, 1965

A classic area stretching over two miles of coast, and studied since 1816; it includes exposures, many of them fossiliferous, of Triassic Dolomitic Conglomerate, Coal Measures Pennant Grit, Carboniferous Limestone Series (lower beds) and Old Red Sandstone.

Structurally the area is very complex and difficult to interpret. There are two master ridges of palaeozoic rocks which influence the grain of the country; one strikes east–west in the Armorican trend, from Portishead Dock Railway Station to Battery Point, forming the district known as Eastwood; and the second strikes north-east to south-west, forming Portishead Down and its northern continuation Wood Hill, and determines the coastline from Battery Point southwards to Clevedon.

The district has been claimed as exhibiting tectonics of the North-west Highland type: folds truncated by overthrusts whereby the southern limbs are driven across the cores of the folds on to their north-western limbs (Reynolds and Greenly, 1924).

The following sites are of particular interest.

ST474774, QUARRY OPPOSITE PORTISHEAD DOCK RAILWAY STATION

At the south end the Dolomitic Conglomerate is seen; at the north end good exposure shows:

Dolomitic conglomerate with a low dip to the east.
----------unconformity----------
Carboniferous Limestone: 50 ft Basal beds, including 18 ft Bryozoa Bed.
Upper Old Red Sandstone   154 ft (seen).

The beds below the Trias dip 65–75° west-north-west.

ST475776, ON THE SHORE BELOW THE ROYAL HOTEL

Dolomitic Conglomerate rests upon nearly unfossiliferous Carboniferous Limestone; a little to the west this is much disturbed and faulted against the Pennant Grit with poor plant remains. Farther westward this grit is sharply folded.

ST464776, BATTERY POINT

Carboniferous Limestone of Horizon Beta and Zone Z_1 are exposed and dip $35°$ north; fossils are plentiful, often beekitised or silicified, with crinoids and *Spirifer clathrata* the commonest. South of the Point the upper K_2 beds are exposed on the shore in a low cliff. There is included a red limestone crowded with bryozoa, crinoids and small gastropods, and including the zone fossil *Vaughania vetus* [=*Cleistopora geometrica*]. Folding and a reversed fault bring the Bryozoa Bed over the K_2 beds, which are thrown into a series of five little anticlines and synclines (Vaughan, 1905).

The following development of the Old Red Sandstone is found in this area:

Upper Old Red Sandstone, 900 ft	Portishead Beds, including the Shirehampton Beds, which are here lenticular and cannot be separately mapped; compact sandstones, conglomerates and marls; the Woodhill Bay Fish Bed lies 50 ft above the basal Woodhill Bay Conglomerate (10 ft thick) of unsorted pebbles and cobbles, partly cemented by irregular masses of cornstone.
	------------unconformity------------
Lower Old Red Sandstone, c. 2000– 3000 ft	Black Nore Sandstone: dingy, purplish-red sandstones, feebly calcareous; subordinate siltstones and conglomerates, and some cornstones in the highest beds.

ST460770, AT THE SOUTH END OF THE PORTISHEAD PARADE

A fine section of the basal 130 feet of the Portishead Beds is exposed, sandstones, conglomerate and marls, with a particularly interesting cornstone development. The beds dip about $20°$ south, and show symmetrical ripple markings, sun-cracks, raindrop impressions, current-bedding and lenticularity.

The Woodhill Bay Fish Bed contains many scales of *Holoptychius, Glyptopomus* and other fishes, and much interesting material is awaiting description. Faulting and the unconformable junction of Trias and Old Red Sandstone are admirably seen.

Just above the Fish Bed, at the southern end of Woodhill Bay, a sandstone bed has yielded many fragmentary specimens of a Eurypterid (animals apparently related to the scorpions and spiders) *Drepanopterus abonensis* described by Simpson (1951). From the reconstructed figure the animal was about 21 inches long.

ST445765, BLACK NORE POINT

The Dolomitic Conglomerate rests upon the Black Nore Sandstone, and many fine unconformable junctions are exposed.

1817, Gilby; 1921a, Reynolds; 1924, Reynolds and Greenly; 1951, Simpson; 1955, Kellaway and Welch; 1964, Pick.

St Cuthbert's Cave, 1 mile SE of Priddy Church, 3 miles N of Wells

---

ST(31)543505	165	Bristol	28SW	C
			(G)	—

A typical phreatic cave system of complex type unusual in Mendip, St Cuthbert's Cave comprises a network of large and small passages in three dimensions. The entrance is in Carboniferous Limestone, apparently of Z zone. Its stream emerges at Wookey Hole, one and three-quarter miles to the south-south-west. Discovered in 1953, it is so far known to be about one mile long and 400 feet deep.

The cave has had an involved history, its features recording a long period of solution while below the water table. Fossils are found projecting from the walls, indicative of such solution. It also contains deposits of gravel (often cemented by stalagmite), sand and mud, which represent an extensive episode of vadose activity probably dating from Pleistocene times.

There are a number of fairly large chambers reaching some 60 feet long and 60 feet high; beautiful formations such as a white stalagmite cascade 70 feet long; rich decorations of stalactites, some as curtains which may have dark-brown banding; and an orange-coloured beehive formation. A number of gours—some now filled with alluvial deposits—are coloured brown by manganese. One nest of cave pearls was noteworthy because of their spherulitic structure.

1956, Bennett et al.; 1957, Barrington.

Saltford Railway Cutting, Saltford, 6 miles SE of Bristol

ST(31)684673	156	Bristol	7SW	P, S
			(–)	1966

The best exposures lie between ST685671 and 681676, at each end of a short tunnel 670 yards north-west of Saltford Railway Station. About 60 feet of strata is exposed, almost the whole thickness of the Blue Lias in this district, dipping gently to the west or north-west.

At the Saltford (south-eastern) end of the tunnel the lower part of the succession (*planorbis* and *liasicus* Zones) is best seen for about 200 yards from the tunnel mouth, on the eastern side of the railway line; and the complete *angulata* sub-zone is preserved in a nearly vertical face on the opposite (west) side of the cutting, 70 yards from the tunnel mouth. Near the top of the cutting, on both sides of the railway, the *calcaria* Bed is exposed.

At the north-western end of the tunnel, 33 yards from its mouth, the base of the *calcaria* Bed lies about 23 feet above the level of the track, but the beds below it are not well exposed. The higher part of the succession is best seen behind a hut on the eastern side of the cutting half-way between the tunnel mouth and the next bridge across the railway. Here is a continuous section above the *calcaria* Bed through the *bucklandi* and most of the *semicostatum* Zones. In places two or three feet of the overlying *sauzeanum* sub-zone are seen on the opposite side of the railway line. Above that horizon the succession is obscured, but apparently consists entirely of shales.

Correlation between the slightly overlapping sections at each end of the tunnel is given by the *calcaria* Bed near the base of the *bucklandi* Zone. The combined succession is approximately as follows (Donovan, 1956).

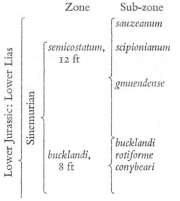

Zone	Sub-zone	
	sauzeanum	3½ ft Shales and limestones, *Euagassiceras spinaries*.
semicostatum, 12 ft	scipionianum	2 ft Limestones and shales, *scipionianum* Bed, with *Arnioceras* sp. group of *bodleyi*, and, below, *Paracoroniceras* sp.
	gmuendense	6½ ft Limestones and shales, *Arietites solarium*, *Coroniceras* group of *reynesi*, *Arnioceras rejectum*, *A.* cf. *mendax;* below, *Coroniceras* sp. group of *lyra*, *Megarietites meridionalis*, *Arietites* aff. *bucklandi*.
	bucklandi	2½ ft Limestones and shales, *Arietites bucklandi*.
bucklandi, 8 ft	rotiforme	3½ ft Limestones and shales.
	conybeari	2 ft Limestones and shales, *Calcirhynchia calcaria* Bed [=*Rhynchonella calcicosta*], *Coroniceras* aff. *caesar*.

Lower Jurassic: Lower Lias — Sinemurian

(continued overleaf)

Hettangian	angulata, 12 ft	angulata	12 ft Limestones with many shale partings, *Schlotheimia* sp.; and, near the base, *S. extranodosum*.
	liasicus, 10 ft	laqueus	10 ft Thick shales and limestones, *Alsatites liasicus* above; *Laqueoceras laqueus* below; *Schlotheimia curviornatum* and *S. megastoma* at base.
	planorbis, 15 ft	johnstoni	15 ft: 10 ft. Mainly thick shales; 5 ft. Mainly limestones, with *Psiloceras* sp.

The thick shale formation with little limestone, five feet at the base of the *laqueus* sub-zone with ten feet at the top of the *johnstoni* sub-zone, is known as the *Saltford Shales*

This Saltford Site of Special Scientific Interest is a replacement for Keeling's Quarry, which lies a mile and a quarter distant to the north-west, and had to be abandoned. Since a more detailed account of the lithological succession and record of some of the fossils other than ammonites were given by Tutcher (1917, 1923) for the similar succession once exposed in Keeling's Quarry, details are given below to provide a more complete picture of the Lower Lias development found in this district.

KEELING'S QUARRY, KEYNSHAM, ST(31)659681

An ancient and famous quarry, disused since about 1927, in which were wrought the Lower Lias Limestones, of indifferent quality for building but also burnt for lime. It included the best local exposure of the *bucklandi* Beds.

'Cainesham' is referred to in the Domesday survey of 1086, and its stone quarries are mentioned by Leland (about 1535–43), who recorded from them 'stones figurid like serpentes wounde into circles'. That these were really fossil sea-shells (ammonites) was concluded by Hooke in 1687 (1705), and independently by Edward Owen in 1754, although they continued to be popularly known as snake-stones. Walcott (1779) states in a footnote (p. 30): 'Formerly the credulous inhabitants of the village believed these snake-stones to have been real serpents, changed into stone by one Keina, a devout British virgin, from whom they likewise denominated the name of the place.'

Large specimens of '*Ammonites bucklandi*' have been obtained here up to 21 inches or more in diameter, sometimes lacking the inner whorls. One such was collected by Dean Buckland, who, thrusting his head and one shoulder through it, as if it were a French horn, so mounted his horse and rode home, dubbed by his friends an Ammon Knight (Sowerby, 1818).

Keeling's Quarry exposed some 26 feet thickness of interbedded limestones and clays, highly fossiliferous, and especially rich in ammonites, with other molluscs, and brachiopods; and foraminifera in the *bucklandi* Zone. Some 127 forms are recorded from 'Keynsham and Stoat's Hill' by Moore (1867).

KEELING'S QUARRY SECTION AFTER TUTCHER (1923)

Zone	Thickness	Sub-zone	
	3½ ft	*sauzeanum*	Clay with four 4-in bands of pale marl; 6 spp. of ammonites with *Agassiceras* cf. *sauzeanum*, *Belemnites infundibulum*, *Chlamys textorius* and *Rhynchonella semicostata*.
	2¼ ft	*scipionianum*	1 ft 3 in. Blue limestone with few phosphatic nodules at base; *Aetornoceras scipionianum* and *Arnioceras geometricum*, *Chlamys textorius*, *Terebratula ovatissima*. 1 ft 0 in. Paper-shale with thin limestone bands at top, 2 spp. ammonites, *Oxytoma* [*Avicula*] *inequivalvis*, *Pecten hehli*, *Anomia pellucida*. 4 in. Sandy shale.
	2½ ft	*gmuendense*	4 in. Blue gritty limestone with badly preserved ammonites. 2 ft 2 in. Four beds blue limestone with shale partings; 5 spp. ammonites, including *Paracoroniceras gmuendense* and *Plagiostoma giganteum*, etc.
	2 ft	*bucklandi*	Three blue limestones with shale partings; 3 spp. ammonites, with *Coroniceras bucklandi*, *C.* cf. *caesar* and *Schlotheimia* cf. *chamassei*.
	3½ ft	*rotiforme*	Limestone and shale with *Coroniceras rotiforme*, *C. rotator* and *Nautilus intermedius*.
	2 ft	*conybeari*	6 in. Shale, *Schlotheimia* cf. *chamassei*, *Plagiostoma giganteum*. 1 ft 4 in. *Calcicosta* Limestone, blue limestone, with *Vermiceras conybeari* and two other ammonites, '*Rhynchonella calcicosta* Dav.' very common.
	8 ft 2 in		Alternating clays and limestones; in the basal 4½ ft the limestones are nodular; with *Schlotheimia angulata*, *S. extranodosa*, *Calcirhynchia calcaria* [= *Rhynchonella calcicosta*], small *Gryphaea*, *Plagiostoma hettangiense*, *P. giganteum*, *Gresslya galathea*.
	4 ft 11 in		Alternating clays and limestones, with *Alsatites liasicus*, *Nautilus intermedius*, *Plagiostoma succinctum*, *Modiolus hillanoides*, *Ornithella sarthacensis*.

Left margin labels: Lower Jurassic: Lower Lias — Sinemurian (semicostatum, bucklandi) — Hettangian (angulata, liasicus)

The Hettangian of the above section was described by Tutcher from his Section A, at ST662692, 1200 yards distant to the north-north-east.

1535–43, Leland; (1687) in 1705, Hooke; 1754, Owen; 1779, Walcott; 1867, Moore; 1917, 1923, Tutcher; 1941, Macfadyen; 1956, Donovan.

Snowdon Hill Quarry, ½ mile NW of Chard

ST(31)312089	177	311	91NE (-)	P, S 1963	

A large quarry, disused before 1900, with an excellent exposure of Upper Greensand, and overlying it a small degraded and overgrown section in the basal Lower Chalk. This is claimed to be the only remaining exposure of the Middle Cenomanian in the area, and one of the most interesting localities in England for its study.

The richly fossiliferous 'Chloritic Marl' hereabouts was noted by De la Beche in 1839, and Jukes-Browne (1903) lists 101 fossil species from it, from Chard and Chardstock. The ammonites of the Chalk Basement Bed correlate it with the upper part of the Middle Cenomanian of the type area of Le Mans, Sarthe, Western France.

The Chalk here contains detrital minerals derived from the West of England granites, including Dartmoor (Groves, 1931).

The dip in the quarry is 5° west and the face shows two parallel north–south faults 30 feet apart, with the block between them upthrown five or six feet. The Chalk was worked for lime-burning, and the Greensand for building stone.

SNOWDON HILL QUARRY (AFTER WIEST, IN DAVIDSON, 1852–5)

Upper Cretaceous Lower Chalk, Middle Cenomanian, *subglobosus* Zone

c. 15 ft Pure white chalk with a few flints; fossils rare.

1 ft 6 in Nodular glauconitic chalk.

2 ft Hard sandy glauconitic chalk, with some ammonites and echinoids. [This 18½ feet of chalk is not now exposed (in 1963)].

1 ft Chalk Basement Bed: 'Chloritic Marl'; hard, slightly glauconitic calcareous sandstone, with scattered brown phosphatic nodules, pebbles of calcareous grit usually having phosphatic skins; and a rich and well-preserved fauna, mostly phosphatised, some fossils water-worn and covered with Serpulae and Polyzoa. Includes some 40 species of Ammonites, with *Eutrephoceras, Scaphites, Turrilites, Schloenbachia, Acanthoceras rhotomagense* and other species, and *Holaster subglobosus, Discoidea cylindrica* and some bivalves and gastropods, etc.

———————— { Well-marked water-worn plane } ————————
{ marking a Cenomanian transgression }

Lower Cretaceous Upper Greensand, Upper Albian

6 in Sandy limestone with a rich fauna of rather poorly preserved crustaceans, and also *Cyclothyris schloenbachi*, 'Rhynchonella' *grasiana, Catopygus columbarius, Discoidea subuculus*, etc.

5 ft Calcareous grit, including calcareous grit boulders.

c. 30 ft (seen). Chert Beds: grey coarse shelly calcareous glauconitic grit, the top 3 in containing quartz pebbles, passing down into finer-grained rocks, with courses of black or brown tabular chert.

I am indebted to Dr J. M. Hancock for information about this site.

1852–1855, Davidson; 1900, 1903, 1904, Jukes-Browne; 1931, Groves; 1960, Tresise; 1962, Smith and Drummond.

Spring Cove, Weston-super-Mare

ST(31)310625	165	OS20	9SE	Ig, P, R
			(G)	PSD, 1965

Spring Cove and Middle Hope (q.v.) are two differing examples of the minor isolated contemporary volcanicity in the Carboniferous Limestone of Somerset.

The volcanic rock of Spring Cove (or Birnbeck Cove) near the western end of Worle Hill was recorded in the 1860s. It is exposed in a coastal cliff section about 150 yards long, and is some 45 feet thick. It is predominantly of basalt, interbedded with Carboniferous Limestone containing volcanic dust and vesicular lapilli, and dips about 40° south-east. The basalt is much decomposed, but contains pseudomorphs after olivine and augite in a microlithic groundmass. The lava immediately overlies the Caninia (Gully) Oolite (C_1 Zone) at a horizon about 350 feet higher than that of Middle Hope and other neighbouring exposures.

At the south-west end, where the lava passes below sea level, the rock is a compact red, slightly amygdaloidal olivine-orthoclase-basalt, mingled with some lumps of limestone. Rather beyond half-way along the exposure it becomes a very coarse tuff or agglomerate, with fluxion structure and with lenticular masses of slaggy basalt up to six feet long, and limestone masses up to 12 feet across. Still farther east the rock again becomes a basalt, very amygdaloidal and often variolitic, with pillow structures; there are some large included limestone masses and some tuff.

It is considered that both lava and tuff flowed out on the sea-bottom in a fragmental condition. The underlying limestone is red-stained and baked at the junction. The overlying limestone contains disseminated ash particles up to at least eight feet above its base, and especially abundant examples of two corals, *Caninia subibicina* and *Campophyllum cylindricum*.

A second feature of Spring Cove is a small patch of Raised Beach, the platform on which the deposits rest ranging up to 50 feet above O.D. It is equivalent to the higher of the two Raised Beaches at Middle Hope, assigned to a late Monastirian date within the Eemian Interglacial.

The section was described as follows (from Day, 1866):

Pleistocene
- 4 ft Head.
- 3 ft Concretionary sandstone over blown sand.
- c. 4 ft Conglomerate of water-worn shingle and flints, and angular fragments of limestone and flint, with sea-shells; also mammalian bones and teeth, mainly of Horse, but some of Fox and Hyaena.

259

1898, Geikie and Strahan; 1904, Boulton; 1905, Sibly; 1907a, 1921a, Reynolds; Raised Beach: 1866, Day; 1868, Mackintosh.

Stoke Lane Slocker, ½ mile NE of Stoke Lane Village, 4 miles NE of Shepton Mallet

---

ST(31)669474	166	OS19	42NW	C, B
			(G)	(1966)

An influent cave system in the upper part of the Carboniferous Limestone, first systematically explored in 1900. In 1947 cave-diving through a water-trap further extended the known part, which now reaches a length of about 2500 feet; the depth is 60 feet. The system with its stream runs north-west along the strike of the strata, and nearly at right angles to the surface drainage feature, a dry valley running slightly east of north. The stream is resurgent 1200 yards from the swallet (which in this part of Mendip is termed a slocker) at the spring called St Dunstan's Well, where the water is thrown out by the relatively impervious Coal Measures and 'Millstone Grit'. All the beds dip steeply at about 50° to the east-north-east.

Stoke Lane Slocker is one of the finest cave systems in the West Country, with six great chambers so far found, and containing a very fine series of cave formations, including cave pearls. The Main Chamber is more than 100 feet high and is adorned with huge flowing curtains and massive banks of stalagmite; the C.B. Chamber beyond is still more beautiful, with stalagmite sheets, bosses and pillars, and many long slender stalactite 'straws'. Next comes the Bone Chamber, where there were found two human skeletons, charcoal and contemporary Roe Deer remains behind a collapsed exit, but nothing to indicate the age. Still farther is the Throne Room, with one pillar 10 feet in diameter, and on the left a series of grottoes leading to the Main Chamber.

The influent stream is well charged with organic matter, and this has resulted in the growth of an alga, *Chroococcus turgidus*, in part of the cave. Of the living cave fauna, at least 14 forms of invertebrates have been collected and identified; a few of them may have been merely washed in, and not be true cave residents.

1930, Welch; 1953, 1962, Cullingford, Ed.; 1956–9, Cave Research Group; 1957, Barrington.

Swildon's Hole, Priddy, 3½ miles NNW of Wells

ST(31)531513	165	Bristol	27NE (G)	C, B (1959)

This swallet is so named because it was on land owned before the fifteenth century by St Swithin's Priory, Winchester, and the name was corrupted by the Mendip folk to Swildon's. Its exploration started in 1901. It is a complex influent swallet in Carboniferous Limestone, the entrance lying in the S Zone, a quarter of a mile north-west of the East Priddy Fault. It has hydrological interests and a very fine series of cave formations. There are both active and abandoned stream-ways, a lower series in part submerged, and an unusual upper series leading from them; and extensive solution features, including 'tubes' from 6 to 30 inches in diameter. Its subterranean stream is thought probably to issue in the Cheddar rising. The altitude of the swallet is 780 feet above O.D., its length (1957) some 10 000 feet, and its depth 460 feet.

Swildon's Hole is described as perhaps the wildest and most magnificent cave in Britain, and one of the most beautiful pieces of underground scenery. In the Old Grotto great pendant masses of stalactite hang from the roof and there are many fine sheets and large bosses on the floor. Pools are lined with crystals of dog-toothed spar in little bunches. In the Shrine and the White Way are found exquisite sheets of stalactite like fabric curtains, gems of Nature's handiwork. In another chamber is a remarkable pillar of great height capped by a square-topped tower. Beyond, in a grotto with wild profusion of decoration, countless long delicate pencil-like stalactites and stalagmites rival in their wealth everything else in Mendip. A few helictites are found in Tratman's Temple.

Extensions of the known limits of the cave system were made by diving through water-traps in 1936, and extensive new passages have been similarly discovered in the years since World War II.

A great variety of living flora and fauna has been reported from this cavern, including fungi, crustacea, springtails, worms, beetles, non-marine gastropods, spiders and mites, a total of at least 39 species of invertebrates.

1907, Baker and Balch; 1948, Balch; 1953, 1962, Cullingford, Ed.; 1956–9, Cave Research Group; 1957, Barrington.

Vallis Vale, 1 mile NW of Frome

ST(31)757491	166	OS19	30SW	U, S, P
			(-)	IGC, PSD, 1966

Vallis Vale is a steep-sided combe eroded by the Mells and Egford streams about 100 feet deep through the Jurassic and Triassic strata and into the Carboniferous Limestone below, in which have been opened quarries now mainly disused. The geological interest is threefold: first, the spectacular unconformity between nearly horizontal Jurassic strata which rest directly upon a fairly level planed-off and bored marine erosion surface of highly inclined Carboniferous Limestone; second, sections in the Upper Inferior Oolite, Lower Lias, and Rhaetic; and third, sections in the Carboniferous Limestone ranging from Zone S_1 to Zone Z_2.

At ST761495, near Hapsford Bridge, a section shows (after Richardson, 1911):

Middle Jurassic: Bajocian
Upper Inferior Oolite: $\left\{ garantiana \text{ Zone} \right.$ 3 ft Upper *Trigonia* Grit, with many brachiopods and echinoids.

----------unconformity----------

Lowest part of the Upper, the whole of the Middle and Lower $\left.\right\}$ (missing)
Inferior Oolite, and of the Upper, Middle and Lower Lias

Upper Rhaetic $\left\{\begin{array}{l} \text{Langport Beds} \\ \text{Cotham Beds} \end{array}\right.$ 2½ ft Well-bored White Lias limestone.
8 ft Conglomerate, marly limestone, shelly limestone, and marls; *Euestheria* and *Lycopodites*.

Lower Rhaetic   Westbury Beds   4 ft Clays and conglomerates, with bored Carboniferous Limestone pebbles, and some fossils.

----------unconformity----------

Carboniferous Limestone   *Seminula* S Beds   c. 10 ft Limestones, with well-planed surface.

At ST757487, by Egford Brook, is found a little Lower Lias, cream-coloured limestone with many bivalves; the presence of *Spiriferina walcotti* shows that the *bucklandi* Zone is represented. This rests upon a planed surface of Carboniferous Limestone.

The whole Carboniferous Limestone succession at Vallis Vale comprises the following (after Bush, 1926):

Lower Carboniferous $\left\{\begin{array}{l} \text{Viséan} \\ \\ \text{Tournaisian} \end{array}\right.$

Viséan $\left\{ \right.$ 324 ft (seen). S_1. Oolite with chert nodules; corals and brachiopods abundant.

Tournaisian $\left\{ \right.$ 535 ft C_1. Massive dark-grey crinoidal limestone, locally dolomitised, with an abundant coral, *Amplexus*, and well-preserved ostracods and foraminifera.
280 ft Hor. Gamma: black dolomitised crinoidal limestone.
240 ft Z_2. Dark crinoidal limestone with much nodular chert.

In the Egford Valley the southernmost exposures are of the Z (*Zaphrentis*) Zone, followed successively to the north by Horizon Gamma and the C (*Syringothyris*) Zone; the C Zone extends into the adjoining part of the Mells valley, followed farther north, at both ends, by the S (*Seminula*) Zone.

The U-shaped borings of the marine worm *Polydora* (?) and the larger borings of the mollusc *Lithophagus* were first noted here by John Phillips in 1829. These borings and the attached oysters relate to the incoming Upper Inferior Oolite fauna at the close of the Upper Bajocian Transgression, and are found on the underlying surface, the borings riddling the top few inches of the limestone that was then exposed, whether this was the planed surface of the Carboniferous Limestone or, as near Hapsford Bridge, the Rhaetic White Lias.

The S_1 Zone at Vallis Vale is the type horizon and locality of the trilobite *Phillipsia scabra*, H. Woodward.

1822, Conybeare and Phillips; 1846, De la Beche (p. 288); 1867, Moore; 1907, 1909, 1911, Richardson; 1921a, Reynolds; 1926, Bush; 1933, Welch.

A Provisional Rough Guide to the British Pleistocene and Recent Succession

More than 200 geological Sites of Special Scientific Interest in the Nature Conservancy's files have at least a part of their interest in the Pleistocene or Recent periods. In the present volume there are 36. The purpose of this rough guide is to provide a general though obviously provisional framework into which such sites may be fitted, to promote a better appreciation of their age and other relationships.

The Recent (Holocene or Post-glacial) succession is generally agreed, and the deposits can be dated in terms of years by carbon-14 measurements. This part is therefore plotted in the diagram on a larger scale. All but the uppermost part of the Pleistocene succession is shown on a much smaller scale, because the extent in time is vastly greater and much less accurately known.

Manifestations of the Pleistocene in Great Britain include boulder clays (till), drumlins, moraines, glacial drift, glacial gravels and sands of all kinds, including kames and eskers; glacial lake deposits, including varves, and lake beaches, glacial overflow channels; glacially sculptured hills and valleys, the latter U-shaped in section, with truncated spurs; glaciated rock pavements, etc., smoothed and striated, with *roches moutonnées*, crag-and-tail, erratic boulders and pebbles, some of them far-travelled; displaced blocks of strata on a considerable scale; river gravels and terraces; marine beaches and platforms, both below and above Ordnance Datum; cave and other deposits, unfossiliferous or with a variety of fossils, including the remains of man, large and small mammals and other animals, mollusca, insects or vegetation; foraminifera, pollen and other microfossils; and man-made implements.

Owing to the difficulties of correlating such scattered, patchy and disparate phenomena, the general succession within the country remains largely un-coordinated, and uncertainly correlated with the Continental successions.

The Pleistocene in western Europe has been divided into four main glacial episodes (for simplicity here ignoring the less well defined earlier indications), first described in the Alps. They are the Günz (which is not known in Great Britain), Mindel, Riss and Würm, separated by three main interglacial (warmer) periods, the Cromerian, Hoxnian and Eemian.

In general, the sea level relative to the land was low during the glacial episodes, much of the water having been withdrawn to form the enlarged ice-caps. During the Interglacials, conversely, the sea level was high, replenished with water from the melted ice-caps. In the Mediterranean area the types of a number of Raised Beaches or Platforms are recorded under the names Calabrian, Sicilian, Milazzian, Tyrrhenian, and the Main, Late and Epi-Monastirian, associated with the interglacial (or interstadial) episodes. The position of the Pliocene–Pleistocene boundary was defined by a Committee of the Nineteenth International Geological Congress in 1952 as the base of the continental Villafranchian beds in Italy, with the appearance therein of abundant Elephant, Horse and Ox. The date in years of that boundary and of the sub-divisions of the Pleistocene are as yet difficult to measure, and are not generally agreed. Various estimates have been made, however, and the results of Ericson, Ewing and Wollin (1964), derived from their studies of foraminifera found in cores taken from the bed of the Atlantic Ocean, have been provisionally adopted here, at the lowest valuation as a convenience in plotting the diagram.

Glossary

ACID   Applied to an igneous rock of whose mass silica forms more than about two-thirds, and usually containing quartz (free silica).

ADIT   A horizontal opening by which a mine is entered or drained.

AGGLOMERATE   A coarse-textured rock consisting of large and small fragments of volcanic rock set in an ashy matrix, resulting from explosive volcanic activity.

ALBITISATION   The replacement of plagioclase (soda-lime feldspar) by albite (soda-feldspar, sodium aluminosilicate) in igneous rocks during the final stage of the cooling of a magma, or during metamorphism.

AMYGDALOIDAL   Almond-like, of a lava in which gas bubbles have been filled with secondary minerals, often oval and light-coloured, suggesting almonds in a cake.

ANGULAR DISCORDANCE   An unconformity (q.v.) in which the bedding planes of the upper strata rest at an angle upon those of the lower.

ANTHODITE   A cluster of radiating masses of long needle-like crystals of aragonite (calcium carbonate) found (rarely) in caves.

ANTICLINE   A structure resulting from strata folded or bent into the form of an arch, and dipping on both sides away from the central line.

APHANITIC   Applied to an igneous rock having a uniform texture showing no distinct grains to the naked eye.

APLITE   A very fine-grained light-coloured form of granite, usually occurring as veins in the parent granite.

ARGILLACEOUS   Clayey.

ARTIFACT   An object made or fashioned by man, especially a stone tool made by Early Man.

AUREOLE (metamorphic)   A series of concentric rings of altered rocks with newly formed minerals, surrounding and baked by an intruded mass of igneous rock, and representing successive belts of decreasing metamorphism; e.g., surrounding the Dartmoor granite.

AVENS   Vertical shafts leading upwards from a cave passage; they may be either connected with a passage above or closed at the top.

'BEEF'   Interbedded layers of fibrous calcite (calcium carbonate), the fibres lying at right angles to the bedding of the strata, usually clay or limestone, e.g. in the Lower Lias.

BEEKITE   Chalcedony (colloidal silica) occurring in the form of sub-spherical, discoid, rosette or doughnut-shaped accretions commonly found on silicified fossils.

BIVALVE   A mollusc whose shell consists of two halves joined by an elastic ligament at the hinge; earlier generally called a lamellibranch or pelecypod.

BOULDER CLAY   Or till; unstratified stiff clay (rock flour) usually with sand, sub-angular or rounded pebbles and scratched boulders embedded in it; a product of glacial action.

BRECCIA   A rock consisting of angular fragments of stone (or sometimes of bone) and cemented by calcium carbonate, etc.

CARTILAGINOUS   Used of those fishes (such as sharks) whose skeletons are of cartilage and not of true bone.

CAVE PEARLS   Sometimes found in caves, smooth rounded concretions of calcite or aragonite (calcium carbonate) formed *in situ* in certain circumstances, by concentric precipitation round a nucleus.

CEMENTSTONE   Thin stone bands of fine-grained clayey limestone, etc., sometimes found interbedded in clays, e.g. in Kimeridge Clay.

CHERT   A siliceous replacement of limestone; practically flint.

COCCOLITHS   Almost ultra-microscopic disc-like plates which built certain marine algae whose fossil remains abound in the Chalk.

COMPETENT ROCKS   Hard and coherent rocks such as limestone and granite, strong enough to transmit pressure without becoming plastic or labile, as opposed to weak unconsolidated gravel, sand or clay strata.

COMPOUND CORAL A kind of coral always consisting of a colony of individuals growing together.

CONCHOIDAL FRACTURE One in which smooth shell-like convex and concave surfaces are presented by the broken rock.

CONE-IN-CONE STRUCTURE A concretionary structure of obscure origin occurring in marls, ironstones, coals, etc., characterised by the development of a succession of small cones, one within another, often of calcite.

CONGLOMERATE A rock formed of rounded pebbles, cobbles or boulders of other rocks, consolidated by some kind of cement; practically a natural concrete; sometimes termed 'pudding-stone'.

CONSEQUENT RIVER A primary main stream that flows directly down the slope of a newly uplifted land surface.

COOMBE-ROCK A tumultuous deposit of local chalk rubble, flints, resorted clays, sands and dirty gravel, etc., in a matrix of chalky paste; a Pleistocene torrential deposit, probably formed as a sludge during summer melting of a frozen ground surface under periglacial conditions (q.v.); it is included as a form of Head (q.v.).

COPROLITE Fossilised animal excrement, generally composed largely of calcium phosphate.

CORNSTONE In the Old Red Sandstone, beds of calcareous concretions embedded in marls, and grading to solid concretionary limestones; their presence increases the fertility of the soil, hence the name.

COUNTRY ROCK A mining term for the surrounding rock through which a vein of ore runs; but used in geology with a wider meaning as the surrounding rock at any site.

CRAG AND TAIL A land form resulting from an obstacle in the path of a moving mass of ice, a prominent rock knob, in the lee of which glacial drift accumulated in a long trail; found especially in the south of Scotland, e.g. Edinburgh Castle Rock.

CRYOTURBATION A disturbance of the strata by severe frost action, likely to be caused in periglacial conditions (q.v.).

CRYPT The hole or burrow made in rock by a boring organism, and in which it then lived.

CYCADS Woody plants resembling both palms and tree-ferns, and allied to the conifers; the male plant is cone-bearing.

DICEY CLAY Hard clay or shale which weathers into small cubes like dice.

DILUVIUM A term no longer in use, meaning the superficial, Pleistocene deposits, or Drift, which was formerly attributed to Noah's Flood.

DIP Or true dip; the inclination of the greatest downward slope of a stratum or vein measured by its angle from the horizontal.

DOGGERS Cement-hardened patches in sandy strata, of rounded outline and sometimes reaching several feet in length, e.g. cowstones.

DOG-TOOTH SPAR A variety of calcite crystals in the form of pointed pyramids.

DOLERITE A medium-grained basic igneous rock composed of plagioclase feldspar and augite, often with olivine, of similar composition to basalt but more coarsely grained; the commonest rock of minor intrusions such as dykes and sills.

DROWNED VALLEY An old river valley now become a sea inlet because of a relative sinking of the land.

DRUMLINS Glacial phenomena, smooth oval mounds or ridges of boulder clay, some reaching 100 feet high, and having their long axes in the direction of the ice motion; they commonly occur in groups, and date from the last glaciation.

DRUSE A crust of small crystals lining a cavity in a rock; also the cavity itself.

DYKE A wall-like sheet of rock, usually igneous, occurring in a more or less vertical fissure in the country rock (q.v.).

ELVAN A Cornish name for the intrusive igneous rock quartz porphyry.

EPIDIORITE A metamorphosed igneous rock such as gabbro, dolerite or basalt, in which the augite has been altered to hornblende.

ERRATICS Boulders, usually of hard rocks, ranging in size from that of a cottage to that of a pebble, usually well-rounded, smoothed, and scratched by glacial action, and transported often over great distances during the Pleistocene by land or sea ice.

ESKER A gravel ridge which may be 30 feet high and several miles long, formed

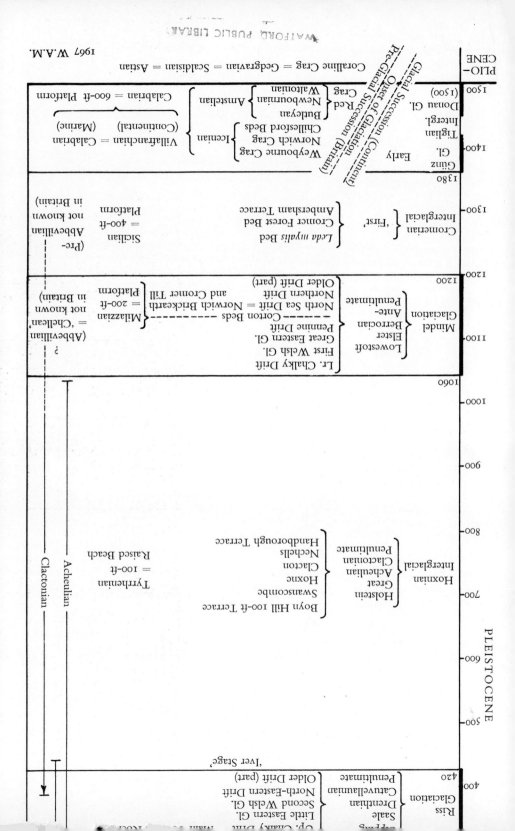

1967 W.A.M.

A PROVISIONAL ROUGH GUIDE TO THE BRITISH PLEISTOCENE AND RECENT SUCCESSION

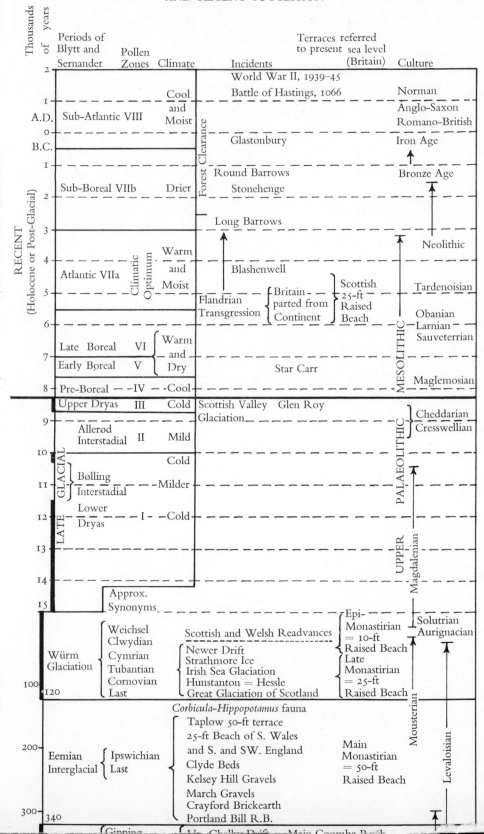

during the melting of an ice sheet, by deposition of material in a sub-glacial stream channel.

EXTRUSIVE Or volcanic; those igneous rocks which have been ejected upon the earth's surface, such as lava flows, and also solid fragmental material such as ash and other detritus.

FACIES Appearance, character; when a certain geological formation of (say) limestones passes laterally into (say) clays or sandstones of contemporary age, these different types of deposit are referred to as different facies of that formation, or the same facies may be found in rocks of different ages, suggesting a common mode of origin.

FAULT A fracture in the earth's crust, along which the rocks on one side have been displaced relative to those on the other side generally in a downward direction. (But see TEAR-FAULT.)

FLOWSTONE A continuous sheet of calcium carbonate on the floor or walls of a cavern, formed by precipitation from a film of slowly flowing water; so called to differentiate it from the same material deposited by dripping water, as a stalactite or stalagmite.

FLUXION STRUCTURE Flow structure; an arrangement of the individual particles of an igneous rock in streaky lines, indicating internal movements in the mass previous to its consolidation.

FRACTURE CLEAVAGE Or strain-slip cleavage; the structure involved where closely spaced tight joints give the rock a capacity to part along parallel surfaces.

FUCOID A term used for fossils thought to resemble seaweeds, specially of the Fucaceae, Brown Algae; or similar markings not clearly identifiable.

GABBRO A dark-grey or black coarse-grained basic igneous rock, essentially a crystalline aggregate of plagioclase (lime-soda feldspar) and augite, sometimes with olivine.

GLACIAL DRIFT A general term embracing all the Pleistocene glacial deposits.

GLACIATION The condition of being covered by an ice-sheet or glacier; glacial action and its results; also one of the main glacial periods of the Pleistocene (in this volume).

GLAUCONITE An amorphous green mineral often occurring abundantly as rounded grains or larger agglomerations; it may give a general green tint to the rock; a hydrated silicate of potassium and iron, it is found only in sedimentary rocks of marine origin.

GOUR A calcareous deposit built up in the form of a dam round the edge of an overflowing natural basin in a cave.

GRANITE The main Cornish and Devon granites are coarsely crystalline acid igneous rocks coming from a molten magma; they consist of orthoclase (potash feldspar), quartz, and mica or hornblende.

GREENSTONE An old term for dark-greenish igneous rocks, mainly altered diorite, dolerite or basalt, and consisting essentially of plagioclase (soda-lime feldspar) and hornblende.

GREISEN An alteration product formed as a selvage a few inches thick lining the walls of open vertical fissures in granite; consisting chiefly of quartz and white mica, sometimes with topaz and fluorspar, it is produced by superheated steam, fluorine and lithium compounds rising from the granite magma at a late stage of its cooling.

GRIKE An open surface reticulation of deep furrows produced by rain-water solution enlarging the joints on a limestone surface; they may reach several inches in width and a foot or more deep.

HEAD A superficial structureless deposit of rubble, etc., often with angular rock fragments of local origin; it is largely the result of solifluxion in periglacial areas during the Pleistocene.

HELICTITE A distorted stalactite, generally slender, which has grown from the roof or wall of a cave at an angle to the vertical; it may have changed the direction of its growth more than once.

HOLOCRYSTALLINE Of igneous rocks, meaning completely crystalline, without any interstitial glass.

HORNFELS A hard, tough rock of exceedingly fine grain, consisting of an intimate aggregate of quartz, feldspar, mica, and usually some metamorphic minerals; it is produced by the thermal metamorphism of clayey strata.

HYDRAULIC CEMENT A cement which sets hard under water, first made by burning limestone with a clay content.

HYPERSALINE Used of water containing a very high concentration of salts.

267

ICE AGE As used in this volume, the approximate equivalent of the Pleistocene.

ICHNITE A fossil footprint.

INCOMPETENT ROCKS Rocks which yield to lateral pressure by plastic adjustment and flow; or soft or unconsolidated strata, such as gravel, sand and clay; in contrast to competent rocks (q.v.).

INFLUENT CAVE A cave into whose mouth a stream runs.

INLIER An exposure of older rocks surrounded by newer rocks.

INTERGLACIAL A lengthy milder period separating two glacial periods.

INTERSTADIAL A minor interglacial, a milder episode within a glacial period.

INTERSTRATIFIED One kind of strata alternating or interbedded with another, e.g. limestone bands in a clay formation.

INTRUSIVE Used generally of igneous rocks which when molten or plastic have been forced among other rocks, e.g. dykes, sills, the Dartmoor Granite.

IRON PAN A hardened substratum of soil containing iron hydroxides, and more or less impervious to water.

IRONSHOT Descriptive of a limestone or clay containing many oolitic grains of limonite (amorphous hydrated oxide of iron).

ISOTOPIC AGE See RADIOMETRIC AGE.

KAME An irregular mound or ridge, often multiple, of unstratified boulders, gravel and sand, deposited at the front of an ice-sheet; they may reach several hundred feet in height.

KAOLIN China clay; a white hydrated silicate of aluminium formed by weathering or other natural chemical decomposition of feldspar *in situ*, e.g. in the Devon and Cornish granites.

KARST Dry limestone country without surface streams, where the rainfall goes deeply underground through fissures and pot-holes.

KILLAS The Cornish name for the usual slates of that county, the Mylor Slates of Downtonian age.

LAPILLI Pieces of lava, usually porous, from the size of a pea to that of a walnut, that have been ejected from a volcano.

LAVA Any rock which issues in a molten state from a volcanic vent or fissure and consolidates on the ground surface; see also PILLOW LAVA.

LIGNITE Brown coal; it may be regarded as an intermediate stage between peat and coal.

LIMB OF A FOLD Flank; a fold is made up of two limbs, one on each side of the hinge of the structure.

LIMONITE Brown iron ore; an amorphous hydrated oxide of iron, of a brown or yellow colour.

LITHOGRAPHIC Descriptive of a very fine-grained uniform limestone such as that employed in lithography.

LITHOLOGY That branch of geology concerned with rocks *qua* rocks; also the general character of a rock.

LITTORAL FACIES Those kinds of deposits laid down near the shore.

LODE In mining, a vein of metal ore, relatively long compared with its thickness.

MAGMA The molten mother-material of an igneous rock, at depth in the earth's crust.

MAGMATIC DIFFERENTIATION Separation of a homogeneous magma into chemically unlike portions, e.g. by gravity, so that rocks erupted or intruded from the magma over a period may differ.

MANDIBLE The lower jaw or jawbone of an animal.

MANDIBULAR RAMUS The process of the lower jawbone, i.e. the rear end which articulates with the face-bones.

METAMORPHISM Change in the character of rocks brought about by heat or pressure or shear, or a combination of these, causing the production of new minerals and, in some cases, new structures, e.g. fissility or schistosity.

METASOMATISM Change in the substance of rocks by the removal of chemical constituents or the introduction of new ones brought about by gases or liquids rising from a cooling magma.

MICROFOSSILS Fossils so small that they can be studied only under the microscope.

MICROLITHIC GROUNDMASS The apparently featureless groundmass of an igneous rock which under the microscope is seen to contain microliths, incipient crystals in the form of minute rods or needles.

MICROZOA Microscopic animals.

MONOCLINE A pronounced bend in a series of sedimentary rocks which, on each side of it, are horizontal, or nearly so; this structure presents only half of a complete fold, so that only one limb is present instead of two.

MOON MILK Sometimes found in caves, a soft cheesy form of calcium carbonate seen as a deposit, as an exudation from joints, or on calcite formations. It is imperfectly understood and probably of several kinds.

MORAINE Detritus or rock waste left by an ice-sheet or glacier, consisting of earthy, sandy or clayey material (boulder clay), more or less charged with stones of all sizes up to blocks weighing many tons.

NAPPE A large-scale flake of the earth's crust which may start as an extreme case of overfolding or as the overlying part of a nearly horizontal overthrust, giving rise to a sheet of rocks perhaps several miles wide that has been pushed, sometimes for many miles, over newer rocks.

NEOLITHIC SUBSIDENCE Relative sinking of the land over parts of England during the Neolithic Period, about 3000 to 2000 B.C.

NEPTUNIAN DYKE In certain places, because of tension due to earth movements, more or less vertical fissures opened in the sea floor and became filled with sediments; these are now found as sedimentary rocks in the form of dykes.

NON-SEQUENCE A temporary stoppage in the continuous deposition of sediment so that the succession of strata is not complete at that place.

OCHRE Limonite (q.v.).

OOLITE Limestone in texture looking like the roe of a fish, composed of small rounded concentrically layered grains (ooliths).

OSSICLES Small plates or joints of chitinous or calcareous substance in the animal framework; e.g. crinoid ossicles, the small disc-like sections of the crinoid stem.

OUTLIER An exposure of newer rocks surrounded by older rocks, the newer rock being an outlying patch separated from the main outcrop.

OVATE A kind of palaeolithic hand-axe, so called from its ovate outline.

OVERSTEP An unconformable bed which passes across the upturned edges of the underlying older rocks is said to overstep each of them in turn.

OVERTURNED BEDS Strata the plane of whose bedding has been tilted by folding or faulting to any degree beyond the vertical.

PALAEOCLIMATE The climate of a past geological age.

PALAEOLITH A man-made stone implement of a palaeolithic culture of the Old Stone Age, roughly equivalent to the Pleistocene.

PALAEOSALINITY The salinity of a body of water, or of the sea, during a past geological age.

PAPER-SHALE A very finely laminated kind of shale which, on weathering, splits into a series of almost paper-thin layers.

PEGMATITE Any coarse-textured igneous rock; the individual crystals may sometimes be measured in feet; it is often associated with granites, and may contain rare minerals.

PERIGLACIAL Bordering an ice-sheet, where a frost climate obtained; during the glacial periods of the Pleistocene in England such conditions were found roughly south of a line joining the Bristol Channel and the mouth of the Thames.

PERMAFROST Perennially frozen ground.

PHENOCRYSTS Large crystals, usually perfect in shape, which occur in the fine-grained matrix of some igneous rocks.

PHREATIC ZONE Below the level of a water-table (see VADOSE ZONE).

PICRITE Very dark-coloured, ultrabasic, deep-seated (plutonic) rock consisting mainly of olivine and augite, with a little plagioclase feldspar.

PILLOW LAVA Or spilite; lava generally basaltic, which was poured out as a submarine flow; the rock consolidated under water as an irregular aggregate of blocks resembling a pile of soft cushions pressed upon and against one another; the pillows range from a few inches to several feet or even yards in diameter.

PINGO The Eskimo name that is used for a sizeable land surface feature found in periglacial regions, a domed hill, an ice-blister or intrusion, or a cone, believed to be formed from a body of unfrozen water under sufficient hydraulic head, trapped between permafrost below and a frozen surface layer, often of peaty lake sediments, above. The water, encroached upon by frost from all directions, turns to ice and exerts pressure upwards, the direction of least resistance, owing to its nine per cent expansion of volume on freezing. Modern pingos have been found of anything up to 140 feet high and 950 feet in diameter at the base; one experimental drilling proved an upper layer of sediments 47 feet thick overlying more than 180 feet of coarsely crystalline ice, without bottoming the ice intrusion. Dykes and sills of ice may be found round the margin.

PISOLITE Limestone similar to oolite but the concretionary grains are larger, the size of peas.

PITCHING FOLD A fold of the rocks, an anticline or a syncline, whose axis is inclined appreciably from the horizontal.

PLATEAU GRAVEL High-level spreads of gravel found in the southern half of England and often of doubtful origin; now mapped under the term Head (q.v.).

PLUTONIC ROCKS Those large and deep-seated masses of igneous rocks which have cooled slowly and are holocrystalline (q.v.), e.g. the Dartmoor Granite.

PORPHYRITIC STRUCTURE Of an igneous rock, having large crystals embedded in a groundmass of much finer texture.

PSEUDOMORPH A mineral (A) which has replaced another (B) and has retained the external crystal form of (B) instead of assuming its own proper crystal form.

PYROMETASOMATISM Metasomatic changes in rocks at or near their contact with an intrusive igneous rock mass, by emanations from the magma, combined with high temperature and pressure.

RADIOMETRIC OR ISOTOPIC AGE The age in years of certain minerals (and so of the rock in which they are found) is estimated by measuring the proportion in them of a certain isotope (A) of an element relative to another isotope (B), from which it is known that (A) must have been derived by radioactive change at a known half-life rate.

RAGSTONE A rough, hard, cavernous type of limestone, such as that formed in coral reefs.

RAISED BEACH See STRAND-LINE.

REGRESSION A retreat of the sea away from the land because of a relatively rising shore-line; the converse of transgression.

ROCHE MOUTONNÉE A prominent knob of rock over which glacial ice has flowed, so that the upstream side is smoothed and striated and the downstream side left rough. From a distance it may resemble a couched sheep.

ROTATIONAL SHEAR SLIP A kind of land-slipping in which the slip-plane is curved, as it were round a sector of an imaginary horizontal tube whose diameter may perhaps reach half a mile; the inner side of the slipped mass descends close against the mother-cliff, towards which the strata now dip, and the distant toe of the slip is forced upwards to form a ridge. Originally described by the Frenchman Alexandre Collin in 1846; see his *Landslides in Clays*, translated by W. R. Schriever, University of Toronto Press, 1956.

SARSENS Or greywethers (from their likeness to couched sheep); residual hard blocks of silicified sandstone which once formed part of a now vanished bed of sand belonging to the Eocene Reading Beds; found scattered over the Chalk of southern England, particularly Wiltshire.

SCHORL-ROCK A quartz-tourmaline rock, the final product of granite altered by magmatic vapours containing fluorine and boron.

SCORIA Rough clinker-like material such as fragments ejected during an explosive volcanic eruption.

SELENITE The transparent crystalline form of gypsum, calcium sulphate.

SEPTARIA Concretionary nodules from two or three inches to more than a foot in diameter, of clayey limestone or clay ironstone and found in clays; the core is of calcite, and the part near the centre has cracked open and the radial fissures have been filled with calcite which is visible on the surface.

SERPENTINE An altered ultrabasic rock now consisting of hydrated magnesium silicate; it is often dark-green or red with a peculiar mottled appearance and greasy lustre; rather a soft rock, it is used for ornamental purposes, taking a good polish.

SERPULITE A limestone or other sedimentary rock containing abundant specimens of *Serpula*, a genus of worms which build firm little calcareous tubes, often clustered together in great numbers.

SHATTER-BELT A belt of shattered but not intensely crushed rock. It may usually be considered as taking the place of a clean-cut fault.

SHELF FACIES The kind of deposits found in shallow open seas overlying the continental shelf.

SILL An intruded sheet of igneous rock which has penetrated for a considerable distance along a bedding plane of the strata, so that its area is great in comparison with its thickness.

SKARN A silicate gangue or matrix with various new minerals produced by contact metamorphism acting particularly upon limestones.

SLICKENSIDES Rubbed, smoothed and striated rock surfaces on the plane of a joint or fault showing that relative movement of the sides has taken place.

SLOCKER The local Mendip name for a swallet (q.v.).

SOLIFLUXION Soil creep; the process by which surface rubble on sloping ground gradually makes its way downhill year after year, largely by alternate freezing and thawing; specially effective in a cold climate, as in periglacial areas.

SPHERULITIC Having an internal structure of minutely fibrous crystals radiating from the centre.

SPILITE See PILLOW LAVA.

STALACTITE An icicle-like formation generally of calcium carbonate depending from the roof or sides of a cavern; it is produced by re-deposition of the material from dripping water which had percolated and dissolved a part of the overlying limestone.

STALACTITE 'STRAWS' Hollow thin-walled stalactites uniform in diameter throughout their length.

STALAGMITE An incrustation or deposit more or less like an inverted stalactite on the floor of a cavern; formed in the same way as a stalactite and of the same material.

STOCKWORK A mining term for metalliferous ore dispersed in an irregular network of veinlets in the country rock.

STRAND-LINE Or shore-line, raised beach, platform of erosion and deposit; conspicuous features at various levels on some coastlines, both above and (less conspicuously) below the present sea level; they are ancient sea beaches, mostly assignable to various dates in the Pleistocene, and important as records of the local sea level relative to the land at those dates; and often for their fossil faunas.

STRATIFIED Disposed in strata or layers.

STRATIGRAPHY The branch of geology that is concerned with the order and relative position of the strata in the earth's crust.

STRIKE The direction of a horizontal line at right angles to the direction of dip.

SUBLITHOGRAPHIC Nearly but not quite lithographic (q.v.).

SWALLET Or swallow-hole, pot-hole or sink: the opening, usually in limestone, through which a stream disappears, or once disappeared, to its underground course.

SYNCLINE A structure resulting from strata folded or bent into the form of a trough, and dipping on both sides towards the centre-line.

TEAR-FAULT A fault (q.v.) in which relative horizontal (rather than vertical) movement of the rocks predominates.

TECTONICS The branch of geology that is concerned with movement of the rocks in the earth's crust, such as folding, faulting, thrusting, and the resulting structures; 'the architecture of the earth's crust'.

THRUST An extreme kind of low-angled reversed fault, in which the rocks on one side of the fault have been driven nearly horizontally over those on the other.

THRUST PLANE The plane of dislocation in a thrust, upon which the movement has taken place.

TILL Boulder clay (q.v.).

TRANSGRESSION The invasion of the land by the sea over a relatively subsiding coast; the converse of regression (q.v.).

TRAP An old geological term not in current use, embracing all kinds of basaltic lavas.

TROGLOBITES Animals which live in the interior parts of caverns and spend their entire life cycle in darkness.

'TUBE' A peculiar kind of cave passage of nearly circular or elliptical section.

TUFA, CALCAREOUS A porous, concretionary or compact form of calcium carbonate deposited from solution in spring-water, sometimes by the agency of special vegetation, near the springhead and along the subsequent course of the stream.

TUFF, VOLCANIC A general term embracing all the finer kinds of detritus from coarse gravelly deposits to volcanic ash, and compact fine-grained rocks formed of dust blown out of a volcano during an explosive eruption, and later consolidated.

ULTRABASIC ROCKS Very dark-coloured igneous rocks such as peridotite (which alters to serpentine), and picrite (q.v.); they consist of ferromagnesian minerals such as olivine, augite and hornblende, and iron ores, with no quartz and little or no feldspar.

UNCONFORMITY A series of strata which follow one another uninterruptedly in sequence without any changes in their general parallel arrangement, is said to be *conformable;* when such a series rests upon an older series of rocks, or one with which continuity of deposition has been broken, such a junction is called an *unconformity*, and the upper series is said to be *unconformable* upon the lower rocks; (see ANGULAR DISCORDANCE and NON-SEQUENCE).

VADOSE ZONE The permeable zone above the water table in which water (vadose water) descends freely and is not under hydrostatic pressure; the water-table thus forms the boundary between the upper, vadose, zone and the lower, phreatic, zone.

VARIOLITE An igneous rock embedded with pea-like spherulites which give it a pock-marked appearance.

VARVES Late glacial and post-glacial laminated clays deposited in lakes during the retreat of the glaciers. They show annual alternations of thick coarse layers and thin finer-grained layers corresponding, respectively, to floods produced by rapid melting of glacier ice in summer, and slower melting in winter. By counting these varves back from an event of known date, the age in years of earlier events so far back as late glacial times have been obtained.

VERTEBRAL CENTRUM The thick, disc-like body of a vertebra.

XENOLITH The fragment of a foreign rock caught up by an igneous magma during its intrusion through the country rock, and found in the marginal part of the intruded mass, e.g. of granite.

References

AGER, D. V. and SMITH, W. E. (1965). *The coast of South Devon and Dorset between Brans-combe and Burton Bradstock.* Colchester; Benham. (Geologists' Association Guides, No. 23.)

ANDERSON, F. W. (1939). Wealden and Purbeck Ostracoda. *Ann. Mag. nat. Hist.* s. 11, **3**, 291–310.

ANDERSON, F. W. (1962). Correlation of the Upper Purbeck Beds of England with the German Wealden. *Lpool Manchr geol. J.* **3**, 21–32.

ANNISS, L. G. (1927). The geology of the Saltern Cove area, Torbay. *Q. Jl geol. Soc. Lond.* **83**, 492–500.

ANNISS, L. G. (1933). The Upper Devonian rocks of the Chudleigh area, south Devon. *Q. Jl geol. Soc. Lond.* **89**, 431–448.

APSIMON, A. M. (1969). 1919–1969: 50 years of archaeological research. *Proc. spelaeol. Soc.* **12**, 31–56.

APSIMON, A. M. and DONOVAN, D. T. (1956). Marine Pleistocene deposits in the Vale of Gordano, Somerset. *Proc. spelaeol. Soc.* **7**, 134–136.

APSIMON, A. M., DONOVAN, D. T. and TAYLOR, H. (1961). The stratigraphy and archaeo-logy of the Late-Glacial and Post-Glacial deposits at Brean Down, Somerset. *Proc. spelaeol. Soc.* **9**, 67–136.

ARBER, M. A. (1940). The coastal landslips of south-east Devon. *Proc. Geol. Ass.* **51**, 257–271.

ARBER, M. A. (1960). Pleistocene sea-levels in north Devon. *Proc. Geol. Ass.* **71**, 169–176. Discussion **72**, 469–471.

ARKELL, W. J. (1933). *The Jurassic System in Great Britain.* Oxford; Clarendon Press.

ARKELL, W. J. (1934). Whitsun field meeting, 1934: The Isle of Purbeck. *Proc. Geol. Ass.* **45**, 412–419.

ARKELL, W. J. (1935). The Portland Beds of the Dorset mainland. *Proc. Geol. Ass.* **46**, 301–347.

ARKELL, W. J. (1938). Three tectonic problems of the Lulworth district: studies on the middle limb of the Purbeck fold. *Q. Jl geol. Soc. Lond.* **94**, 1–54.

ARKELL, W. J. (1939). The ammonite succession at the Woodham Brick Company's pit, Akeman Street Station, Buckinghamshire, and its bearing on the classification of the Oxford Clay. *Q. Jl geol. Soc. Lond.* **95**, 135–222.

ARKELL, W. J. (1941). The Gastropods of the Purbeck Beds. *Q. Jl geol. Soc. Lond.* **97**, 79–128.

ARKELL, W. J. (1943). The Pleistocene rocks at Trebetherick Point, north Cornwall. *Proc. Geol. Ass.* **54**, 141–170.

ARKELL, W. J. (1947a). The geology of the country around Weymouth, Swanage, Corfe and Lulworth. Expl. Sheets 341, 342, 343 etc. *Mem. geol. Surv. U.K.*

ARKELL, W. J. (1947b). *Geology of Oxford.* Oxford; Clarendon Press.

AUSTEN, A. C. (1835). An account of the raised beach near Hope's Nose in Devonshire. *Proc. geol. Soc. Lond.* **2**, 102–103.

AUSTEN, J. H. (1852). *A guide to the geology of the Isle of Purbeck, and the south-west coast of Hampshire.* Blandford; Shipp.

AUSTEN, J. H. (1856). The Agglestone. *Pap. Purbeck Soc.* 79–81.

AUSTEN, J. H. (1857a). On the marine shells in the Blashenwell deposit. *Pap. Purbeck Soc.* 124–129.

AUSTEN, J. H. (1857b). On the discovery of fossil mammalian remains in Durlston Bay. *Pap. Purbeck Soc.* 172–174.

BADEN-POWELL, D. F. W. (1930a). On the geological evolution of the Chesil Bank. *Geol. Mag.* **67**, 499–513.

BADEN-POWELL, D. F. W. (1930b). Notes on raised beach mollusca from the Isle of Portland. *Proc. malac. Soc. Lond.* **19**, 67–76.

BADEN-POWELL, D. F. W. (1942). On the marine mollusca of Studland Bay, Dorset, and the supply of lime to the sand dunes. *J. Anim. Ecol.* **11**, 82–95.

BADEN-POWELL, D. F. W. (1953). Correlation of Pliocene and Pleistocene marine beds. *Nature, Lond.* **172**, 762–763.

BAKER, E. A. and BALCH, H. E. (1907). *The netherworld of Mendip*. Clifton, Bristol; Baker.

BALCH, H. E. (1914). *Wookey Hole: its caves and cave dwellers*. London; Oxford University Press.

BALCH, H. E. (1947a). *Mendip: The great cave of Wookey Hole*. 3rd ed., London; Simpkin, Marshall.

BALCH, H. E. (1947b). *Mendip: Cheddar its gorge and caves*. 2nd ed., London; Simpkin, Marshall.

BALCH, H. E. (1948). *Mendip: its swallet caves and rock shelters*. 2nd ed., London; Simpkin, Marshall.

BAMBER, A. E. (1924). The Avonian of the western Mendips from the Cheddar valley to the sea. *Proc. Bristol Nat. Soc.* s. 4, **6**, 75–91.

BARNARD, T. (1949). Foraminifera from the Lower Lias of the Dorset coast. *Q. Jl geol. Soc. Lond.* **105**, 347–391.

BARNARD, T. (1953). Foraminifera from the Upper Oxford Clay (Jurassic) of Redcliff Point, near Weymouth, England. *Proc. Geol. Ass.* **64**, 183–197.

BARRINGTON, N. (1957). *The caves of Mendip*. Clapham, Yorks; Dalesman Publications.

BARTON, R. M. (1964). *An introduction to the geology of Cornwall*. Truro; Truro Bookshop.

BEAUMONT, J. (1681). An account of Okey-hole, and several other subterraneous grottos and caverns in Mendip-hills, Somersetshire. *Phil. Collect. R. Soc.* No. 2, 1. (*Abridged Phil. Trans. R. Soc.* **2**, 1672–1683, 487–488.)

BELL, A. (1898). On the Pliocene shell beds at St Erth. *Trans. R. geol. Soc. Cornwall.* **12**, 111–166.

BENNETT, R. and others. (1956). A preliminary report on St Cuthbert's Swallet. *Bristol Explor. Club; Caving Rep.* No. 2, 1–17.

BLYTH, F. G. H. (1957). The Lustleigh fault in north-east Dartmoor. *Geol. Mag.* **94**, 291–296.

BONNEY, T. G. (1877). On the serpentine and associated rocks of the Lizard district. *Q. Jl geol. Soc. Lond.* **33**, 884–924.

BONNEY, T. G. (1888). Note on a picrite from the Liskeard district. *Mineralog. Mag.* **8**, 108–111.

BORLASE, W. (1758). *The natural history of Cornwall*. 2 vols. Oxford; private.

BOSWELL, P. G. H. (1923). The petrography of the Cretaceous and Tertiary outliers of the west of England. *Q. Jl geol. Soc. Lond.* **79**, 205–230.

BOTT, M. P. H., DAY, A. A. and MASSON-SMITH, D. (1958). The geological interpretation of gravity and magnetic surveys in Devon and Cornwall. *Phil. Trans. R. Soc.* **251**, A.992, 161–191.

BOULTON, W. S. (1904). On the igneous rocks at Spring Cove, near Weston-super-Mare. *Q. Jl geol. Soc. Lond.* **60**, 158–169.

BOWMAN, H. L. (1904) On the occurrence of bertrandite at the Cheesewring Quarry, near Liskeard, Cornwall. *Mineralog. Mag.* **16**, 47–50.

BRADSHAW, R. (1966). The Avon Gorge. *Proc. Bristol Nat. Soc.* **31**, 203–220.

BRIGHT, R. (1817). On the strata in the neighbourhood of Bristol. *Trans. geol. Soc. Lond.* s. 1, **4**, 193–205.

BRISTOW, H. W. and FISHER, O. (1857). Comparative vertical sections of the Purbeck strata of Dorsetshire. *Geol. Surv. vert. sects.* 22.

BRISTOW, H. W. and WHITAKER, W. (1869). On the formation of the Chesil Bank, Dorset. *Geol. Mag.* **6**, 433–438.

BRODIE, P. B. (1845). *A history of the fossil insects in the Secondary rocks of Britain*. London; John van Voorst.

BRODIE, P. B. (1850). On certain beds in the Inferior Oolite, near Cheltenham. *Q. Jl geol. Soc. Lond.* **6**, 239–249.

BROWN, P. A. (1966). A cist burial at Blashenwell tufa pit, Corfe Castle. *Proc. Dorset nat. Hist. archaeol. Soc.* **87**, 97, 98.

BROWN, P. R. (1963). Algal limestones and associated sediments in the basal Purbeck of Dorset. *Geol. Mag.* **100**, 565–573.

BUCKLAND, W. and CONYBEARE, W. D. (1824). Observations on the south-western coal district of England. *Trans. geol. Soc. Lond.* s. 2, **1**, pt. 2, 210–316.

BUCKLAND, W. and DE LA BECHE, H. T. (1836). On the geology of the neighbourhood of Weymouth and the adjacent parts of the coast of Dorset. *Trans. geol. Soc. Lond.* s. 2, **4**, 1–46.

BUCKMAN, S. S. (1889). The relations of Dundry with the Dorset-Somerset and Cotteswold areas during the Jurassic period. *Proc. Cotteswold Nat. Fld Club.* **9**, 374–387.

BUCKMAN, S. S. (1892). Inferior Oolite ammonites. *Palaeontogr. Soc. Monogn.* **40**, 293.

BUCKMAN, S. S. (1893). The Bajocian of the Sherborne district; its relation to subjacent and superjacent strata. *Q. Jl geol. Soc. Lond.* **49**, 479–522.

BUCKMAN, S. S. (1897). Deposits of the Bajocian age in the northern Cotteswolds: the Cleeve Hill plateau. *Q. Jl geol. Soc. Lond.* **53**, 607–629.

BUCKMAN, S. S. (1901). Excursion to Dundry Hill. *Proc. Geol. Ass.* **17**, 152–158.

BUCKMAN, S. S. (1910). Certain Jurassic (Lias—Oolite) strata of south Dorset, and their correlation. *Q. Jl geol. Soc. Lond.* **66**, 52–89.

BUCKMANN, S. S. (1922). Jurassic chronology II. Preliminary studies. Certain Jurassic strata near Eypesmouth (Dorset); the Junction-Bed of Watton Cliff and associated rocks. *Q. Jl geol. Soc. Lond.* **78**, 378–457.

BUCKMAN, S. S. and WILSON, E. (1896). On Dundry Hill. . . . *Q. Jl geol. Soc. Lond.* **52**, 669–720.

BUCKMAN, S. S. and WILSON, E. (1897). The geological structure of the upper portion of Dundry Hill. *Proc. Bristol Nat. Soc.* n.s., **8**, 188–231.

BURTON, E. ST. J. (1937). The origin of Lulworth Cove. *Geol. Mag.* **74**, 377–383.

BURY, H. (1950). Blashenwell tufa. *Proc. Bournemouth Nat. Sci. Soc.* **39**, 67–75.

BUSH, G. E. (1926). The Avonian of Spring Gardens and Vallis Vale, Frome, Somerset. *Proc. Bristol Nat. Soc.* s. 4, **6**, 250–259.

BUSH, G. E. (1930). Carboniferous Limestone (Avonian) succession in the Woodspring promontory, Weston-super-Mare. *Proc. Bristol Nat. Soc.* s. 4, **7**, 138–141.

CALKIN, J. B. (1953). Kimmeridge Coal—Money. *Proc. Dorset nat. Hist. archaeol. Soc.* **75**, 45–71.

CANTRILL, T. C., SHERLOCK, R. L. and DEWEY, H. (1919). Iron Ores: Sundry unbedded ores of Durham . . . Devon and Cornwall. *Spec. Rep. Min. Resour. G.B.* **9**, *Mem. geol. Surv. U.K.*

CARRECK, J. N. (1960). Whitsun field meeting to Weymouth, Abbotsbury and Dorchester, Dorset. *Proc. Geol. Ass.* **71**, 341–347.

CARRECK, J. N. and DAVIES, A. G. (1955). The Quaternary deposits of Bowleaze Cove, near Weymouth, Dorset. *Proc. Geol. Ass.* **66**, 74–100.

CASEY, R. (1961). The stratigraphical palaeontology of the Lower Greensand. *Palaeontology.* **3**, 487–621.

CASEY, R. (1963). The dawn of the Cretaceous period in Britain. *Bull. S.-East Un. scient. Socs.* **117**, 1–15.

CAVE RES. GP. GT. BR. (1956–1959). *Biol. Suppl.* 1–4. Berkhamsted, Herts. (cyclostyled).

CHANDLER, M. E. J. (1957). The Oligocene flora of the Bovey Tracey lake basin. *Bull. Br. Mus. nat. Hist. (Geology).* **3**, 73–123.

CHANDLER, M. E. J. (1962). *The Lower Tertiary floras of southern England. II. Flora of the Pipe-Clay series of Dorset (Lower Bagshot).* London; British Museum (Nat. Hist.).

CHANDLER, M. E. J. (1964). *A summary and survey of findings in the light of recent botanical observations. Lower Tertiary Floras of Southern England.* Vol. 4. London; B.M. (Nat. Hist.).

CHARIG, A. J. and NEWMAN, B. H. (1962). Footprints in the Purbeck. *New Scient.* **14**, 234–235.

CIFELLI, R. (1959). Bathonian foraminifera of England. *Bull. Mus. comp. Zool. Harvard.* **121**, 265–368.

CLARK, J. G. D. (1938). Microlithic industries from tufa deposits at Prestatyn, Flintshire, and Blashenwell, Dorset. *Proc. prehist. Soc.* n.s., **4**, 330–332.

CLEMENS, W. A. (1963). Late Jurassic mammalian fossils in the Sedgwick Museum, Cambridge. *Palaeontology.* **6**, 373–377.

275

REFERENCES

COLLINS, J. H. (1912). Observations on the west of England mining region. *Trans. R. geol. Soc. Cornwall.* **14,** i–xxiv, 1–683.

COLLINSON, J. (1791). *The history and antiquities of the county of Somerset.* 3 vols. Bath; Cruttwell.

CONYBEARE, W. D. and BUCKLAND, W. (1840). *Ten plates . . . representing the changes produced on the coast of east Devon between Axmouth and Lyme Regis by the subsidence of the land . . . on the 26th December 1839, and 3rd February 1840 . . .* London.

CONYBEARE, W. D. and PHILLIPS, W. (1822). *Outline of the geology of England and Wales.* London; W. Phillips.

COODE, J. (1853). Description of the Chesil Bank. . . . *Minut. Proc. Instn. civ. Engrs.* **12,** 520–559.

COPPER, P. (1965). Unusual structures in Devonian Atrypidae from England. *Palaeontology,* **8,** 358–373.

CORNISH, V. (1898a). On sea-beaches and sandbanks. Sec. 11. On the Chesil Beach, a local study in the grading of beach shingle. *Geogr. J.* **11,** 628–634.

CORNISH, V. (1898b). On the grading of the Chesil Beach shingle. *Proc. Dorset nat. Hist. antiq. Fld Club.* **19,** 113–121.

COSGROVE, M. E. and HEARN, E. W. (1966). Structures in the Upper Purbeck Beds at Peveril Point, Swanage, Dorset. *Geol. Mag.* **103,** 498–507.

COTT, H. B. (1961). Scientific results of an inquiry into the ecology and economic status of the Nile Crocodile (*Crocodilus niloticus*) in Uganda and Northern Rhodesia. *Trans. zool. Soc. Lond.* **29,** 211–356.

COX, L. R. (1925). The fauna of the Basal Shell-Bed of the Portland Stone, Isle of Portland. *Proc. Dorset nat. Hist. antiq. Fld Club.* **46,** 113–172.

COX, L. R. (1936). The gastropoda and lamellibranchia of the Green Ammonite Beds of Dorset. *Q. Jl geol. Soc. Lond.* **92,** 456–471.

COYSH, A. W. (1931) U-shaped burrows in the Lower Lias of Somerset and Dorset. *Geol. Mag.* **68,** 13–15.

CULLINGFORD, C. H. D. (Ed.) (1953). *British Caving.* London; Routledge. (2nd ed., 1962.)

CUMBERLAND, G. (1821). On the limestone beds of the River Avon, near Bristol. . . . *Trans. geol. Soc. Lond.* s. 1, **5,** 95–113.

CUMMINGS, R. H. (1958). The faunal analysis and stratigraphic application of Upper Palaeozoic smaller foraminifera. *Micropaleontology.* **4,** 1–24.

DAMON, R. (1884). *Geology of Weymouth, Portland, and coast of Dorsetshire from Swanage to Bridport-on-the-sea.* 3rd ed. Weymouth; R. F. Damon. London; Stanford.

DAVIDSON, T. (1825–1855). A monograph of British Brachiopoda. Part 2: Cretaceous Brachiopoda. *Palaeontogr. Soc. Monogr.*

DAVIDSON, T. (1870). Notes on the Brachiopoda hitherto obtained from the 'Pebble-Bed' of Budleigh Salterton near Exmouth in Devonshire. *Q. Jl geol. Soc. Lond.* **26,** 70–90.

DAVIDSON, T. (1881). On the Devonian and Silurian Brachiopoda that occur in the Triassic pebble-bed of Budleigh Salterton near Exmouth in Devonshire. *Palaeontogr. Soc. Monogr.* **4,** 317–368.

DAVIES, A. T. and KITTO, B. K. (1878). On some beds of sand and clay in the parish of St Agnes, Cornwall. *Trans. R. geol. Soc. Cornwall.* **9,** 196–203.

DAVIS, A. G. and CARRECK, J. N. (1958). Further observations on a Quaternary deposit at Bowleaze Cove, near Weymouth, Dorset. *Proc. Geol. Ass.* **69,** 120–122.

DAVIS, J. W. (1881). Notes on the fish remains of the bone bed at Aust, near Bristol. . . . *Q. Jl geol. Soc. Lond.* **37,** 414–426.

DAWKINS, W. B. (1862). On a hyaena den at Wookey Hole, near Wells. *Q. Jl geol. Soc. Lond.* **18,** 115–125.

DAWKINS, W. B. (1863a). On a hyaena den at Wookey Hole, near Wells. Part 2. *Q. Jl geol. Soc. Lond.* **19,** 260–274.

DAWKINS, W. B. (1863b). Wookey Hole hyaena den. II. *Proc. Somerset archaeol. nat. Hist. Soc.* **11,** 197–219.

DAWKINS, W. B. (1864). Rhaetic and White Lias of west and central Somerset. . . . *Q. Jl geol. Soc. Lond.* **20,** 396–412.

DAWKINS, W. B. (1865). *On the caverns of Burrington Combe explored in 1864 by Messrs. W. Ayshford Sanford and W. Boyd Dawkins.* Taunton, 1–16.

276

DAY, E. C. H. (1863). On the Middle and Upper Lias of the Dorsetshire coast. *Q. Jl geol. Soc. Lond.* **19,** 278–297.

DAY, E. C. H. (1866). On a raised beach and other Recent formations near Weston-super-Mare. *Geol. Mag.* **3,** 115–119.

DEARMAN, W. R. (1959). The structure of the Culm Measures at Meldon, near Oke-hampton, north Devon. *Q. Jl geol. Soc. Lond.* **115,** 65–106.

DEARMAN, W. R. (1962). *Dartmoor.* Colchester; Benham. (Geologists' Association Guides, No. 23.)

DEARMAN, W. R. (1963). Wrench-faulting in Cornwall and south Devon. *Proc. Geol. Ass.* **74,** 265–287.

DEARMAN, W. R. and BUTCHER, N. E. (1959). The geology of the Devonian and Car-boniferous rocks of the north-west border of the Dartmoor granite, Devonshire. *Proc. Geol. Ass.* **70,** 51–92.

DEARMAN, W. R. and CLARINGBULL, G. F. (1960). Bavenite from the Meldon aplite quarries, Okehampton, Devon. *Mineralog. Mag.* **32,** 577–578.

DE LA BECHE, H. T. (1822). Remarks on the geology of the south coast of England from Bridport Harbour, Dorset, to Babbacombe Bay, Devon. *Trans. geol. Soc. Lond.* s. 2, **I,** 40–47.

DE LA BECHE, H. T. (1826a). On the Lias of the coast in the vicinity of Lyme Regis, Dorset. *Trans. geol. Soc. Lond.* s. 2, **2,** 21–30.

DE LA BECHE, H. T. (1826b). On the Chalk and sands beneath it (usually termed Green-Sand) in the vicinity of Lyme Regis, Dorset, and Beer, Devon. *Trans. geol. Soc. Lond.* s. 2, **2,** 109–118.

DE LA BECHE, H. T. (1835). Note on the Trappean rocks associated with the (New) Red Sandstone of Devonshire. *Proc. geol. Soc. Lond.* **2,** 196–198.

DE LA BECHE, H. T. (1839). Report on the geology of Cornwall, Devon, and west Somerset. *Mem. geol. Surv. U.K.*

DE LA BECHE, H. T. (1846). On the formation of the rocks of south Wales and south western England. *Mem. geol. Surv. U.K.* **1,** 253.

DELAIR, J. B. (1958–1960). The Mesozoic reptiles of Dorset. *Proc. Dorset nat. Hist. archaeol. Soc.* I. 1958, **79,** 47–72; II. 1959, **80,** 52–90; III. 1960, **81,** 59–85.

DELAIR, J. B. (1966). New records of Dinosaurs and other fossil reptiles from Dorset. *Proc. Dorset nat. Hist. archaeol. Soc.* **87,** 57–66.

DE LUC, J. A. (1811). *Geological travels.* Vols 2 and 3. London.

DEWEY, H. (1910). Notes on some igneous rocks from north Devon. *Proc. Geol. Ass.* **21,** 457–472.

DEWEY, H. (1913). The raised beach of north Devon. *Geol. Mag.* **50,** 154–163.

DEWEY, H. (1914). The geology of north Cornwall. *Proc. Geol. Ass.* **25,** 154–179.

DEWEY, H. (1916). On the origin of some river-gorges in Cornwall and Devon. *Q. Jl geol. Soc. Lond.* **72,** 63–76.

DEWEY, H. (1948). *South-west England.* 2nd ed. Brit. reg. Geol., geol. Surv. U.K.

DEWEY, H. (1949). The geology of south Devon. *Trans. Proc. Torquay nat. Hist. Soc.* **10,** 59–69.

DEWEY, H. and FLETT, J. S. (1911). British pillow lavas and the rocks associated with them. *Geol. Mag.* 202–209; 241–248.

DINELEY, D. L. (1961). The Devonian System in south Devonshire. *Fld Stud.* **I,** 121–140.

DINELEY, D. L. and RHODES, F. H. T. (1956). Conodont horizons in the west and south-west of England. *Geol. Mag.* **93,** 242–248.

DINELEY, D. L. (1963). Contortions in the Bovey Beds (Oligocene), S.W. England. *Biuletin Peryglacjalny.* Lodz. No. 12, 151–160.

DINES, H. G. (1956). The metalliferous mining region of south-west England. *Mem. geol. Surv. U.K.*

DIVER, C. (1933). The physiography of South Haven peninsula, Studland Heath, Dorset. *Geogr. J.* **81,** 404–427.

DODSON, M. H. (1961). Isotopic ages from the Lizard peninsula, south Cornwall. *Proc. geol. Soc. Lond.* 133–136.

DONOVAN, D. T. (1955). The Pleistocene deposits at Gough's Cave, Cheddar. . . . *Proc. spelaeol. Soc.* **7,** 76–104.

277

DONOVAN, D. T. (1956). The zonal stratigraphy of the Blue Lias around Keynsham, Somerset. *Proc. Geol. Ass.* **66,** 182–212.

DONOVAN, D. T. (1960). Gravels below the flood plain of the Bristol Avon at Keynsham. *Proc. Bristol Nat. Soc.* **30,** 55–66.

DONOVAN, D. T. (1969). Geomorphology and hydrology of the central Mendips. *Proc. spelaeol. Soc.* **12,** 63–74.

DOWNES, W. (1882). The zones of the Blackdown Beds and their correlation with those at Haldon. *Q. Jl geol. Soc. Lond.* **38,** 75–94.

DOWNIE, C. (1956). Microplankton from the Kimeridge Clay. *Q. Jl geol. Soc. Lond.* **112,** 413–434.

DREGHORN, W. (1965). *Leckhampton Hill.* Cheltenham; Author, St Paul's College.

DREGHORN, W. (1966). *Geology and the Severn Bridge.* Cheltenham; Author, St Paul's College.

DUNCAN, P. M. (1879). On the Upper Greensand coral fauna of Haldon, Devonshire. *Q. Jl geol. Soc. Lond.* **35,** 89–96.

EDGELL, A. W. (1874). Notes on some lamellibranchs of the Budleigh Salterton pebbles. *Q. Jl geol. Soc. Lond.* **30,** 45–49.

EDMONDS, E. A. *et al.* (1968). Geology of the country around Okehampton. Expl. Sheet 324. *Mem. geol. Surv. U.K.*

EDMUNDS, F. H. (1938). (In: discussion, in W. V. Lewis: The evolution of shoreline curves. *Proc. Geol. Ass.* **49,** 107–127).

EGERTON, P. DE M. G. (1841). A catalogue of fossil fish in the collections of the Earl of Enniskillen and Sir Philip Grey Egerton. *Ann. Mag. nat. Hist.* **7,** 487–498.

EGERTON, P. DE M. G. (1879). The Lias fishes of Lyme Regis. *In,* T. Wright, 1878–1886, Monograph of the Lias ammonites of the British Islands, II. *Palaeontogr. Soc. Monogr.* **33,** 61–64.

ELLIOTT, G. F. (1961). A new British Devonian alga, *Palaeoporella lummatonensis,* and the brachiopod evidence of the age of the Lummaton Shell-Bed. *Proc. Geol. Ass.* **72,** 251–260.

ELLIOTT, G. F. (1963). Whitsun field meeting report: Devonian of Torbay. *Proc. Geol. Ass.* **74,** 81–85.

ERICSON, D. B., EWING, M. and WOLLIN, G. (1964). The Pleistocene epoch in deep-sea sediments. *Science, N.Y.,* **146,** 723–732.

EVANS, J. (1897). *The ancient stone implements, weapons and ornaments of Great Britain.* 2nd ed. London; Longmans.

EVENS, E. D. (1958). Origin of picrite blocks near Wells. *Geol. Mag.* **95,** 511.

EXLEY, C. S. (1958). Magmatic differentiation and alteration in the St Austell granite. *Q. Jl geol. Soc. Lond.* **114,** 197–230.

FITTON, W. H. (1824). Inquiries respecting the geological relations of the beds between the Chalk and the Purbeck limestone in the south-east of England. *Ann. Phil.* **24,** 365–458.

FITTON, W. H. (1836). Observations on some of the strata between the Chalk and the Oxford Oolite in the south-east of England. *Trans. geol. Soc. Lond.* s. 2, **4,** 103–388.

FLETT, J. S. (1913). The geology of the Lizard. *Proc. Geol. Ass.* **24,** 118–133.

FLETT, J. S. (1946). Geology of the Lizard and Meneage. 2nd ed. Expl. sheet 359. *Mem. geol. Surv. U.K.*

FLETT, J. S. and HILL, J. B. (1913). Report of an excursion to the Lizard, Cornwall. *Proc. Geol. Ass.* **24,** 313–327.

FORBES, E. (1850). On the succession of strata and distribution of organic remains in the Dorsetshire Purbecks. *Rep. Br. Ass. Advmt. Sci. Trans. Sect.* 79–81.

FORBES, J. (1822). On the geology of the Land's End district. *Trans. R. geol. Soc. Cornwall.* **2,** 242–280.

FORD, D. C. and STANTON, W. I. (1968). The geomorphology of the south-central Mendip Hills. *Proc. Geol. Ass.* **79,** 401–427.

FOSTER, C. LE N. (1875). Notes on Haytor Iron Mine. *Q. Jl geol. Soc. Lond.* **31,** 628–630.

FROST, G. A. (1926). Otoliths of fishes from the Jurassic of Buckinghamshire and Dorset. *Ann. Mag. nat. Hist.* **18,** 81–85.

FRYER, G. (1960). Evolution of the land form of Kerrier. *Trans. R. geol. Soc. Cornwall.* **19,** 122–153.

GARDINER, C. I. and Others. (1934). The geology of the Gloucester district. *Proc. Geol. Ass. Lond.* **45,** 109–144; excursion, 445–450.

GEIKIE, A. and STRAHAN, A. (1898). Volcanic group in the Carboniferous Limestone of north Somerset. *Mem. Geol. Surv. U.K. Summ. Prog.* 104–111.

GEORGE, T. N. (1958). Lower Carboniferous palaeogeography of the British Isles. *Proc. Yorks. geol. Soc.* **31,** 227–318.

GHOSH, P. K. (1927). The petrology of the Bodmin Moor granite (eastern part), Cornwall. *Mineralog. Mag.* **21,** 285–309.

GILBERT, E. B. (1963). An account of recent developments in G. B. Cave, Charterhouse-on-Mendip, Somerset. *Proc. spelaeol. Soc.* **10,** 58–64.

GILBY, W. H. (1817). On the Magnesian Limestone and the Red Marl or sandstone in the neighbourhood of Bristol. *Trans. geol. Soc. Lond.* s. 1, **4,** 210–215.

GLENNIE, E. A. (1957). The two Plumleys of Burrington Combe. *Newsl. Cave Res. Grp Gt Br.* No. 68/69, 4–5.

GOOD, R. (1946). *Weyland: the story of Weymouth and its countryside.* Dorchester; Longmans.

GORDON, W. A. (1965). Foraminifera from the Corallian beds, Upper Jurassic, of Dorset, England. *J. Paleont.* **39,** 828–863.

GRAY, J. W. (1920). Leckhampton Hill. *Trans. Worcs. Nat. Club.* **7,** 255–264.

GREEN, G. W. and WELCH, F. B. A. (1965). Geology of the country around Wells and Cheddar. Expl. Sheet 280. *Mem. Geol. Surv. U.K.*

GREEN, J. F. N. (1947). Some gravels and gravel pits in Hampshire and Dorset. *Proc. Geol. Ass.* **58,** 128–143.

GROVES, A. W. (1931). The unroofing of the Dartmoor granite and the distribution of its detritus in the sediments of southern England. *Q. Jl geol. Soc. Lond.* **87,** 62–94.

HALLAM, A. (1956). The Rhaetic and Lias beds at Tolcis, near Axminster. *Proc. Dorset nat. Hist. archaeol. Soc.* **78,** 58–63.

HALLAM, A. (1960). A sedimentary and faunal study of the Blue Lias of Dorset and Glamorgan. *Phil. Trans. R. Soc.* **243,** B.698, 1–44.

HALLAM, A. (1961). Cyclothems, transgressions and faunal changes in the Lias of north-west Europe. *Trans. Edinb. geol. Soc.* **18,** 124–174.

HALLAM, A. and PAYNE, K. W. (1958). Germanium enrichment in lignites from the Lower Lias of Dorset. *Nature, Lond.* **181,** 1008–1009.

HAMPTON, J. S. (1957). Some Holothurian spicules from the Upper Bathonian of the Dorset coast. *Geol. Mag.* **94,** 507–510.

HANDLIRSCH, A. (1906–1908). *Die Fossilen Insekten* . . . Leipzig; Engelmann.

HARMER, F. W. (1907). On the origin of certain canyon-like valleys associated with lake-like areas of depression. *Q. Jl geol. Soc. Lond.* **63,** 470–514.

HARMER, F. W. (1914–1925). The Pliocene Mollusca of Great Britain. *Palaeontogr. Soc. Monogr.*

HARRIS, T. M. (1939). *British Purbeck Charophyta.* London; British Museum. (Nat. Hist.) cat.

HAWKES, C. F. C. (1943). Two palaeoliths from Broom, Dorset. *Proc. prehist. Soc.* n.s., **9,** 48–52.

HAWKINS, A. B. (1962). The buried channel of the Bristol Avon. *Geol. Mag.* **99,** 369–374.

HAWKINS, J. (1832). On a very singular deposit of alluvial matter on St Agnes Beacon and on the granitical rock which occurs in the same situation. *Trans. R. geol. Soc. Cornwall.* **4,** 135–144.

HEAP, W. (1958). The Mammal Bed of Durlston Bay. *Yearb. Soc. Dorset Men in Lond.* 1957/58, 83–85.

HEER, O. (1862). On the fossil flora of Bovey Tracey. *Phil. Trans. R. Soc.* **152,** 1039–1066.

HENDRIKS, E. M. L. (1937). Rock succession and structure in south Cornwall. *Q. Jl geol. Soc. Lond.* **93,** 322–367.

HENDRIKS, E. M. L. (1959). A summary of present views on the structure of Cornwall and Devon. *Geol. Mag.* **96,** 253–257.

HENWOOD, W. J. (1858). Notice of the submarine forest near Padstow. *Roy. Inst. Cornwall.* 40th Ann. Rep., Appendix I, 17–19.

HEPWORTH, J. V. and STRIDE, A. H. (1950). A sequence from the Old Red Sandstone to Lower Carboniferous, near Burrington, Somerset. *Proc. Bristol Nat. Soc.* **28**, 135–138.

HINDE, G. J. (1887–1912). A Monograph of the British fossil sponges. 1893, vol. I, part III: Sponges of Jurassic strata. *Palaeontogr. Soc. Monogr.* **45**, 189–254.

HODSON, F., HARRIS, B. and LAWSON, L. (1956). Holothurian spicules from the Oxford Clay of Redcliff, near Weymouth (Dorset). *Geol. Mag.* **93**, 336–344.

HOOKE, R. (1705). *The posthumous Works of Robert Hooke.* V. Discourses on earthquakes (p. 286). London; S. Smith and B. Walford.

HOOPER, J. H. D. (1939). Baker's Pit Cave, Buckfastleigh, Devon. *J. Mendip Explor. Soc.* **4**, 87–92.

HOOPER, J. H. D. (1947). The caverns at Buckfastleigh. *Rep. Trans. Devon Ass. Advmt Sci.* **79**, 113–116.

HOOPER, J. H. D. (1950). Reed's Cave, Buckfastleigh. *Rep. Trans. Devon Ass. Advmt Sci.* **82**, 291–294.

HORNER, L. (1816). Sketch of the geology of the south western part of Somersetshire. *Trans. geol. Soc. Lond.* s. 1, **3**, 228–384.

HOUSE, M. R. (1956). Devonian goniatites from north Cornwall. *Geol. Mag.* **93**, 257–262.

HOUSE, M. R. (1958). *The Dorset coast from Poole to the Chesil Beach.* Colchester; Benham. (Geologists' Association Guides, No. 22).

HOUSE, M. R. (1961). The structure of the Weymouth anticline. *Proc. Geol. Ass.* **72**, 221–238.

HOUSE, M. R. (1963a). Devonian ammonoid successions and facies in Devon and Cornwall. *Q. Jl geol. Soc. Lond.* **119**, 1–28.

HOUSE, M. R. (1963b). Dorset geology. *Proc. Dorset nat. Hist. archaeol. Soc.* **84**, 77–91.

HOWARTH, M. K. (1957). The Middle Lias of the Dorset coast. *Q. Jl geol. Soc. Lond.* **113**, 185–204.

HOWE, J. A. (1914). A handbook to the collection of Kaolin, China-clay and China stone in the Museum of Practical Geology. *Mem. geol. Surv. U.K.*

HUDLESTON, W. H. (1887). Monograph of the British Jurassic Gasteropoda. Part I, Gasteropoda of the Inferior Oolite. *Palaeontogr. Soc. Monogr.* **40**, 56.

HUGHES, T. M. (1887). On the ancient beaches and boulders near Braunton and Croyde. *Q. Jl geol. Soc. Lond.* **43**, 657–670.

HULL, E. (1857). The geology of the country around Cheltenham. Expl. Sheet 44. *Mem. geol. Surv. U.K.*

HULL, E. (1903). The Cheesewring, Cornwall, and its teachings. *J. Trans. Vict. Inst.* **35**, 140–148.

HUTCHINS, J. (1774). *The history and antiquities of the county of Dorset.* 3rd ed., rev., 4 vols. London; J. B. Nichols (1861).

ILCHESTER v. RAISHLEY (1888). *Transcripts of plaintiff's documents of title to Abbotsbury Estate.* Dorchester, Archive Office. [Unpublished.]

JACKSON, J. F. (1922). Appendix I: Sections of the Junction-Bed and contiguous deposits (*In:* S. S. Buckman: Jurassic Chronology II. *Q. Jl geol. Soc. Lond.* **78**, 436–448).

JACKSON, J. F. (1926). The Junction-Bed of the Middle and Upper Lias on the Dorset coast. *Q. Jl geol. Soc. Lond.* **82**, 490–525.

JENKIN, A. K. H. (1961). *Mines and miners of Cornwall.* 1. Around St Ives. Truro; Truro Bookshop.

JOFFE, J. (1967). The 'dwarf' crocodiles of the Purbeck formation, Dorset: a reappraisal. *Palaeontology.* **10**, 629–639.

JONES, T. R. (1885). On the Ostracods of the Purbeck formation; with notes on the Wealden species. *Q. Jl geol. Soc. Lond.* **41**, 311–355.

JUDD, J. W. (1871). On the Punfield formation. *Q. Jl geol. Soc. Lond.* **27**, 207–227.

JUKES-BROWNE, A. J. (1898). On an outlier of Cenomanian and Turonian near Honiton. *Q. Jl geol. Soc. Lond.* **54**, 239–250.

JUKES-BROWNE, A. J. (1900). The Cretaceous Rocks of Britain. 1. The Gault and Upper Greensand of England. *Mem. geol. Surv. U.K.*

JUKES-BROWNE, A. J. (1903). The Cretaceous Rocks of Britain. 2. The Lower and Middle Chalk of England. *Mem. geol. Surv. U.K.*

JUKES-BROWNE, A. J. (1904). The geology of the country round Chard. *Proc. Somerset archaeol. nat. Hist. Soc.* **49**, 12–22.

JUKES-BROWNE, A. J. (1906). The Devonian limestones of Lummaton Hill, near Torquay. *Proc. Geol. Ass.* **19,** 291–302.

JUKES-BROWNE, A. J. (1911). *The Building of the British Isles.* 3rd ed. London; Stanford.

KAYSER, E. (1889). Ueber das Devon in Devonshire und im Boulonnais. *Neues Jb. Miner.* **I,** 179–190.

KEELING, P. S. (1961). Cornish Stone. *Trans. Br. Ceram. Soc.* **60,** 390–426.

KELLAWAY, G. A. and WELCH, F. B. A. (1948). *Bristol and Gloucester District.* 2nd ed. Brit. reg. Geol., Geol Surv. U.K.

KELLAWAY, G. A. and WELCH, F. B. A. (1955). The Upper Old Red Sandstone and Lower Carboniferous rocks of Bristol and the Mendips compared with those of Chepstow and the Forest of Dean. *Bull. geol. Surv. Gt Br.* **9,** 1–21.

KENDALL, P. F. and BELL, R. G. (1886). On the Pliocene beds of St Erth. *Q. Jl geol. Soc. Lond.* **42,** 201–215.

KENNARD, A. S. (1945). The early digs in Kent's Hole, Torquay, and Mrs Cazalet. *Proc. Geol. Ass.* **56,** 156–213.

KENNARD, A. S. and WOODWARD, B. B. (1901). The Post-Pliocene non-marine mollusca of the south of England. *Proc. Geol. Ass.* **17,** 213–268.

KING, W. B. R. (1954). The geological history of the English Channel. *Q. Jl geol. Soc. Lond.* **110,** 77–101.

KINGSBURY, A. W. G. (1941). Mineral localities on the Mendip Hills, Somerset. *Mineralog. Mag.* **26,** 67–80.

KINGSBURY, A. W. G. (1954). New occurrences of rare copper and other minerals in Devon and Cornwall. *Trans. R. geol. Soc. Cornwall.* **18,** 386–406.

KINGSBURY, A. W. G. (1958). Two beryllium minerals new to Britain: euclase and herderite. *Mineralog. Mag.* **31,** 815–817.

KINGSBURY, A. W. G. (1961). Beryllium minerals in Cornwall and Devon: helvine, genthelvite, and danalite. *Mineralog. Mag.* **32,** 921–940.

KINGSBURY, A. W. G. and HARTLEY, J. (1956). Atacamite from Cumberland and Cornwall. *Mineralog. Mag.* **31,** 349–350.

KINGSTON, J. T. (1828). Account of the iron mine at Haytor. *Phil Mag.* **3,** 359–365.

KOWALSKI, K. (1966). Cave palaeontology. *Newsl. Assoc. William Pengelly Cave. Res. Centre.* **6,** 9–12.

KÜHNE, W. G. (1947). The geology of the fissure-filling 'Holwell 2' . . . *Proc. zool. Soc. Lond.* **116,** 729–733.

LANG, W. D. (1904). The zone of *Hoplites interruptus* (Bruguière) at Black Ven, Charmouth. *Geol. Mag.* 124–131.

LANG, W. D. (1907). The Selbornian of Stonebarrow Cliff, Charmouth. *Geol. Mag.* 150–156.

LANG, W. D. (1914). The geology of the Charmouth cliffs, beach and foreshore. *Proc. Geol. Ass.* **25,** 293–360.

LANG, W. D. (1924). The Blue Lias of the Devon and Dorset coasts. *Proc. Geol. Ass.* **35,** 169–185.

LANG, W. D. (1928). The Belemnite Marls of Charmouth, a series in the Lias of the Dorset coast. *Q. Jl geol. Soc. Lond.* **84,** 179–257.

LANG, W. D. (1932). The Lower Lias of Charmouth and the Vale of Marshwood. *Proc. Geol. Ass.* **43,** 97–126.

LANG, W. D. (1936). The Green Ammonite Beds of the Dorset Lias. *Q. Jl geol. Soc. Lond.* **92,** 423–455.

LANG, W. D. (1946). Geological notes. *Proc. Dorset nat. Hist. archaeol. Soc.* **68,** 90.

LANG, W. D. (1957). Report on Dorset natural history for 1957. Geology. *Proc. Dorset nat. Hist. archaeol. Soc.* **78,** 23.

LANG, W. D. and SPATH, L. F. (1926). The Black Marl of Black Ven and Stonebarrow in the Lias of the Dorset coast. *Q. Jl geol. Soc. Lond.* **82,** 144–187.

LANG, W. D., SPATH, L. F. and RICHARDSON, W. A. (1923). 'Shales-with-Beef', a sequence in the Lower Lias of the Dorset coast. *Q. Jl geol. Soc. Lond.* **79,** 47–99.

LANG, W. D. and THOMAS, H. D. (1936). Whitsun Field Meeting, 1936. The Lyme Regis District. *Proc. Geol. Ass.* **47,** 301–315.

REFERENCES

LAWSON, J. D. (1955). The geology of the May Hill Inlier. Q. Jl geol. Soc. Lond. 111, 85–116.
LEE, J. E. (1877). Notice of the discovery of Upper Devonian fossils in the shales of Torbay. Geol. Mag. 100–102.
LEECH, J. G. C. (1929). St Austell detritals. Proc. Geol. Ass. 40, 139–146.
LELAND, J. (1907). The itinerary of John Leland in or about the years 1535–1543. Ed. L. T. Smith, 5 vols London; Bell.
LLOYD, A. J. (1959). Arenaceous foraminifera from the type Kimeridgian (Upper Jurassic). Palaeontology. 1, 298–320.
LLOYD, A. J. (1962). Polymorphinid, miliolid and rotaliform foraminifera from the type Kimmeridgian. Micropaleontology. 8, 369–383.
LLOYD, W. (1933). Geology of the country around Torquay. 2nd ed. Expl. Sheet 350. Mem. geol. Surv. U.K.
LONG, C. D. F. (1932). Holwell Cave Survey. Wells nat. Hist. Arch. Soc., pub. Mendip nat. Research Ctte.
LOVE, L. G. (1962). Further studies on micro-organisms and the presence of syngenetic pyrite. Palaeontology. 5, 444–459.
LYDEKKER, R. (1899). Note on a fossil crocodile from Chickerell. Proc. Dorset nat. Hist. antiq. Fld Club. 20, 171–173.
LYDEKKER, R. (1906). Palaeontology. Victoria Hist. Counties of England: Somersetshire. 1, 35–39.
MACALISTER, D. A. (1912). In: Reid, C. Geology of Dartmoor. Expl. Sheet 338. Mem. geol. Surv. U.K.
MACFADYEN, W. A. (1941). Foraminifera from the Green Ammonite Beds, Lower Lias, of Dorset. Phil. Trans. R. Soc. 231, B.576, 1–73.
MACKINTOSH, D. (1868). On the mode and extent of encroachment of the sea on some parts of the shores of the Bristol Channel. Q. Jl geol. Soc. Lond. 24, 279–283.
MANSEL, J. C. (1857). On the Pleistocene tufaceous deposit at Blashenwell. Pap. Purbeck Soc. 120–123.
MANSELL-PLEYDELL, J. C. (1886). On a tufaceous deposit at Blashenwell, Isle of Purbeck. Proc. Dorset nat. Hist. antiq. Fld Club. 7, 109–113.
MANSELL-PLEYDELL, J. C. (1895). Anniversary address of the President. Proc. Dorset nat, Hist. antiq. Fld Club. 16, lxvi.
MARTIN, A. J. (1967) Bathonian sedimentation in southern England. Proc. Geol. Ass. 78, 473–488.
MATON, W. G. (1797). Observations relative chiefly to the natural history, picturesque scenery, and antiquities of the western counties of England made in the years 1794 and 1796. 2 vols. Salisbury; Eaton.
MEYER, C. J. A. (1872). On the Wealden as a fluvio-lacustrian formation and on the relations of the so called Punfield formation to the Wealden and Neocomian. Q. Jl geol. Soc. Lond. 28, 243–255.
MEYER, C. J. A. (1873). Further notes on the Punfield formation. Q. Jl geol. Soc. Lond. 29, 70–76.
MIDDLETON, G. V. (1960). Spilitic rocks in south-east Devonshire. Geol. Mag. 97, 192–207.
MIDDLETON, J. (1812). Continuation of the account of the British mineral strata. Monthly Mag., 1 December 1812, 395
MILLER, J. A. and GREEN, D. H. (1961). Preliminary age-determinations in the Lizard area. Nature, Lond. 191, 159–160.
MILLES, J. (1760). Remarks on the Bovey coal. Phil. Trans. R. Soc. 51, 534–553.
MILLETT, F. W. (1887–1902). Notes on the fossil foraminifera of the St Erth clay pits. Trans. R. geol. Soc. Cornwall 10, 213–216; 222–226; 11, 655–661; 12, 43–46; 174–176; 719–720.
MILNE-EDWARDS, H. and HAIME, J. (1853). British fossil corals. Pt. 4: Devonian. Palaeontogr. Soc. Monogr.
MILNER, H. B. (1922). The nature and origin of the Pliocene deposits of the county of Cornwall . . . Q. Jl geol. Soc. Lond. 78, 348–377.

MITCHELL, G. F. (1965). The St Erth Beds—an alternative explanation. *Proc. Geol. Ass.* **76**, 345–366.

MOIR, J. R. (1936). Ancient man in Devon. *Proc. Devon Archaeol. Explor. Soc.* **2**, 264–281.

MOORE, C. (1858). On Triassic beds near Frome and their organic remains. *Rep. Br. Ass. Advmt. Sci.* pt. 2, 93–94.

MOORE, C. (1860). On the contents of three square yards of Triassic Drift. *Rep. Br. Ass. Advmt. Sci., Trans. Sects.* 87–88.

MOORE, C. (1867). On abnormal conditions of secondary deposits when connected with the Somersetshire and South Wales coal-basins, and on the age of the Sutton and Southerndown Series. *Q. Jl geol. Soc. Lond.* **23**, 440–568.

MOORE, G. W. (1956). Aragonite Speleothems as indicators of paleotemperatures. *Am. J. Sci.* **254**, 746–753.

MORGAN, C. L. (1890). Mendip notes. *Proc. Bristol Nat. Soc.* s. 3, **6**, 169–182.

MORGAN, C. L. and REYNOLDS, S. H. (1901). The igneous rocks and associated sedimentary beds of the Tortworth inlier. *Q. Jl geol. Soc. Lond.* **57**, 267–284.

MORGAN, C. L. and REYNOLDS, S. H. (1903). The field relations of the Carboniferous volcanic rocks of Somerset. *Proc. Bristol Nat. Soc.* s. 3, **10**, 188–212.

MORGAN, C. L. and REYNOLDS, S. H. (1904). The igneous rocks associated with the Carboniferous Limestone of the Bristol district. *Q. Jl geol. Soc. Lond.* **60**, 137–157.

MUIR-WOOD, H. M. (1936). Brachiopoda from the Lower Lias, Green Ammonite Beds of Dorset. *Q. Jl geol. Soc. Lond.* **92**, 472–485.

MURCHISON, R. I. (1834). *Outline of the geology of the neighbourhood of Cheltenham.* Cheltenham, n.p., 1–40.

MURCHISON, R. I. (1845). *Outline of the geology of the neighbourhood of Cheltenham.* New ed. rev. by J. Buckman and H. E. Strickland. London, 1–109.

NEATE, D. J. M. (1967). Underwater pebble grading of Chesil Bank. *Proc. Geol. Ass.* **78**, 419–426.

OAKLEY, K. P. (1949). *Man the Tool-Maker.* London; British Museum (Nat. Hist.).

OAKLEY, K. P. (1964). *Frameworks for dating Fossil Man.* London; Weidenfeld & Nicolson.

OGILVIE, A. H. (1941). Kent's Cavern. *Sch. Nat. Study*, October, 120–125.

ORME, A. R. (1960). The raised beaches and strandlines of south Devon. *Fld Stud.* **1**, 109–130.

OWEN, E. (1754). *Observations on the earth, rocks, stones and minerals for some miles about Bristol and on the nature of the Hot Well and the virtues of its water.* London; W. Johnston.

OWEN, R. (1860). On some small fossil vertebrae from near Frome, Somerset. *Q. Jl geol. Soc. Lond.* **16**, 492–497.

PALMER, C. P. (1966a). Notes on the fauna of the Margaritatus Clay (Blue Band) in the Domerian of the Dorset Coast. *Proc. Dorset nat. Hist. archaeol. Soc.* **87**, 67, 68.

PALMER, C. P. (1966b). The fauna of Day's Shell Bed in the Middle Lias of the Dorset Coast. *Proc. Dorset nat. Hist. archaeol. Soc.* **87**, 69–80.

PALMER, L. S. (1931). On the Pleistocene succession of the Bristol district. *Proc. Geol. Ass.* **42**, 345–361.

PALMER, L. S. (1934). Some Pleistocene breccias near the Severn estuary. *Proc. Geol. Ass.* **45**, 145–161.

PALMER, L. S. (1958). Plumley's Hole—a sequel. *Newsl. Cave Res. Gr Gt Br.* 72/77, 6–7.

PARRINGTON, F. R. (1947). On a collection of Rhaetic mammalian teeth. *Proc. zool. Soc. Lond.* **116**, 707–728.

PENGELLY, W. (1862). The lignites and clays of Bovey Tracey. *Phil. Trans. R. Soc.* **152**, 1019–1038.

PENGELLY, W. (1868–84). The literature of Kent's Cavern. *Rep. Trans. Devon Ass. Advmt. Sci.* 1, 1868, **1**, 469; 2, 1869, **3**, 191; 3, 1871, **4**, 467; 4, 1878, **10**, 141; 5, 1884, **16**, 189.

PENGELLY, W. (1873a). The ossiferous caverns and fissures in the neighbourhood of Chudleigh, Devonshire. *Rep. Trans. Devon Ass. Advmt Sci.* **6**, 46–60.

PENGELLY, W. (1873b). Literature of the caverns at Buckfastleigh, Devonshire. *Rep. Trans. Devon Ass. Advmt Sci.* **6**, 70–72.

PHILLIPS, W. J. (1964). The structures in the Jurassic and Cretaceous rocks on the Dorset coast between White Nothe and Mupe Bay. *Proc. Geol. Ass.* **75**, 373–405

PICK, M. C. (1964). The stratigraphy and sedimentary features of the Old Red Sandstone, Portishead coastal section, north-east Somerset. *Proc. Geol. Ass.* **75**, 199–221.

POLWHELE, R. (1797–1806). *The history of Devonshire.* 3 vols. London and Exeter; Cadell, Johnson & Dilly.

PRESTWICH, J. (1875a). On the origin of the Chesil Bank. . . . *Min. Proc. Instn. civ. Engrs.* **40**, 61–79.

PRESTWICH, J. (1875b). Notes on the phenomena of the Quaternary period in the Isle of Portland and around Weymouth. *Q. Jl geol. Soc. Lond.* **31**, 29–54.

PRESTWICH, J. (1892). The raised beaches and 'Head' or rubble drift of the south of England. *Q. Jl geol. Soc. Lond.* **48**, 263–343.

PRINGLE, J. (1936). (Stubblefield in) *Mem. geol. Surv. U.K. Summ. Prog.* pt. 1, 80.

PUGH, M. E. (1968). Algae from the Lower Purbeck limestones of Dorset. *Proc. Geol. Ass.* **79**, 513–523.

PUGH, M. E. and SHEARMAN, D. J. (1967). Cryoturbation structures at the south end of the Isle of Portland. *Proc. Geol. Ass.* **78**, 463–471.

RANKINE, W. F. (1961). The Mesolithic age in Dorset and adjacent area. *Proc. Dorset nat. Hist. archaeol. Soc.* **83**, 91–99.

REED, F. R. C. and REYNOLDS, S. H. (1908). On the fossiliferous Silurian rocks of the southern half of the Tortworth Inlier. *Q. Jl geol. Soc. Lond.* **64**, 512–543.

REID, C. (1890). The Pliocene deposits of Britain. *Mem. geol. Surv. U.K.*

REID, C. (1896). An early Neolithic kitchen midden and tufaceous deposit atBlashenwell, near Corfe Castle. *Proc. Dorset nat. Hist. antiq. Fld Club.* **17**, 67–75.

REID, C. (1909). Cornwall and Devonshire. *Mem. geol. Surv. U.K. Summ. Prog.* 14–18.

REID, C. (1912). The geology of Dartmoor. Expl. Sheet 338. *Mem. geol. Surv. U.K.*

REID, C. and FLETT, J. S. (1907). The geology of the Land's End district. Expl. Sheets 351, 358. *Mem. geol. Surv. U.K.*

REID, C. and SCRIVENOR, J. B. (1906). The geology of the country near Newquay. Expl. Sheet 346. *Mem. geol. Surv. U.K.*

REID, C., BARROW, G. and DEWEY, H. (1910). The geology of the country around Padstow and Camelford. Expl. Sheets 335 and 336. *Mem. geol. Surv. U.K.*

REID, C. and Others (1911). The geology of the country around Tavistock and Launceston. Expl. Sheet 337. *Mem. geol. Surv. U.K.*

REYNOLDS, S. H. (1907a). The igneous rocks of the Bristol district. *Proc. Geol. Ass.* **20**, 59–65.

REYNOLDS, S. H. (1907b). Excursion to Bristol. *Proc. Geol. Ass.* **20**, 150–156.

REYNOLDS, S. H. (1921a). *A geological excursion handbook for the Bristol district.* 2nd ed. Bristol; Arrowsmith.

REYNOLDS, S. H. (1921b). The lithological succession of the Carboniferous Limestone (Avonian) of the Avon Section at Clifton. *Q. Jl geol. Soc. Lond.* **77**, 213–245.

REYNOLDS, S. H. (1921c). On the rocks of the Avon Section, Clifton. *Geol. Mag.* 543–548.

REYNOLDS, S. H. (1924). The igneous rocks of the Tortworth Inlier. *Q. Jl geol. Soc. Lond.* **80**, 106–112.

REYNOLDS, S. H. (1938). A section of Rhaetic and associated strata at Chipping Sodbury, Glos. *Geol. Mag.* **75**, 97–102.

REYNOLDS, S. H. (1946). The Aust Section. *Proc. Cotteswold Nat. Fld Club.* **29**, 29–39.

REYNOLDS, S. H. and GREENLY, E. (1924). The Old Red Sandstone and Carboniferous Limestone of the Portishead–Clevedon area. *Proc. Bristol Nat. Soc.* s. 4, **6**, 92–97.

REYNOLDS, S. H. and VAUGHAN, A. (1911). Faunal and lithological sequence in the Carboniferous Limestone series (Avonian) of Burrington Combe (Somerset). *Q. Jl geol. Soc. Lond.* **67**, 342–392.

RICHARDSON, L. (1903). The Rhaetic rocks of north-west Gloucestershire. *Proc. Cotteswold Nat. Fld Club.* **14**, 127–174.

RICHARDSON, L. (1904). *A handbook to the geology of Cheltenham and Neighbourhood.* Cheltenham; Sawyer.

RICHARDSON, L. (1906a). Half-day excursion to Leckhampton Hill, Cheltenham. *Proc. Cotteswold Nat. Fld Club.* **15**, 182–189.

RICHARDSON, L. (1906b). On the occurrence of *Ceratodus* in the Rhaetic at Garden Cliff, Westbury-on-Severn, Gloucestershire. *Proc. Cotteswold Nat. Fld Club.* **15**, 267–271.

RICHARDSON, L. (1906c). On the Rhaetic and contiguous deposits of Devon and Dorset. *Proc. Geol. Ass.* **19**, 401–409.

RICHARDSON, L. (1907). The Inferior Oolite and contiguous deposits of the Bath—Doulting district. *Q. Jl geol. Soc. Lond.* **63**, 383–423.

RICHARDSON, L. (1909). Excursion to the Frome district, Somerset. *Proc. Geol. Ass.* **21**, 209–228.

RICHARDSON, L. (1911). The Rhaetic and contiguous deposits of west, mid, and part of east Somerset. *Q. Jl geol. Soc. Lond.* **67**, 1–72.

RICHARDSON, L. (1928). The Inferior Oolite and contiguous deposits of the Burton Bradstock—Broadwindorr district. *Proc. Cotteswold Nat. Fld Club.* **23**, pt. 2, 149–185.

RICHARDSON, L. (1930). The Inferior Oolite and contiguous deposits of the Sherborne district, Dorset. *Proc. Cotteswold Nat. Fld Club.* **24**, pt. 1, 35–85.

RICHARDSON, L. (1933). The country around Cirencester. Expl. Sheet 235. *Mem. Geol. Surv. U.K.*

RICHARDSON, L. (1947). The upper limit of the Rhaetic series and the relationship of the Rhaetic and Liassic series. *Proc. Cotteswold Nat. Fld Club.* **29**, 143–144.

RICHARDSON, L. and THACKER, A. G. (1920). On the stratigraphical and geographical distribution of the sponges of the Inferior Oolite of the west of England. *Proc. Geol. Ass.* **31**, 161–186.

RICHARDSON, W. A. (1923). A micrometric study of the St Austell granite (Cornwall). *Q. Jl geol. Soc. Lond.* **79**, 546–576.

RICHTER, D. (1966). On the New Red Sandstone Neptunian Dykes of the Tor Bay Area (Devonshire). *Proc. Geol. Ass.* **77**, 173–186.

ROBINSON, P. L. (1957). The Mesozoic fissures of the Bristol Channel area and their vertebrate faunas. *J. Linn. Soc (Zool.).* **43**, 260–282.

ROEMER, F. (1880). Notice on the occurrence of Upper Devonian Goniatite Limestone in Devonshire. *Geol. Mag.* 145–147.

ROGERS, E. H. (1956). Stratification of the cave earth in Kent's Cavern. *Proc. Devon Archaeol. Explor. Soc.* **5**, 68–92.

ROWE, A. W. (1901). The zones of the White Chalk of the English coast. 2. Dorset. *Proc. Geol. Ass.* **17**, 1–80.

ROWE, A. W. and SHERBORN, C. D. (1903). The zones of the White Chalk of the English coast. 3. Devon. *Proc. Geol. Ass.* **18**, 1–51.

RUSSELL, A. (1911). On the occurrence of phenacite in Cornwall. *Mineralog. Mag.* **16**, 55–62.

RUTTER, J. (1829). *Delineations of the north-west division of Somersetshire.* Shaftesbury; author. London; Longmans.

SALTER, D. L. and WEST, I. M. (1965). Calciostrontianite in the basal Purbeck beds of Durlston Head, Dorset. *Mineralog. Mag.* **35**, 146–150.

SALTER, J. W. (1864). On some points of ancient physical geography illustrated by fossils from a pebble-bed at Budleigh Salterton, Devonshire. *Geol. Mag.* **1**, 5–12.

SANDERS, W. (1840). Account of a raised sea-beach at Woodspring Hill, near Bristol. *Rep. Br Ass. Advmt Sci., Trans. Sects.* 102–103.

SARJEANT, W. A. S. (1960). New Hystrichospheres from the Upper Jurassic of Dorset. *Geol. Mag.* **97**, 137–144.

SAVAGE, R. J. G. (1969). Pleistocene mammal faunas. *Proc. Spelaeol. Soc.* **12**, 57–62.

SAVAGE, R. J. G. and LARGE, N. F. (1966). On *Birgeria acuminata* and the absence of labyrinthodonts from the Rhaetic. *Palaeontology.* **9**, 135–141.

SAVAGE, R. J. G. and WALDMAN, M. (1966). *Oligokyphus* from Holwell Quarry, Somerset. *Proc. Bristol Nat. Soc.* **31**, 185–194.

SCRIVENOR, J. B. (1903). The granite and greisen of Cligga Head, western Cornwall. *Q. Jl geol. Soc. Lond.* **59**, 142–159.

SEDGWICK, A. (1821). On the physical structure of those formations which are immediately associated with the primitive ridge of Devonshire and Cornwall. *Trans. Camb. phil. Soc.* **1**, 89–146.

SEDGWICK, A. (1822). On the physical structure of the Lizard district in the county of Cornwall. *Trans. Camb. phil. Soc.* **1**, 291–330.

SEDGWICK, A. and MURCHISON, R. I. (1840). Description of a raised beach in Barnstaple or Bideford Bay on the north-west coast of Devonshire. *Trans. geol. Soc. Lond.* s. 2, **5**, 279–286.

SELWOOD, E. B. (1960). Ammonoids and trilobites from the Upper Devonian and lowest Carboniferous of the Launceston area of Cornwall. *Palaeontology*, **3**, 153–185.

SERNANDER, R. (1915–1916). Svenska Kalktuffer. *Geol. Fören. Stockh. Förh.* **37**, 521–554; **38**, 127–190.

SEWARD, A. C. (1897). On *Cycadeoidea gigantea*, a new cycadean stem from the Purbeck beds of Portland. *Q. Jl geol. Soc. Lond.* **53**, 22–39.

SEWARD, A. C. (1904). The Jurassic Flora. II. Liassic and Oolitic floras of England. Catalogue of the Mesozoic plants in the Department of Geology. London; *British Museum (Nat. Hist.).*

SHANNON, W. G. (1928). Geology of the Torquay district . . . *Proc. Geol. Ass.* **39**, 103–136.

SHAW, T. R. (1949a). Pixie's Hole, Chudleigh. *Br. Caver.* **19**, 70–77.

SHAW, T. R. (1949b). The caves at Chudleigh. *Rep. Trans. Devon Ass. Advmt Sci.* **81**, 341–345.

SHAW, T. R. (1962). Lamb Leer in the 17th Century. *Proc. spelaeol. Soc.* **9**, 183–187.

SHORT, A. R. (1903). A new theory of the Cotham Marble. *Proc. Bristol Nat. Soc.* s. 3, **10**, 135–149.

SHRUBSOLE, O. A. (1903). On the probable source of some of the pebbles of the Triassic pebble-beds of south Devon and the Midland counties. *Q. Jl geol. Soc. Lond.* **59**, 311–333.

SIBLY, T. F. (1905). The Carboniferous Limestone of the Weston-super-Mare district. *Q. Jl geol. Soc. Lond.* **61**, 548–561.

SIBLY, T. F. (1907). On the Carboniferous Limestone (Avonian) of Burrington Combe and Cheddar. *Proc. Geol. Ass.* **20**, 66–69.

SIBLY, T. F. and REYNOLDS, S. H. (1937). The Carboniferous Limestone of the Mitcheldean area, Gloucestershire. *Q. Jl geol. Soc. Lond.* **93**, 23–51.

SIMPSON, G. G. (1928). *A Catalogue of the Mesozoic mammalia in the Geological Department of the British Museum (Nat. Hist.).* London.

SIMPSON, G. G. (1933). Paleobiology of Jurassic mammals. *Palaeobiologica.* **5**, 127–158.

SIMPSON, S. (1951). A new Eurypterid from the Upper Old Red Sandstone of Portishead. *Ann. Mag. nat. Hist.* s. 12, **4**, 849–861.

SIMPSON, S. (1957). On the trace-fossil *Chondrites. Q. Jl geol. Soc. Lond.* **112**, 475–499.

SMITH, S. (1934). The Tortworth Inlier. *Proc. Geol. Ass.* **45**, 114–120.

SMITH, W. E. (1957a). The Cenomanian Limestone of the Beer district, south Devon. *Proc. Geol. Ass.* **68**, 115–135.

SMITH, W. E. (1957b). Summer field meeting in south Devon and Dorset. *Proc. Geol. Ass.* **68**, 136–152.

SMITH, W. E. (1961). The Cenomanian deposits of south-east Devonshire: The Cenomanian Limestone and contiguous deposits west of Beer. *Proc. Geol. Ass.* **72**, 91–134.

SMITH, W. E. (1965). The Cenomanian deposits of south-east Devonshire: The Cenomanian Limestone east of Seaton. *Proc. Geol. Ass.* **76**, 121–136.

SMITH, W. E. and DRUMMOND, P. V. O. (1962). Easter field meeting: The Upper Albian and Cenomanian deposits of Wessex. *Proc. Geol. Ass.* **73**, 335–352.

SOWERBY, J. (later SOWERBY, J. DE C.). (1812–1846). *The Mineral Conchology of Great Britain. . . .* 7 vols. London.

SPEED, J. (1611). *The theatre of the Empire of Great Britaine* . . . London; Sudbury & Humble.

STODDART, W. W. (1867). Geology of Dundry Hill. *Proc. Bristol Nat. Soc.* s. 2, **2**, 29–33.

STODDART, W. W. (1876). Geology of the Bristol Coal-field. Part 3: Carboniferous. *Proc. Bristol Nat. Soc.* s. 3, **1**, 313–334.

STODDART, W. W. (1879). Geology of the Bristol Coal-field. Part 5: Jurassic strata. *Proc. Bristol Nat. Soc.* s. 3, **2**, 279–291.

STRAHAN, A. (1898). The geology of the Isle of Purbeck and Weymouth. *Mem. geol. Surv. U.K.*

STRICKLAND, H. E. (1842). On the occurrence of the Bristol Bone-Bed in the Lower Lias near Tewkesbury. *Proc. geol. Soc. Lond.* **3**, 585–588. [For full text of paper, see *Memoirs of Hugh Edwin Strickland* by Sir W. Jardine, 1858, 154–160. London; Van Voorst.]

STRICKLAND, H. E. (1846). On two species of microscopic shells found in the Lias. *Q. Jl geol. Soc. Lond.* **2**, 30–31.

STRICKLAND, H. E. (1853). On the distribution and organic contents of the Ludlow Bone-Bed in the districts of Woolhope, May Hill, etc. *Q. Jl geol. Soc. Lond.* **9**, 8–12.

STUBBLEFIELD, C. J. (1960). Trilobites of south-west England. *Trans. R. geol. Soc. Cornwall.* **19**, 101–112.

SUTCLIFFE, A. J. (1960). Joint Mitnor Cave, Buckfastleigh. *Trans. Proc. Torquay nat. Hist. Soc.* **13**, 1–28 (separate).

SUTCLIFFE, A. J. (1962). A note on some late Pleistocene mammalian remains from Lummaton Quarry, Torquay, Devon. *Trans. Proc. Torquay nat. Hist. Soc.* **13**, 4–7.

SUTCLIFFE, A. J. (1965). Planning England's cave studies centre. *Stud. Speleol.* **1**, 106–124.

SUTCLIFFE, A. J. and ZEUNER, F. E. (1962). Excavations in the Torbryan Caves, Devonshire. I. Tornewton Cave. *Proc. Devon archaeol. Explor. Soc.* **5**, 127–145.

SWINTON, W. E. (1939). A new Triassic Rhynchocephalian from Gloucestershire. *Ann. Mag. nat. Hist.* s. 11, **4**, 591–594.

SWINTON, W. E. (1958). *Fossil Amphibians and Reptiles.* 2nd ed. London; British Museum (Nat. Hist.).

SYLVESTER-BRADLEY, P. C. (1939). Clay in Devon. *Trans. Proc. Torquay nat. Hist. Soc.* **8**, 3–9.

SYLVESTER-BRADLEY, P. C. (1941). The shell structure of the Ostracoda and its application to their palaeontological investigation. *Ann. Mag. nat. Hist.* s. 11, **8**, 1–33.

SYLVESTER-BRADLEY, P. C. (1948). Bathonian ostracods from the *Boueti* Bed of Langton Herring, Dorset. *Geol. Mag.* **85**, 185–204.

SYLVESTER-BRADLEY, P. C. (1949). The Ostracod genus *Cypridea* and the zones of the Upper and Middle Purbeckian. *Proc. Geol. Ass.* **60**, 125–153.

TAWNEY, E. B. (1874). Museum notes: Dundry Gasteropoda. *Proc. Bristol Nat. Soc.* s. 2, **1**, 9–59.

TAWNEY, E. B. (1875). Notes on the Lias in the neighbourhood of Radstock. *Proc. Bristol Nat. Soc.* s. 2, **1**, 167–189.

TAYLOR, C. W. (1956). Erratics of the Saunton and Fremington areas. *Rep. Trans. Devon Ass. Advmt Sci.* **88**, 52–64.

TEALL, J. J. H. (1887). On granite containing andalusite from the Cheesewring, Cornwall. *Mineralog. Mag.* **7**, 161–163.

TEALL, J. J. H. (1888). *British Petrography.* London; Dulau.

THOMAS, H. H. (1902). The mineralogical constitution of the finer material of the Bunter pebble-bed in the west of England. *Q. Jl geol. Soc. Lond.* **58**, 620–632.

TIDMARSH, W. G. (1932). The Permian lavas of Devon. *Q. Jl geol. Soc. Lond.* **88**, 712–775.

TILLEY, C. E. (1935). Metasomatism associated with the Greenstone-Hornfelses of Kenidjack and Botallack, Cornwall. *Mineralog. Mag.* **24**, 181–202.

TILLEY, C. E. and FLETT, J. S. (1929). Hornfelses from Kenidjack, Cornwall. *Mem. geol. Surv. U.K. Summ. Prog.* pt. 2, 24–41.

TORRENS, H. S. (Ed.) (1969). *International Field Symposium on the British Jurassic.* A. Guide for Dorset and South Somerset. B. Guide for North Somerset and Gloucestershire. Geology Dept., Keele University.

TOWNSEND, J. (1813). *The character of Moses established for veracity as an historian recording events from the Creation to the Deluge.* Bath and London.

TRATMAN, E. K. (1963a). Sun Hole Cave, Cheddar, Somerset: Pleistocene fauna. *Proc. speleol. Soc.* **10**, 16–17.

TRATMAN, E. K. (1963b). The hydrology of the Burrington area, Somerset. *Proc. speleol. Soc.* **10**, 22–57.

TRESISE, G. R. (1960). Aspects of the lithology of the Wessex Upper Greensand. *Proc. Geol. Ass.* **71**, 316–339.

TROMELIN, G. DE (1877). Étude de la faune du Grès Silurien de May, Campandré, Mont-Robert, etc. (Calvados). *Bull. Soc. Linn. Normandie*, s. 3, **1**, 5–83.

TROTTER, F. M. (1942). Geology of the Forest of Dean coal and iron-ore field. *Mem geol. Surv. U.K.*

TRUEMAN, A. E. (1938). Erosion levels in the Bristol district and their relation to the development of the scenery. *Proc. Bristol Nat. Soc.* s. 4, **8**, 402–428.

TUCK, M. C. (1925). The Avonian succession between Wickwar and Chipping Sodbury, (Glos.). *Proc. Bristol Nat. Soc.* s. 4, **6**, 237–249.

TUTCHER, J. W. (1917). The zonal sequence of the Lower Lias (lower part). *Q. Jl geol. Soc. Lond.* **73**, 278–281.

TUTCHER, J. W. (1923). Some recent exposures of the Lias (Sinemurian and Hettangian) and Rhaetic about Keynsham. *Proc. Bristol Nat. Soc.* s. 4, **5**, 268–278.

TUTCHER, J. W. and TRUEMAN, A. E. (1925). The Liassic rocks of the Radstock district (Somerset). *Q. Jl geol. Soc. Lond.* **81**, 595–666.

USSHER, W. A. E. (1877). A chapter on the Budleigh pebbles. *Rep. Trans. Devon Ass. Advmt Sci.* **9**, 222–226.

USSHER, W. A. E. (1879). *The Post-Tertiary geology of Cornwall.* Hertford. (Distinct from the paper of the same author, date and title, published in *Geol. Mag.*)

USSHER, W. A. E. (1902). The geology of the country around Exeter. Expl. Sheet 325. *Mem. geol. Surv. U.K.*

USSHER, W. A. E. (1907). The geology of the country around Plymouth and Liskeard. Expl. Sheet 348. *Mem. geol. Surv. U.K.*

USSHER, W. A. E. (1913). The geology of the country around Newton Abbot. Expl. Sheet 339. *Mem. geol. Surv. U.K.*

USSHER, W. A. E., BARROW, G. and MACALISTER, D. A. (1909). The geology of the country around Bodmin and St Austell. Expl. Sheet 347. *Mem. geol. Surv. U.K.*

VACHELL, E. T. (1953). Kent's Cavern, its origin and history. *Trans. Proc. Torquay nat. Hist. Soc.* **11**, 51–73.

VAUGHAN, A. (1905). On the palaeontological sequence in the Carboniferous Limestone of the Bristol area. *Q. Jl geol. Soc. Lond.* **61**, 181–305.

VAUGHAN, A. (1906). The Avonian of the Avon Gorge. *Proc. Bristol Nat. Soc.* s. 4, **1**, 87–100.

VAUGHAN, A. and REYNOLDS, S. H. (1935). The Carboniferous Limestone series (Avonian) of the Avon Gorge. *Proc. Bristol Nat. Soc.* s. 4, **8**, 29–90.

VICARY, W. and SALTER, J. W. (1864). On the Pebble-Bed of Budleigh Salterton. *Q. Jl geol. Soc. Lond.* **20**, 283–302.

WALCOTT, J. (1779). *Descriptions and figures of petrifactions found in the quarries, gravel pits etc. near Bath.* Bath.

WALFORD, E. A. (1889). On some Bryozoa from the Inferior Oolite of Shipton Gorge, Dorset. *Q. Jl geol. Soc. Lond.* **45**, 561–574.

WALFORD, E. A. (1894). On some Bryozoa from the Inferior Oolite of Shipton Gorge, Dorset. Part II. *Q. Jl geol. Soc. Lond.* **50**, 72–78.

WALKER, C. T. (1964). Depositional environment of Purbeck formation. *Geol. Mag.* **101**, 189.

WALKER, H. H. and SUTCLIFFE, A. J. (1967). James Lyon Widger 1823–1892 and the Torbryan caves. *Rep. Trans. Devon Ass. Advmt. Sci.* **99**, 49–110.

WALLIS, F. S. (1924). The Avonian of the Tytherington—Tortworth—Wickwar Ridge, Gloucestershire. *Proc. Bristol Nat. Soc.* s. 4, **6**, 57–74.

WARWICK, G. T. (1958). Aragonite clusters in caves. *New Scient.* **3**, 41–42.

WEAVER, T. (1824). Geological observations on part of Gloucestershire and Somerset. *Trans. geol. Soc. Lond.* s. 2, **1**, 317–368.

WEBSTER, T. (1816). Observations on the strata of the island and their continuation in the adjacent parts of Dorsetshire. *In* H. C. C. Englefield: *A description of the principal picturesque beauties, antiquities, and geological phaenomena of the Isle of Wight.* London; Payne & Fosse.

WEBSTER, T. (1826). Observations on the Purbeck and Portland Beds. *Trans. geol. Soc. Lond.* s. 2, **2**, 37–44.

288

Welch, F. B. A. (1929). The geological structure of the central Mendips. Q. Jl geol. Soc. Lond. **85,** 45–76.

Welch, F. B. A. (1930). The hydrology of the Stoke Lane area (Somerset). Proc. Cotteswold Nat. Fld Club. **24,** 87–96.

Welch, F. B. A. (1933). The geological structure of the eastern Mendips. Q. Jl geol. Soc. Lond. **89,** 14–52.

West, I. (1960). On the occurrence of celestine in the Caps and Broken Beds at Durlston Head, Dorset. Proc. Geol. Ass. **71,** 391–401.

West, R. G. (1968). Pleistocene geology and biology with especial reference to the British Isles. London; Longmans.

Wethered, E. (1886). On the structure and organisms of the Lower Limestone Shales, Carboniferous Limestone, and Upper Limestones of the Forest of Dean. Geol. Mag. 529–540.

Whidborne, G. F. (1883). Notes on some fossils, chiefly Mollusca, from the Inferior Oolite. Q. Jl geol. Soc. Lond. **39,** 487–540.

Whidborne, G. F. (1889–1898). A monograph of the Devonian fauna of the south of England. Palaeontogr. Soc. Monogr. **42.**

Whitley, N. (1849). On the remains of ancient volcanoes on the north coast of Cornwall in the parish of St Minver. 30th Ann. Rep. R. Instn Cornwall. 60.

Whitley, N. (1887). The evidence of glacial action in Cornwall and Devon. Trans. R. geol. Soc. Cornwall. **10,** 132–141.

Whittard, W. F. (1949). Geology of the Aust—Beachley district, Gloucestershire. Geol. Mag. **86,** 365–376.

Wilson, V. and Others (1958). Geology of the country around Bridport and Yeovil. Expl. Sheets 327 and 312. Mem. geol. Surv. U.K.

Wood, S. V. (Jr.). (1885). A new deposit of Pliocene age at St Erth. Q. Jl geol. Soc. Lond. **41,** 65–73.

Woodward, A. S. (1886). The history of the fossil crocodiles. Proc. Geol. Ass. **9,** 288–344.

Woodward, A. S. (1897). On a new specimen of the Mesozoic ganoid fish Pholidophorus from the Oxford Clay of Weymouth. Proc. Dorset nat. Hist. antiq. Fld Club. **18,** 150–152.

Woodward, A. S. (1916–1919). The fossil fishes of the English Wealden and Purbeck formations. Palaeontogr. Soc. Monogr. **69–71.**

Woodward, H. B. (1893, 1895). The Jurassic rocks of Britain (Yorkshire excepted). Mem. geol. Surv. U.K. **3,** Lias of England and Wales. **5,** Middle and Upper Oolitic rocks.

Woodward, H. B., Ussher, W. A. E. and Jukes-Browne, A. J. (1911). The geology of the country near Sidmouth and Lyme Regis. Expl. Sheets 326 and 340. Mem. geol. Surv. U.K.

Woodward, J. (1728–1729). An attempt towards a natural history of the fossils of England. In: A catalogue of the English fossils in the collection of J. Woodward. London.

Worth, R. H. (1920). The geology of the Meldon valleys near Okehampton on the northern verge of Dartmoor. Q. Jl geol. Soc. Lond. **75,** 77–118.

Worth, R. N. (1876). William Cookworthy and the Plymouth china factory. Rep. Trans. Devon Ass. Advmt. Sci. **8,** 480–496.

Wright, C. W. and Wright, E. V. (1949). The Cretaceous Ammonite genera Discohoplites Spath and Hyphoplites Spath. Q. Jl geol. Soc. Lond. **104,** 477–497.

Wright, T. (1856). On palaeontological and stratigraphical relations of the so-called sands of the Inferior Oolite. Q. Jl geol. Soc. Lond. **12,** 292–325.

Wright, T. (1860). The zone of Avicula contorta and the Lower Lias of the south of England. Q. Jl geol. Soc. Lond. **16,** 374–411.

Wright, T. (1878–1886). Monograph on the Lias ammonites of the British Islands. Palaeontogr. Soc. Monogr. **32–39.**

Yeldham, D. (1937). The Pridhamsleigh caves. Caves Caving. **1,** 73–75.

Zeuner, F. E. (1961). Fossil insects from the Lower Lias of Charmouth, Dorset. Bull. Br. Mus. nat. Hist. (Geology). **7,** 155–171.

Index of Sites by Counties

General Index*

* Names of Sites of Special Scientific Interest are shown THUS.

Chesil Beach, 180
Chester's Hill, 176
China Stone ('Cornish Stone'), 36–38
CHUDLEIGH CAVES, 51
Clay workings, 29, 31, 80, 191
CLICKER TOR QUARRY, 21
CLIGGA HEAD, 22
Coccoliths, 128
Code for visitors to geological sites, 15
Coins:
  Roman, 86, 238, 244
  Saxon, 210
Conifers:
  Lr. Lias, 107
  Purbeck, 120, 150, 175
Conodonts, 60, 222
Corallian, 114, 129, 176, 178–179
Corals:
  Lr. Lias, 106
  Lr. Cretaceous, 53
Cornbrash, 129, 176
COVERACK COVE, 26
Cretaceous:
  Upper, 41, 46, 69, 72, 87, 110, 185, 257
  Lower, 41, 52, 72, 91, 110, 152, 185, 192, 257
Crinoids:
  Lr. Lias, 105
  Kimeridgian (Saccocoma), 127, 174
Cromhall Quarry, see SLICKSTONES QUARRY
CROCK HILL BRICK PIT, 133
Crustacea:
  Lr. Lias, 101
  Inf. Oolite, 166, 209
  Portlandian, 130
CULLIMORE'S QUARRY, 205
Culm Measures, 54, 59, 61, 62, 64, 76
Cycads:
  Lr. Lias, 107
  Purbeck, 120, 150, 175

De Luc, J. A., 50, 52, 65, 83, 95, 176, 180, 190
Devonian:
  Upper, 55, 59, 70
  Middle, 55, 60, 70
  Lower, 29, 70
Dolomitic Conglomerate (Trias), 232–233,
  237, 241–242, 252–254
Downtonian, 19, 23, 30, 215
DUNDRY MAIN ROAD SOUTH QUARRY, 239
Dungy Head, 111
Durdle Door, 111, 115, 118
DURLSTON BAY, 134

EBBOR GORGE and WOOKEY HOLE, 240
Echinoids:
  Lr. Carboniferous, 222

Echinoids: (contd)
  Lr. Lias, 104
  Inf. Oolite, 208–209
  Purbeck, 149
  Upper Cretaceous, 43, 69, 73–74, 87, 117, 258
Emmit Hill, 111, 121, 131
Eocene, 52, 165, 191
Epidiorite, 19, 26, 27
Erratics, 35, 71, 168
Estheria, 199, 204, 207, 212, 262
Eurypterid, 253
Explanation of symbols used, 12
EYPE COAST, 152

Fishes:
  Old Red Sandstone, 253
  Lr. Carboniferous, 201, 214, 222–223
  Rhaetic, 199, 201, 203, 204, 207–208, 212,
    226, 247
  Lr. Lias, 100, 106
  Oxford Clay, 133
  Kimeridgian, 126
  Purbeck, 145
Fleet, 175, 182, 184
Flora:
  Old Red Sandstone/Lr. Carboniferous,
    214–215
  Up. Carboniferous, 253
  Lr. Lias, 107
  Purbeck, 119, 120, 139, 150, 173, 175
  Eocene, 191
  Oligocene, 78
  ? Pliocene or Pleistocene, 32
  Recent, 108
Fly ash, 251
Foraminifera:
  Lr. Carboniferous, 214, 222–223, 233, 263
  Lr. Lias, 92, 106, 213, 228, 256
  Mid. Jurassic, 157–158, 167
  Kimeridgian, 127–128
  Lr. Cretaceous, 53, 75, 91
  Up. Cretaceous, 73
  ? Pliocene or Pleistocene, 31–34
  Pleistocene, 168
  Recent, 27
Forest Marble, 155–157, 176–177
FROGDEN QUARRY, 158
Fuller's Earth, 155–157, 176–177

Gabbro, 22, 25–27
Gad Cliff, 110–111, 121
GARDEN CLIFF, 206
Gastropods:
  Lr. Lias, 103
  Purbeck non-marine, 148
  ? Pliocene or Pleistocene, 31–32
G. B. CAVERN, 245